Archaeological Landscape Evolution

Mike T. Carson

Archaeological Landscape Evolution

The Mariana Islands in the Asia-Pacific Region

Springer

Mike T. Carson
Micronesian Area Research Ctr
University of Guam
Mangilao, Guam

ISBN 978-3-319-31399-3 ISBN 978-3-319-31400-6 (eBook)
DOI 10.1007/978-3-319-31400-6

Library of Congress Control Number: 2016939450

Printed on acid-free paper

This Springer imprint is published by Springer Nature
The registered company is Springer International Publishing AG Switzerland

This book is dedicated to the memory of William R. Dickinson, 1931–2015, for his inspiring studies of geoarchaeology and coastal landscapes of Pacific Oceania. His friendly and professional guidance will be missed.

Preface and Acknowledgements

This book presents my views of landscapes as complex natural-cultural systems that have evolved over long periods of time. Over several years of research, eventually I compiled enough information to feel confident about portraying landscape evolution in liberal interdisciplinary terms, centred around archaeology, while drawing on diverse lines of evidence. Surely more could be added as investigations will continue, and I hope to encourage further work.

While preparing this book, I wanted to focus on one geographic area as a convenient illustrative example of the general-global applicability of landscape evolution studies. My research in the Mariana Islands very clearly offered the best candidate for developing this book's main narrative. The Mariana Islands case explicitly is situated in a larger Asia-Pacific regional context, and furthermore the key research principles are presented with the purpose of broader relevance.

Several people have improved my work as presented here. Regarding general approaches to environmental studies linked with archaeology, I am grateful to Steve Athens, Bill Dickinson, Patrick Nunn and John Peterson for their advice and collaborations. Peter Bellwood, Hsiao-chun Hung and Glenn Summerhayes broadened my cross-regional knowledge. Jim Bayman, Hiro Kurashina, Barry Rolett and Rebecca Stephenson encouraged for me to enhance my skills in research and writing. Numerous individuals, too many to mention, have supported my research over the years in Guam and in the Commonwealth of the Northern Mariana Islands, but I want especially to express my gratitude to Vic April, Lufo Babauta, Rosanna Barcinas, Laura Beauregard, Lon Bulgrin, Matt Brown, Diego Camacho, John Castro, Jeremy Cepeda, Gabe Cruz, Tommy Deleon Guerrero, Boyd Dixon, Don Farrell, LaVonne Guerrero Meno, Leonard Iriarte, Mertie Kani, Andy Laguana, Brian Leon Guerrero, Patrick Lujan, Kelly Marsh Taitano, Al Masga, Rita Nauta, John Palacios, Joe Quinata, Dick Randall, Spencer Rearden, Ronnie Rogers, Scott Russell, Emily Sablan, Carmen Sanchez, Joe Schwagerl, Monique Storie, Herman Tudela, Robert Underwood and Eric West. I wish that I could name everyone who has played a role in the developments toward making this book, and I hope that I can be forgiven for this abbreviated list.

Contents

About the Author

Mike T. Carson (Ph.D. in Anthropology, University of Hawaii, 2002) investigates archaeological and palaeo-landscapes throughout the Asia-Pacific region. He currently is associate professor of archaeology at the Richard F. Taitano Micronesian Area Research Center (MARC), University of Guam, and he is co-editor of *Asian Perspectives: The Journal of Archaeology for Asia and the Pacific*, published by the University of Hawaii Press.

Part I
Parameters of Study

Chapter 1
Landscape Evolution as Natural–Cultural History

Landscape studies strike at the core experience of how human beings relate with the natural world, fundamentally shared in one way or another by all people around the globe and throughout human evolutionary history. We all conduct our daily activities within landscapes that very much shape our lives, just as much as we shape and re-shape them. Living in or with a landscape constitutes an essential part of what it means to be a human being. As experienced today worldwide, landscapes embody the results of long-term interactions between people and their environments of the material world, socially constructed world, or hybrids of these perspectives that all are considered equally relevant for understanding how our landscapes operate as complex systems.

This study begins with a simple premise: the landscapes that we all inhabit today have evolved through long-term processes of complex natural–cultural histories, so we can understand our place in the world through unravelling how our landscapes have evolved. This book draws primarily on archaeology as a way to obtain information directly from the past, but landscapes must be understood in the broadest sense as inhabited social-ecological environments that can be studied in multiple ways. Archaeology provides a central focus, while geological records, preserved botanical and faunal remains, ethnohistories, language histories, and other lines of evidence all are pertinent for a full comprehension of landscape evolution.

What can we as human beings learn about our role in shaping our landscapes and being shaped by them? How and why have we evolved or co-evolved with our landscapes? In order to address these grand questions, we first need to answer another: how did the different natural and cultural elements of a landscape affect one another through time? Only by knowing a detailed long-term record of a landscape can we hope to answer these questions. We can begin by making sense of at least one concrete example of coordinating diverse lines of evidence about how a landscape has evolved through changing climate, sea level, geological formations, sediment regimes, vegetation communities, animal life, and of course the richly layered cultural behaviours within this dynamic setting.

M.T. Carson, *Archaeological Landscape Evolution*,
DOI 10.1007/978-3-319-31400-6_1

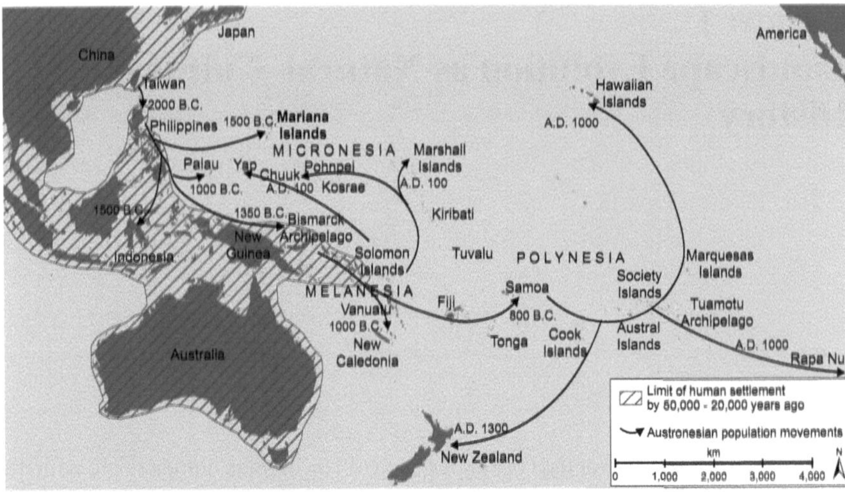

Fig. 1.1 Asia-Pacific region, showing major patterns in settlement chronology

The proposed approach to study landscape evolution could be applied anywhere in the world in principle, but a practical example is necessary to demonstrate this general utility. The "proof of concept" example needs to be presented thoroughly, with a strong chronological basis, numbers of identifiable time periods, and diverse records of evidence that cross through periods of changing landscape conditions. Brief introductory sketches of landscape evolution can highlight the potential for studies in representative areas of the world, such as coastal China, California, and the Hawaiian Islands each with their own unique circumstances and time scales. A most thorough treatment of landscape evolution, however, is presented in for an unsuspecting corner of the Pacific Ocean, where the small and remote Mariana Islands have yielded an instructively data-rich and long-term chronology (Figs. 1.1 and 1.2). Humanity's first contact with the Remote Oceanic environment occurred here about 3500 years ago. Since then, the Marianas landscape has evolved in complex ways, as may be said of any place in the world, but here we have an opportunity to study an exceptionally informative case.

Perhaps, most intriguing in global perspective, Marianas archaeology offers a rare documentation of people living in a truly untouched natural setting where no other human beings ever had lived previously. The oldest Marianas sites attest to the first human settlement of the Remote Oceanic region about 1500 B.C., some centuries earlier than any other inhabitation of the Remote Oceanic environment. What happened during those centuries of first contact between people and a unique natural environment? What did people do in these habitats as they inevitably changed or evolved over time? What does the surviving material evidence imply about the evolution of landscapes as inter-active, symbiotic entities?

The Marianas sites offer unique potential for examining topics that quite simply are not possible (at least not to the same extent) anywhere else. The island setting enables great confidence about the data archives of ancient sites accurately representing the past, more convincingly than in larger land masses and continents

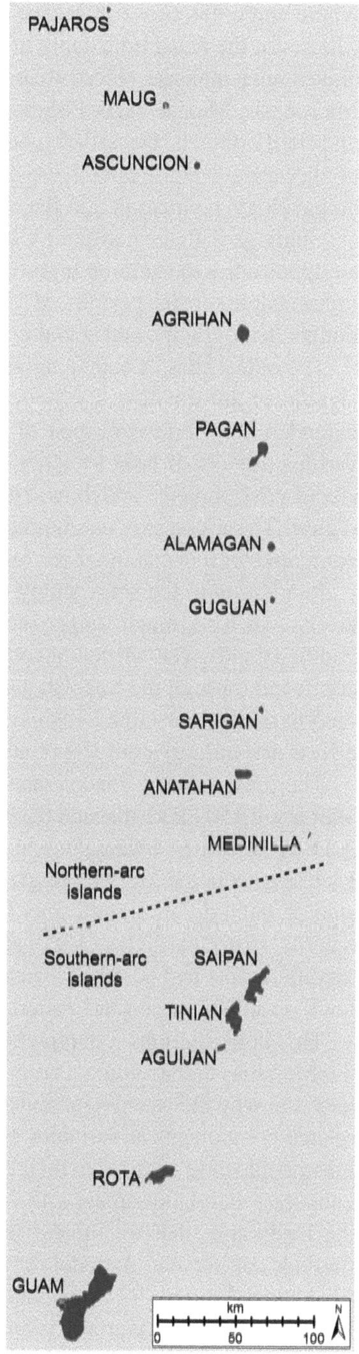

Fig. 1.2 Mariana Islands, noting northern-arc and southern-arc groups

where numerous sites can escape discovery despite decades of intensive searching. Moreover, the dated time depth of human settlement in the Marianas gives a noticeably longer material record than for any other of the Remote Oceanic islands in Micronesia, Melanesia, or Polynesia. Discoveries of distinctive pottery styles, stone and shell artefacts, discarded food refuse, remnants of house structures, and relicts of the ancient landscapes have changed the baseline chronology of Asia-Pacific archaeology as illustrated in the chapters of this book. When combining the oldest site findings with the complete chronology of the last 3500 years, we can maximise an opportunity to examine high-resolution chronicles of long-term adaptations that transcended several periods of change in climate, sea level, forest composition, human demography, and social-cultural behaviours.

The oldest sites of course attract special attention, but later sites contain equally significant information for understanding the past and for comprehending the extended long-term evolution of landscapes. The sites of all time periods reveal various aspects of how societies developed, how they interfaced with a dynamic island environment, and how they related with other groups in the Asia-Pacific region. These contexts can be reconstructed for any time period, and they can be compared with one another for fuller comprehension of chronological change.

In addition to the surprisingly old archaeological evidence, the Mariana Islands are known for a much longer record of written history than is the case for other Pacific Islands. The written history began with Magellan's arrival in A.D. 1521 and continued through the Spanish galleon trade and into modern times. The historical records gained sustained regularity in the late 1600s, stimulated by missionary efforts and military conflicts of an imposed foreign imperial regime.

The present study concentrates on the archaeological sequence from the oldest sites about 1500 B.C. through the historically known Spanish encounters. After about A.D. 1700, written documents and oral traditions substantiate numerous detailed studies of events in the Mariana Islands. Archaeology contributes to these later historical studies, but it differs significantly from research of the earlier periods without written records. Historical archaeology can focus on the circumstances and contingencies of specific events and people, whereas "prehistoric" archaeology (the emphasis of this book) examines large-scale patterns and long-term processes of human societies.

This book outlines a comprehensive natural–cultural history, re-animated from the Marianas archaeological record but with general-global applicability to explore how and why the world's inhabited landscapes have evolved in their multiple interrelated components at variables scales, paces, and magnitudes. New studies reveal how people have adapted to their changing environmental and social conditions that constitute their landscapes, with variable degrees of success and failure, resulting in the landscapes that we inhabit today and with ongoing challenges. The detailed findings can serve as a model for global human–environment relational studies and fundamental queries of human nature; hence, this book aims for a broad relevance beyond the isolated Mariana Islands.

When viewed as complex systems with multiple internal components, natural–cultural landscapes have evolved constantly in their individual components, but occasionally a deep systemic transformation has occurred when multiple components have changed in concert. In order to trace both of these kinds of change through time, archaeological sites of each measured time interval are situated within their broader landscapes, definable as natural environmental and socially mediated settings, resulting in a chronological sequence of the long-term trends and patterns comprising the total landscape system. The complex social-ecological systems and their interacting components have changed or evolved at different rates, paces, and scales that can be understood through long-term chronological sequences as depicted in the chapters of this book.

The landscapes that we inhabit today inevitably will transform through time, just as has been true for countless centuries in the past. The world's landscapes will continue to exist, but they certainly will evolve and in some cases become almost unrecognisable after change in climate and sea level, coupled with re-configuration of coastlines, agriculturally productive lands, and human population demography. Modes of human life, especially in terms of perceiving and interacting with landscapes, will need to change in order to sustain increasing population levels with the world's limited natural resources. In this regard, a long-term view of landscape evolution in its broadest natural–cultural sense will prove most useful.

Evolution of an Inhabited Landscape

Landscapes mean many things to different people, including archaeologists who have no shortage of definitions and nuanced viewpoints. The word "landscape" has been employed in so many ways that its meaning has become obscured in the literature, often concerned with justifying the use of the term. The word has its roots in objectively viewing the land or surrounding environment from a detached perspective as if through a timeless window of a framed painting, yet most people today recognise that they live actively within landscapes (Thomas 2012). The broad appeal of landscapes for archaeologists can be appreciated in a collection of more than 60 essays and case studies (David and Thomas 2008), each making use of slightly different perspectives about landscapes. Many of the key social anthropological issues were discussed in another collection of essays by Tim Ingold (2000), developing notions of how people live in the world and interact with landscapes. Although now landscapes may be accepted as a fundamental part of human experience, scholars sometimes disagree about how to conceptualise landscapes as objective realities of the natural world, as subjective human cognitive creations, or as both.

In the present study, landscapes are most important in terms of how they create inter-dependent linkages between nature and human beings. The natural world certainly has influenced much of human history, just as human actions have affected much of the natural world, inter-linked to such as degree as to create inter-dependence. The process is similar to what happens when people domesticate dogs or other animals, upon closer inspection forcing questions about the humans or the dogs actually controlling the domestication process. Perhaps, no single operator really is in control, but more realistically a symbiotic or co-dependent relationship creates synergistic outcomes greater than the sum of its parts. Furthermore, the beginning point of domestication is nearly impossible to specify, while no ending point at all is evident among species that continue to evolve independently or co-evolve together (Rindos 1984).

A morphological change in a species (such as enlarged or more durable portions of an edible plant) signifies that domestication already has created a physical response, but it does not clarify when, how, or why the process started. Likewise, the long-term evolution of a fully operating natural–cultural landscape involves much more than a morphological change in a single species, in a set or artefacts, or in the configuration of geological landforms. Rather, it involves a set of dynamic interactions among numerous natural and cultural variables, each changing at different magnitudes and time scales that may not necessarily coincide with one another. A liberal view of a landscape is necessary for understanding how it functions, and a long-term perspective is essential for comprehending its transformations through time.

Natural and cultural histories are inherently connected, manifest in records of landscapes found in archaeological sites, faunal and botanical remains, geological formation layers, and other lines of evidence considered in this book. This stance avoids extremist acceptance or rejection of environmental determinism or its counterpart cultural determinism. It meanwhile escapes a trite juxtaposition of nature versus nurture as opposing forces, instead more productively viewed as mutual contributors in a dynamically unified natural–cultural system.

Landscapes may be understood as environments that have evolved over time, with and without human groups, although any archaeological study necessarily maintains focus on the period of time when people inhabited the landscape in question. In the Mariana Islands, this period began at least as early as 1500 B.C., and it continues today. During these several centuries, the landscape underwent numerous transformations of its internal components, due to changing climate, sea level, geological formation processes, forest composition, animal populations, and of course human activities. Much longer time sequences and different kinds of landscape change are known in other regions of the world, as outlined for cases of coastal China and California in Chap. 2. Considerably shorter sequences of landscape evolution have occurred in other Pacific Islands, such as in Hawaii where accessible ethnohistory plays a stronger role than in many other settings, also discussed in Chap. 2 before other chapters consider details of the Marianas example. Several other cases could be mentioned, but this book is intended as a general approach for others to apply.

In whatever illustrative example is chosen, landscapes here are viewed as consisting of multiple inter-related components of the natural environment and cultural setting that mutually affect each other in a complex and dynamic system. This sense of inter-connectedness or inter-dependence is a central principle of landscape ecology (Turner et al. 2001), wherein change in one part can result in change in other parts, often subtle or temporary but sometimes at a larger magnitude and long-lasting. The process of change must be understood as occurring within a defined place or landscape (Cumming 2011), but it also necessarily unfolds over a period of time that in many cases is best documented through long-term records such as in archaeology (Dearing et al. 2006a, b). Given the number of variables involved, multiple time scales, and different rhythms of possible response to any change within the complex landscape system, direct cause and effect are not always obvious when viewed in short time intervals of a few decades or even of a few centuries. A significantly long-term view, however, reveals more useful information about chronologically detectable change within different aspects of a landscape system.

In a chronological perspective, any landscape has evolved through a series of events and processes that can be identified through archaeological and palaeoenvironmental datasets, among other ways of learning about the past. This scope of research shares much in common with geoarchaeological studies of sites in their ecological contexts (Butzer 1982; Wilson 2011), here expanded to the scale of a landscape and in a continuous chronological sequence. It similarly shares much in common with efforts to use archaeological evidence to address urgent questions of human–environment relations in a changing world (Briggs et al. 2006; Crumley 1994, 2001; Hornborg et al. 2007; van der Leeuw and Redman 2002). Towards these goals, one notably high-resolution study of a long-term evolving inhabited landscape was completed for the Ritidian Site in northern Guam (Carson 2012), and the research scope was expanded for the Mariana Islands overall (Carson 2014a), now developed fully in this book.

A chronology-based view of landscape evolution can interface productively with notions of niche construction theory (or NCT). According to NCT, people live in niches, adapt to them, and modify them in ways that affect future generations (Olding-Smee et al. 2003). The Marianas case in this book, along with other examples in Chap. 2, refers to groups of people in the past who targeted specific ecological niches, adapted their lifestyles according to their unique settings, and created new conditions for themselves and with continuing effects on later groups of people who lived there. As the examples in this book illustrate, people at any single point in time inhabit landscapes that are inherited from prior periods of natural and cultural historical change, yet people continue to modify these landscapes and to be affected by these landscapes with ongoing effects.

The record of human habitation in the Marianas exceeds the temporal scope in other remote Pacific Islands, so it necessarily offers more opportunities to study chronological change. In fact, it holds a unique record of the first time when human beings ever lived in the Remote Oceanic region of the world (Carson 2014b; Carson and Kurashina 2012), and hence it also contains the longest known material sequence of an inhabited landscape in this region. Accordingly, the long-term chronological dimension in evolution or co-evolution of the Marianas landscape comprises the central focus of the present study.

This approach to natural–cultural landscape evolution could be applied in any setting, but it is most instructive in a place with a substantive archaeological record that extends through successive periods of changing environmental and social conditions. These qualities are found in the Mariana Islands, but they could be found elsewhere. The Marianas case happens to involve useful datasets within a solid chronological sequence, and moreover the small and remote island setting offers opportunities for thorough investigation of a complete landscape system. Based on this example, researchers may develop landscape evolution studies more productively for the circumstances of other regions, and potentially new cross-regional comparisons will become possible.

The Marianas Landscape as a Model System

What makes the Mariana Islands especially informative about landscape evolution? These and other small islands offer reasonable control of their environmental parameters, but they are not completely closed laboratory-like systems sealed from the outside world. After all, people settled in the Mariana Islands at least as early as 1500 B.C., and overseas contacts occurred periodically ever since that time. Nevertheless, islands inherently involve a degree of isolation and discernible boundary that together can enhance observations of complex processes at work in geology, biology, and cultural systems. Perhaps over-stating the case, islands have been described as microcosms of global conditions (Kirch 2004). No island is completely secluded, due to connectivity across the surrounding ocean water, but greater degree of remoteness implies a stronger boundary and in theory magnifies the interpretive power of observations. Few island groups are more isolated than the Marianas in a distant northwest corner of the Remote Oceanic region.

To the extent that an island setting is isolated or bounded, it may be portrayed as having controllable parameters for studying processes that otherwise are vastly complex and difficult to specify. Vitousek et al. (2010) neatly expressed this point about islands as model systems by drawing an analogy with the use of small and short-lived insects as exemplary models for learning about biological evolution. The insects in this case do not represent the totality of all biological evolution, but they can facilitate observations of the most essential components of the evolutionary process. Likewise, island ecosystems do not necessarily represent the full range of variability in the world, but they offer unique case studies that potentially offer insights into the detailed workings of ecosystems in a general sense.

Regarding the present study of landscape evolution, few topics could be so complicated, involving multiple concurrent processes throughout the world, often interconnected across space and through time. These complexities are profound and varied geographically, culturally, and chronologically to such an extent that arguably they cannot be discussed properly without plunging into the full details of the entire global system of landscapes. More manageable, though, are landscape systems with clear boundaries, within which the most influential key processes can be identified, for example in the Mariana Islands.

The notion of a model system is encouraging for island-based research, but a logical problem arises from not knowing for sure if the particular island in question accurately represents the subject of study. In one point of view, the small scales and skewed natures of islands pose unfortunate obstacles against drawing conclusions of relevance beyond their isolated shores. Additionally, islands tend to have shorter time spans of human occupation than can be found in continents, so many islands do not in fact hold useful records of long-term chronological change. In another point of view, these exact same characteristics of islands offer special opportunities for close examination of details that are compressed in time and space, as in the analogy of examining tiny and short-lived insects for learning about the basic operations of biological evolution. These insects, however, are probably not useful towards learning about processes of interactive agents in complicated natural–cultural systems, just like they are probably not useful towards learning about longer scales of chronological change far beyond their life spans.

What are the key variables involved in the evolution of a landscape, and are they discoverable in a meaningful way in the Mariana Islands? In general terms, landscapes evolve in islands according to the same processes of natural and cultural history that occur anywhere else, but the results are more easily visible and detectable in the model systems of islands. The only remaining concern, therefore, is to find an island setting with substantial datasets about how the inhabited landscape has evolved over an appreciably long period of time, accounting for significant chronological change in the natural and cultural environment.

Structure and Content of This Book

This book works towards the goal of learning about long-term evolution of landscapes as unified natural–cultural systems. Although intended for general applicability, the research approach is presented through real-life examples of archaeology and other avenues of research in the Mariana Islands. This book should not be mistaken as the final word of Marianas archaeology, but rather it facilitates new discussions about this region's archaeology and more generally about landscapes in any part of the world.

This study is based on situating hard data of archaeological materials within dated time intervals and larger contextual landscapes of natural environments and social settings. The book draws on several years of research in the Mariana Islands, most intensively since 2005. The general strategy involves situating sites within their ancient terrain and habitat setting during a series of time periods, together composing a full chronological sequence. Interdisciplinary investigations are integrated for a comprehensive view of the past landscape in terms of inter-acting elements of terrain structure, climate, plant and animal communities, and human activities within specific ecological zones as these variables changed through time.

Towards the goal of examining unified natural–cultural landscape evolution, the book is organised in three parts. Part One (Chaps. 1–6) introduces key concepts of general use and illustrates how to coordinate the multitude of changing landscape elements in a cohesive sequence. Part Two (Chaps. 7–14) presents a chronological narrative of the Marianas example, with attention to the dynamic conditions of landforms, resource zones, material culture, and larger Asia-Pacific regional context in each identifiable time period. Part Three (Chaps. 15–17) applies the datasets in a synthesis of how the Marianas landscape has evolved, building a larger understanding of landscape evolution in general.

References

Briggs, J. M., Spielmann, K. A., Schaafsma, H., Kintigh, K. W., Kruse, M., Morehouse, K., & Schollmeyer, K. (2006). Why ecology needs archaeologists and archaeology needs ecologists. *Frontiers in Ecology and the Environment, 4*, 180–188.

Butzer, K. W. (1982). *Archaeology as human ecology*. Cambridge: Cambridge University Press.

Carson, M. T. (2012). Evolution of an Austronesian landscape: The Ritidian site in Guam. *Journal of Austronesian Studies, 3*, 55–86.

Carson, M. T. (2014a). Paleoterrain research: Finding the first settlement sites of Remote Oceania. *Geoarchaeology, 29*, 268–275.

Carson, M. T. (2014b). *First settlement of Remote Oceania: Earliest sites in the Mariana Islands*. New York: Springer.

Carson, M. T., & Kurashina, H. (2012). Re-envisioning long-distance Oceanic migration: Early dates in the Mariana Islands. *World Archaeology, 44*, 409–435.

Crumley, C. (Ed.). (1994). *Historical ecology: Cultural knowledge and changing landscapes* (School of American research advanced seminar series). Santa Fe: School of American Research Press.

Crumley, C. (Ed.). (2001). *New directions in anthropology and environment: Intersections*. Walnut Creek, CA: Altamira Press.

Cumming, G. S. (2011). Spatial resilience: Integrating landscape ecology, resilience, and sustainability. *Landscape Ecology, 26*, 899–909.

David, B., & Thomas, J. (Eds.). (2008). *Handbook of landscape archaeology*. Walnut Creek, CA: Left Coast Press.

Dearing, J. A., Battarbee, R. W., Dikau, R., Larocque, I., & Oldfield, F. (2006a). Human-environment interactions: Learning from the past. *Regional Environmental Change, 6*, 1–16.

Dearing, J. A., Battarbee, R. W., Dikau, R., Larocque, I., & Oldfield, F. (2006b). Human-environment interactions: Toward synthesis and simulation. *Regional Environmental Change, 6*, 115–123.

Hornborg, A., McNeill, J. R., & Martinez-Alier, J. (Eds.). (2007). *Rethinking environmental history: World-system history and global environmental change*. Lanham, MD: Altamira Press.

Ingold, T. (2000). *The perception of the environment: Essays on livelihood, dwelling and skill*. London: Routledge.

Kirch, P. V. (2004). Oceanic islands: Microcosms of global change. In C. L. Redman, S. R. James, P. R. Fish, & J. Daniel Rogers (Eds.), *The archaeology of global change: The impact of humans on their environment* (pp. 13–27). Washington, DC: Smithsonian Books.

Olding-Smee, F. J., Laland, K. N., & Feldman, M. W. (2003). *Niche construction* (Monographs in Population Biology 37). Princeton, NJ: Princeton University Press.

Rindos, D. (1984). *The origins of agriculture: An evolutionary perspective*. New York: Academic.

Thomas, J. (2012). Archaeologies of place and landscape. In I. Hodder (Ed.), *Archaeological theory today* (2nd ed., pp. 167–187). Cambridge: Polity Press.

Turner, M. G., Gardner, R. H., & O'Neill, R. V. (2001). *Landscape ecology in theory and practice: Pattern and process*. New York: Springer.

van der Leeuw, S., & Redman, C. L. (2002). Placing archaeology at the center of socio-natural studies. *American Antiquity, 67*, 597–605.

Vitousek, P. M., Chadwick, O. A., Hartshorn, A. S., & Hotchkiss, S. C. (2010). Intensive agriculture in Hawai'i: The model system approach. In P. V. Kirch (Ed.), *Roots of conflict: Soils, agriculture, and sociopolitical complexity in ancient Hawai'i* (pp. 31–44). Santa Fe: School for Advanced Research Press.

Wilson, L. (2011). The role of geoarchaeology in extending our perspective. In L. Wilson (Ed.), *Human interactions with the geosphere: The geoarchaeological perspective* (pp. 1–9). London: Geological Society of London. Special Publication Number 352.

Chapter 2
Global Applicability of Landscape Evolution

Beyond the Asia-Pacific context of this book with focus on the Mariana Islands, studies of landscape evolution can be applied in any region and time period, as well as accommodating a diverse range of information about what landscapes are and how they have changed through time. The issues of studying an ethnohistorically defined landscape as it is experienced today are substantially different from the issues of studying the cultural interface with the environment during the Last Glacial Maximum about 24,000–18,000 B.C. Studies of the last 1000 years of dynamic coastlines work with many of the same principles yet at different scales of the last 10,000 years or 100,000 years of changing sea level and effects on coastal ecologies. Landscapes certainly are configured differently, experienced individually, and have undergone unique transformations in coastal zones, inland mountains, river valleys, and other settings.

Each region presents its own concerns about landscape evolution, so brief accounts of coastal China, California, and the Hawaiian Islands here serve as a prelude to illustrate the research approach in longer and shorter time scales and with qualitatively different datasets. Research in coastal China can address transitions from hunter-gatherer to sedentary agriculturalist modes of landscape experience, over a long time scale including *Homo erectus* ancestors, drowning of the former Pleistocene coastlines after the Last Glacial Maximum, and origins of some of the world's oldest complex societies. Landscape studies in California similarly must contend with conditions of remarkably different coastlines and associated ecological zones during the Pleistocene more than 10,000 years ago, in this case during the ancient migrations of people into the North American Continent, followed by a series of changing conditions during the Holocene leading to the historically known patterns of hunter-gatherer land use. The Hawaiian Islands offer an opportunity to examine a richly textured ethnohistoric perspective of a unique landscape system in comparison to geoarchaeological and other material records of how this system has evolved over the last 1000 years.

M.T. Carson, *Archaeological Landscape Evolution*,
DOI 10.1007/978-3-319-31400-6_2

Coastal China

Archaeology in China has proceeded at a rapid pace over the last two decades; yet, a landscape approach has not yet been formally applied except in a few cases. Detailed culture history sequences have been formulated from thousands of site excavations. The amount of information is so vast that scholars may feel lost in a sea of names of archaeological cultures, periods, phases, and areas spanning hundreds of thousands of years and more than 9 million km^2 of diverse terrain. The time range incorporates several thousands of years of *Homo sapiens* presence and much longer if considering evidence of *Homo erectus*, while the geographic scope includes staggering diversity of wet tropical lowlands, river valleys, grassland steppes, desert plateau, and dry and freezing high-altitude mountains.

Focusing just on the coastal region of China, today's coastline extends over more than 14,000 km of linear distance (Fig. 2.1). Coastal China includes zones of humid tropical, subtropical, and temperate climates with variable landforms of sandy beaches, alluvial coastal plains, deeply incised river drainages, broad river deltas, rocky bluffs, colluvial slopes, and steep mountainsides. Additionally, more than 6000 offshore islands contribute to the coastal landscape. All of these settings contain evidence of human occupation at least over the last few thousands of years and often much longer, crossing multiple periods of changing natural-cultural landscape contexts.

The physical shoreline and associated ecological zones have undergone substantial transformations during periods of higher and lower sea level. These alterations were driven primarily by change in global climate and the amount of ocean water trapped in ice sheets, but additional contributing factors included the effects of slope erosion and re-deposition into coastal lowlands and the changing river courses and amounts of alluvial sediment over time. Other considerations involve the roles of human groups, whether intentionally or not, in altering sedimentation rates through forest-clearing, crop cultivation, and other activities. Furthermore as shown in archaeological records, people have caused impacts on the compositions of plant and animal communities and even on geological mineral deposits differentially in some cases more than others, effectively reconfiguring the ecological balance of their inhabited landscapes.

Although ever-changing in a long-term perspective, China's coastal zones consistently have supported some of the world's densest populations. Reasons for the long-term reliability and intensity of coastal habitation are not difficult to imagine and in fact much the same as for any coastal zone, beginning with essential access to fresh water in streams, rivers, and seeps. Coastal communities benefit from the reliability of shellfish and seaweeds in intertidal and shallow sub-tidal zones, fisheries positioned safely near the shore, and naturally healthy plant growth close to the water table and especially near stream and river drainages. These same settings offer good prospects for hunting or trapping animals attracted to the water sources and vegetation. Moreover, the ocean water could facilitate transport of people and supplies. Later developments of sea-crossing vessels expanded the abilities of deep-sea fishing, broadened trade and commerce, and magnified the potential for long-distance migrations.

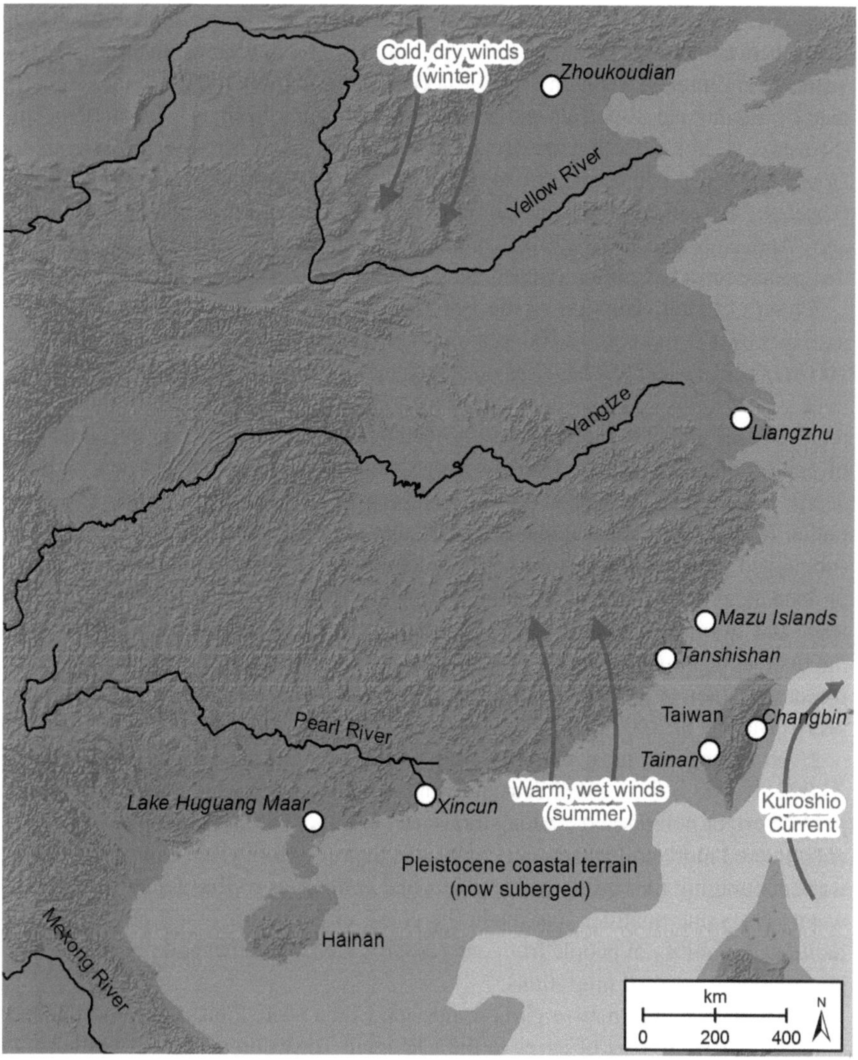

Fig. 2.1 Coastal China, showing major landscape features and sites mentioned in the text

Coasts arguably have been essential for the evolution of the human species and global dispersals of populations, and coastal China evidently played a key role in these events. This region is among the longest-inhabited places on earth, supporting populations of ancient human ancestors such as *Homo erectus*, famously found at Zhoukoudian near Beijing and dated in the range of 700,000–400,000 years old (Shen et al. 2009), while additional fossils of *Homo erectus* have been found on the seafloor of the Taiwan Strait that would have comprised a lowland terrain at the time (Chang et al. 2015). Remains of anatomically moderns humans or *Homo sapiens* have been found beneath and pre-dating a geological layer that could be as much as 100,000 years old at Zhiren in southern China (Liu et al. 2012), closely following the estimated date of first human dispersals and genetic diversification from an African homeland (Soares et al. 2012). Even when accepting a younger age of perhaps 60,000 years for *Homo sapiens* in China, the arrival of our species in Australia by 40,000 years ago (Hiscock 2008) and eventual migrations into the American continents by 14,000 years ago (Anderson and Bissett 2015) cannot be explained without first acknowledging that people must have been living already in other regions such as in coastal China.

China's coastal ecologies of the last glacial period of the Pleistocene, approximately 110,000 through 12,000 years ago, today are drowned beneath more than 100 m of ocean water that has risen due to the melting of polar ice sheets. The approximate shapes of ancient coastlines can be mapped by tracing the depths of the seafloor relative to the global sea-level history, refined through location-specific accountability of geotectonic movements, river discharge patterns, and geological cores showing depths of sedimentary deposits of different ages. Other landscape indictors such as pollen records and faunal remains so far have not been obtained directly from the submerged Pleistocene landforms, but general information is available from today's on-land areas where deep sedimentary coring is possible, especially in the bottoms of lakes and swamps, such as at Lake Huguang Maar where sedimentary records reveal the fluctuating climate conditions of the last 16,000 years (Yancheva et al. 2007) and preserved pollen of the vegetation communities that have changed over the last 13,000 years (Wang et al. 2007).

Even during the most extreme cold period of the Last Glacial Maximum at 24,000–18,000 B.C., most of coastal China was an ice-free zone without ice sheets, permafrost, or polar deserts. Along thousands of km of coastline, people would have encountered more arid and cooler conditions than seen today, but overall the setting was encouraging for finding edible plants and animals as well as reliable sources of water. The same qualities that made the coast attractive for living in general also facilitated mobility of people from one resource area to another and potentially supporting cross-regional migrations.

A particularly informative glimpse into the China's late Pleistocene coastal landscape comes from a set of caves in the Changbin Township on the eastern coast of Taiwan, occupied as early as 25,000 B.C. (Tsang et al. 2009, 2011), at a time when the island of Taiwan was connected to mainland China and the east-facing coastline was positioned just outside these caves. The mountain range of eastern Taiwan happens to have risen dramatically among the world's most rapidly uplifting geological formations, measured at nearly 1 cm per year at its southern end and 5–7 mm per year in the northern end associated with the caves in the Changbin Township (Liew et al. 1993). The ancient living floors of those caves now at 150–170 m elevation would have been scarcely above sea level at 25,000 B.C. (Figs. 2.2 and 2.3). Outside the eastward-facing coastal caves, a steep rocky slope descended to the tides and a broad expanse of coastal resources.

Named after the Changbin Township, archaeological materials in these caves locally are described as the Changbinian Culture (Sung 1969), regarded as a generic term for hunter-gatherer groups who lived in Taiwan for several thousands of years

Fig. 2.2 Deep excavation at one of the Changbinian Cave Sites in eastern Taiwan, September 2014, with Dr. Tsang Cheng-hwa (wearing *dark glasses*) and Dr. Peter Bellwood (holding camera)

prior to the horizon of pottery-bearing sites about 4000 B.C. Tsang et al. (2009, 2011) documented a stone-tool industry of chipped pebbles and cobbles, as well as several small stone flakes. The larger chipped stone stools likely were used for general-purpose tasks, perhaps for chopping through wood, meat, and bone. The small flakes likely were used for finer cutting and slicing tasks. Extremely few animal bone fragments and shells confirm the expectation of coastal foraging, but the poor preservation of material disallows more precise statements.

The coastal zone must have been the major attraction of reliable natural resources for the people engaged in a hunter-gatherer economy at 25,000 B.C., but the limited faunal and botanical records do not yet provide a clear definition of those ancient resources. The ocean water itself may have been somewhat warm due to the passage of the Kuroshio Current bringing generally warmer tropical water from the western-central Pacific, but it may not have been warm enough to support growth of coral reefs prior to the post-glacial conditions of the Holocene. So far, the oldest dated corals embedded in uplifted terrain of eastern Taiwan have produced results of about 4000 B.C. (Chen et al. 1991; Liew et al. 1993). The Pleistocene coastal ecology still needs to be addressed, for example through studying the kinds of shellfish that are preserved even in small numbers in the cave deposits and embedded in the uplifted land mass.

After 10,000–9000 B.C., coastal China underwent massive transformation due to the melting of polar ice sheets and glaciers, when low-lying terrain throughout the world became flooded under more than 100 m of rising oceans. People necessarily adjusted to these changing circumstances, along with new opportunities of plant

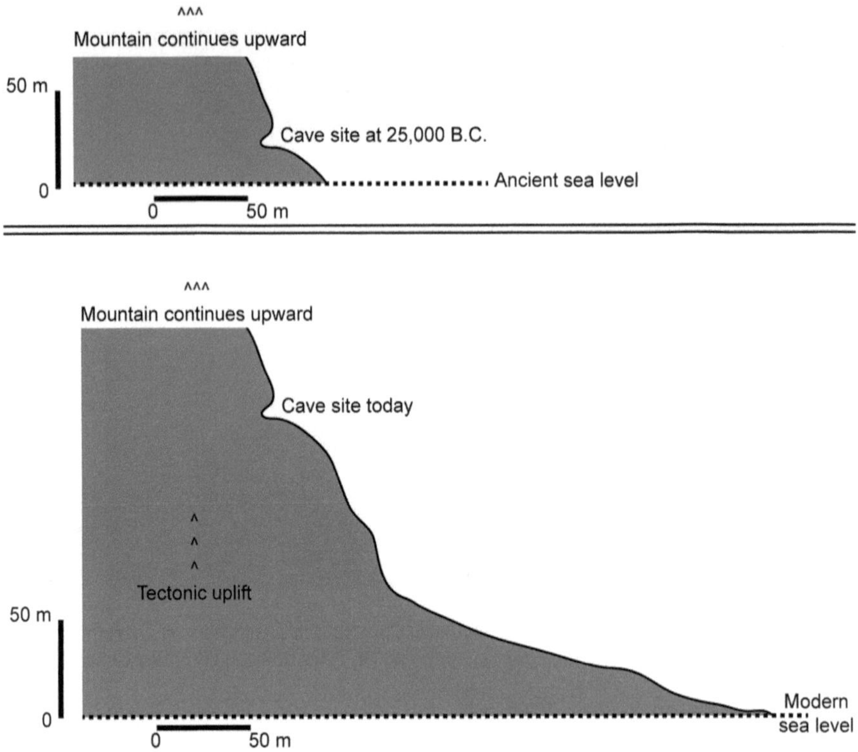

Fig. 2.3 Schematic section view of Chanbginian cave site occupation, view to north, based on information from Tsang et al. (2009; 2011)

growth and different ecological niches of the warmer Holocene conditions. The East Asian monsoon weather patterns have been steady throughout the Holocene: summer winds from the south bring warm moist air and rain, and then winter winds from the north bring cold and dry air. These predictable conditions definitely have influenced the human interface with the landscape, not only in practical terms of growing rice and other crops but also in terms of organising annual activities suitable for each season and developing a long-term relationship with the natural world. Although people were manipulating rice, millet, and other plants in China prior to 10,000 B.C. (Zhang and Hung 2008, 2010), the key turning point in developing a formally domesticated crop-dependent agricultural system began after 7000 B.C. (Fuller et al. 2009; Liu et al. 2007; Zhang and Hung 2013) and very clearly within a context of the overall stability and predictability of the Holocene.

The earliest sedentary landscape systems in China had developed most strongly not in the coastal region but rather farther inland along river valleys and terraces, in close association with the land-use patterns and annual predictability of rice and millet agriculture. By 6000 B.C., sedentary farming economies were spreading along the major river valleys such as the Yangtze and Yellow River that provided the most suitable landforms and reliable water sources. Similar kinds of landforms did not exist at that time in coastal zones, where instead the available terrain for horizontally expansive agriculture was far less than was already being used in the inland river valley terraces. Archaeological evidence reveals a chronology of gradual expansion of the geographic range of sedentary agriculturalists over the course of some millennia. By 5000–4000 B.C., farming communities were established at the coasts associated with some but not all of the major river valleys, while many coastal zones continued to support groups engaged in mixed hunting, gathering, and fishing economies (Liu and Chen 2012:169–212). Until about 3000 B.C., hunter-gatherer groups continued to inhabit most of the southeast coast (Zhang and Hung 2012).

From 6000 to 3000 B.C., coastal communities in southeast China developed a very different kind of sedentary landscape than had emerged among their inland neighbours of land-dependent rice-farming villagers. These coastal groups lived in small settlements and relied primarily on mixed foraging and fishing over broad catchment areas. At the Xincun Site near the inland side of the Pearl River Delta, occupied at 3350 through 2470 B.C., preserved palaeobotanical remains have yielded no evidence of domesticated rice or millet, but rather the plant foods are represented in starches and phytoliths of sago palms, banana, Job's tear, acorns, and other taxa (Yang et al. 2013) indicative of low labour-input subsistence economies of foraging, managed forests, and perhaps limited horticulture. Coastal settlements at this time mostly covered less than 10,000 m^2, although a few exceeded 20,000 m^2, in comparison to their contemporary rice-farming villages along the middle Yangtze River averaging 20,000–30,000 m^2 (Zhang and Hung 2008).

By 6000 B.C., the same warming conditions that encouraged rice-farming along the river valleys had resulted in global rise of sea level reaching approximately its modern level, directly affecting China's coastal landscapes. The lifestyle of coastal people at that time evidently involved sea crossings to several small offshore islands (Fig. 2.4), such as in the Mazu Islands where people created mounds of shells and other debris from mixed foraging and fishing (Chen 2013). People were buried in these mounds in Liang Island or Liangdao (*dao* = "island") as early as 6000 B.C. according to direct radiocarbon dating of the human bones (Ko et al. 2014).

Coastlines in China and elsewhere have been affected by a number of minor sea-level fluctuations during the last few thousand years. These later fluctuations have been generally 2 m or less, indeed minor compared to the post-glacial sea-level rise of more than 100 m but nonetheless bringing significant effects to low-lying coastal zones. The net effects on coastal landforms and ecologies are even greater when accounting for increased rates of slope erosion within the last few thousand years.

A highstand of sea level occurred approximately at 3000 through 1000 B.C., about 1.5–2.5 m higher than the present level and in most cases prior to the accumulations of sediments composing the lowland plains and terraces that now characterise much of coastal China (Zong 2004; Zong et al. 2009). Large expanses of alluvial plains and flat tablelands simply did not exist in low-elevation areas near sea level, so that the coastal settings did not include the same opportunities for agricultural land-use patterns that have been possible only more recently. As rice and millet farming gained more popularity in coastal areas after 3000 B.C., the rate of slope erosion increased and began to

Fig. 2.4 View of ancient shell midden site dated at 6000 B.C., on ridge of Liangdao, Mazu Islands

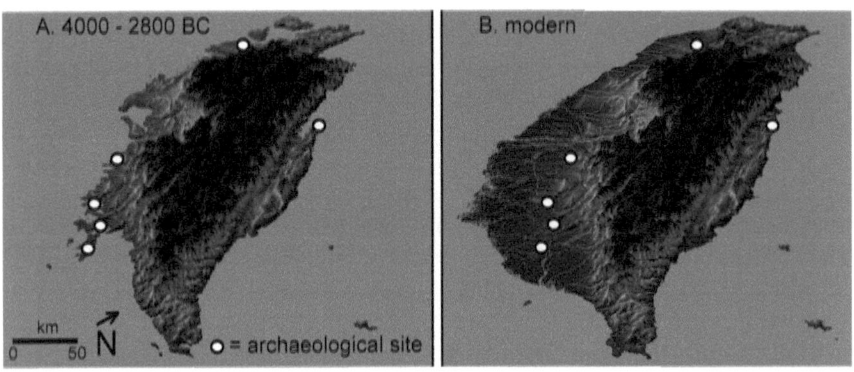

Fig. 2.5 Early Neolithic (4000–2800 B.C.) versus modern landscape of Taiwan. Palaeoterrain model is modified from Carson and Hung (2014)

generate thicker alluvial and colluvial sediments in lowlands of coastal zones and riverside terraces. Due to a higher sea level at that time, however, the lowland sedimentary accumulations in some cases had settled beneath water levels, only later exposed as dry landform surfaces after a period of sea-level drawdown following 1000 B.C.

The western coast of Taiwan offers one example of the landscapes changing with sea level, alluvial deposition, and placement of human settlements that occurred not only there but also along several coastal areas of mainland China (Fig. 2.5). Prior to 3000 B.C., residential sites were situated on low hills and ridges overlooking coastal waters (Hung and Carson 2014). After alluvial sediments began to accumulate, likely due to increased slope erosion prompted by forest-clearing, residential sites began to develop in the newly forming coastal plain, but these sites of 2800 B.C. such as at the Tainan Science Park now are buried under 5–7 m of more recent alluvial sediments and stranded more than 20 km inland from today's coastline of western Taiwan (Tsang 2005). Later, alluvial deposits continued to accumulate over the abandoned site, and a drawdown of sea level after 1000 B.C. magnified the extent of the alluvial plain exposed above sea level (Chen et al. 2004). These combined factors created a substantially different coastal landscape today than had been inhabited at 2800 B.C. and even more dramatically different from the landscape prior to 3000 B.C.

On a larger scale than can be seen in Taiwan, the mouths of major rivers in China present broad deltas and adjacent marshlands, at 3000 B.C. covering much more extensive zones than experienced today. The Yangtze Delta was a centre of one of China's classic early complex societies known as the Liangzhu Culture at 3300–2200 B.C. (Liu and Chen 2012:240), renowned for the production of ornately carved jade artworks, beautifully polished discs, finely made pottery wares, cemeteries with elaborate grave offerings and associated religious rites, and extensive stonework structural remains of formalised village complexes (Qin 2013). The Liangzhu Culture emerged at a time when people populated a low-elevation coastal strip between the river delta and an extensive marshland (Stanley et al. 1999), and much of the habitation complex involved tall stonework structures with living surfaces elevated above the threats of floods (Fig. 2.6). Most impressively, the structures within the walled city of Maojiaoshan covered 290 ha, larger than the Forbidden City in Beijing, regarded as the capital city of the Liangzhu Kingdom.

By 2200 B.C., the Liangzhu residential sites were abandoned apparently all at once, suggestive of a major catastrophe. Among the hypothesised reasons, a failure of the rice-farming complex must be considered, likely related to the brief but sharp aridity that caused collapse of several farming economies all across Asia, known as the "4200 ybp Event" (Yasuda 2008). Following a severe drought and lowered water table, the dry soils along riverbanks and hill slopes became more vulnerable to erosion, so the return of heavy rains brought massive lowland sedimentation in areas such as the Yangtze Delta. Given the drought followed by sediment-filled flooding and landslides, the once thriving Liangzhu Culture was no longer sustainable (Xu et al. 2011; Zhang et al. 2004, 2005). Following later episodes of lowland sedimentation and coastal progradation, today's large expanse of contiguous terrace land around the Yangtze Delta supports an entirely different land-use pattern near modern-day Shanghai.

Approximately at the same time of the Liangzhu Culture at the Yangtze Delta, sedentary landscape systems developed all along coastal China, for example as seen in the Tanshishan Culture in the Fuzhou Basin of the Fujian Province. At approximately 3000–2300 B.C., communities of the Tanshishan Culture lived on low hills and promontories along an estuary that reached nearly 80 km farther inland from today's shoreline (Rolett et al. 2011). The inhabited landscape of that time accommodated small groups scattered on the available landforms, in a much more watery world than later would be the case after 1000 B.C. when a drawdown of sea level and increase of sedimentation created broad coastal plains and extensive river terraces.

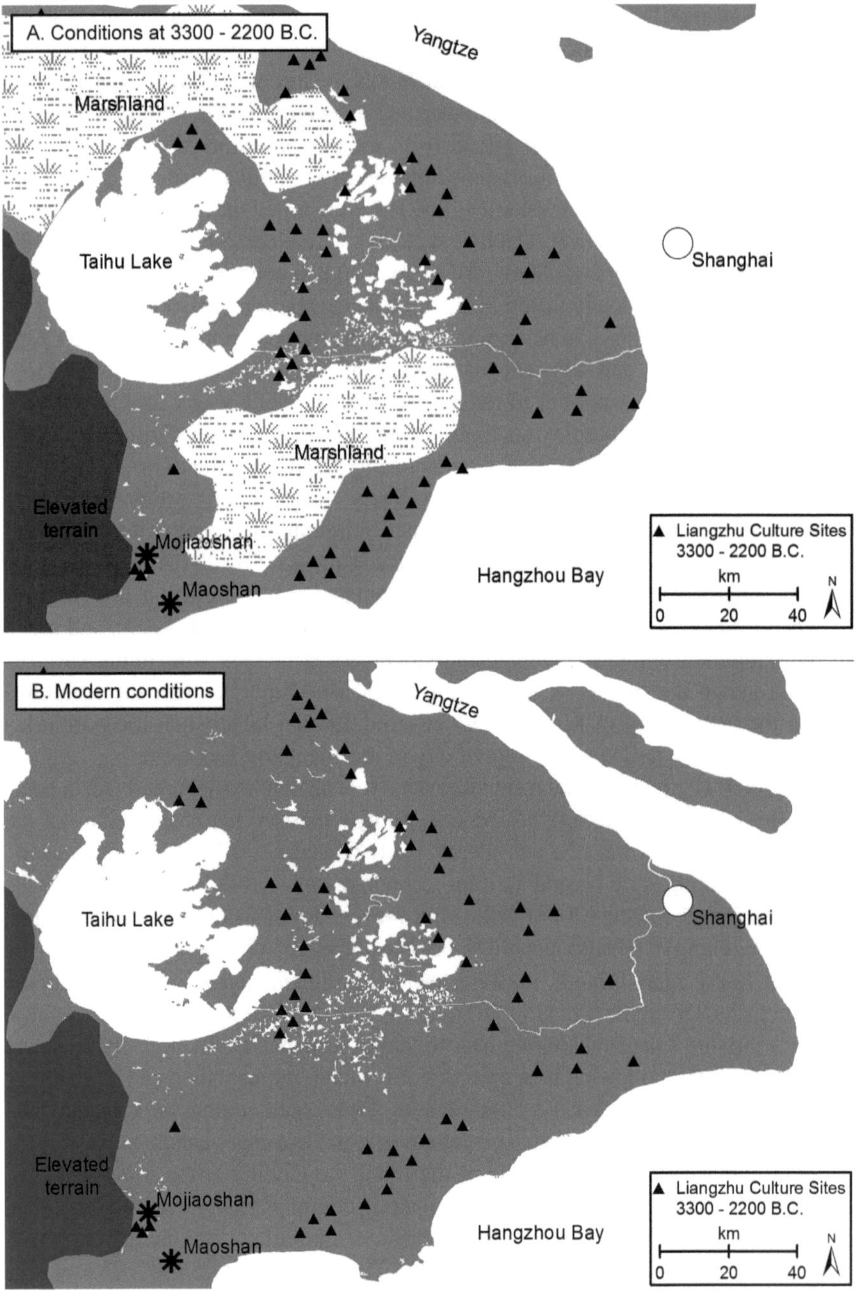

Fig. 2.6 Landscape of the Liangzhu Culture Period, 3300–2200 B.C., as compared to the modern setting. Information follows Stanley et al. (1999), Qin (2013), Xu et al. (2011), and Zhang et al. (2004, 2005)

By the time of the Han Dynasty expansion and imperial regime at 206 B.C.–A.D. 220, China's coastal landforms and ecologies were very much similar to today's conditions. Dense population centres relied for successive generations on widespread rice-farming, but coastal life of course continued to involve use of the sea and related resources. Written records allow vastly more detailed understanding of the natural-cultural landscape, although the Han writers clearly were biased when writing about the conquest of indigenous populations. In southeast coastal China, Han records describe the people of the Yue State or Bai Yue ("Hundred Yue") as barbarians with tattoos who lived in bamboo groves without proper villages (Brindley 2015). The Yue were rebellious from time to time, and eventually they were forced to retreat into marginal inland hills outside their preferred coastal zones and outside the primary concern of Han and later Empires.

This greatly condensed review leaves no doubt about the dynamism of China's coastal landscapes, constantly undergoing change in multiple factors at different but concurrent paces. Archaeological records reveal how groups of people engaged in certain modes of life and traditions of how to interact with their landscapes for periods of time, but eventually the inhabited landscape transformed to such an extent that qualitatively different lifestyles were needed. Archaeologists can refer to the landscape systems of Changbinian hunter-gatherers using caves along the eastern coast of Taiwan at 25,000 B.C., coastal-marine foragers ranging through a watery world of estuaries and offshore islands at 6000–3000 B.C., a thriving complex society emerging on the edge of the Yangzte Delta at Liangzhu about 3300 B.C. yet unsustainable by 2200 B.C., and historical developments of intensive agricultural land-use patterns since the Han Dynasty of 206 B.C.–A.D. 220. Many other examples could be added in a more thorough review not attempted here, and of course a number of minor fluctuations could be discussed within each identifiable time period that did not always cause deep restructuring episodes as highlighted for the landscape system overall.

California

Landscape evolution in California (Fig. 2.7) involves issues of first human migration into the American Continents, long-term development of hunter-gatherer economies in varied ecological niches and patchworks, emergence of complex social structures among semi-sedentary groups, and contributions of history and ethnography in building a sense of place in a landscape. As is the case for the Americas generally, the beginning of human presence is unclear but at least as early as 10,000 B.C. during the last centuries of the Pleistocene, so the origins of an inhabited landscape are not yet understood along the coasts that now are submerged beneath the Pacific Ocean. In today's more accessible terrain, post-glacial Holocene sites have yielded information about how established populations adjusted to their changing environmental settings over the last several thousands of years. People evidently sustained hunter-gatherer economies and lifestyles, despite the demands of dense residential communities and complicated social, economic, and political systems as documented at the time of European written histories in the A.D. 1500s.

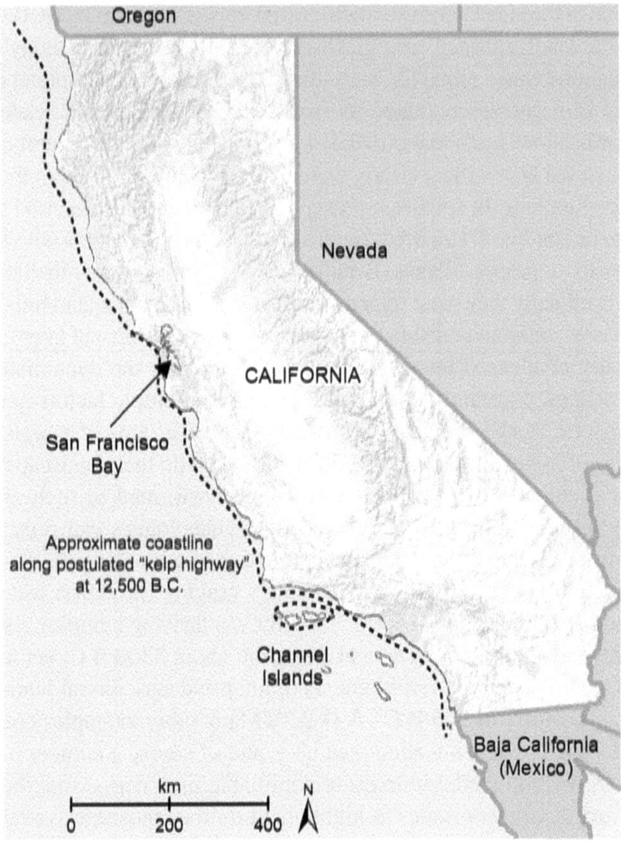

Fig. 2.7 California landscape today, with approximation of Pleistocene shoreline

Historical and modern traditions convey a strong sense of place in the landscape as it is experienced today, with elements of both long-term continuity and disjuncture when considering more than 12,000 years of an inhabited landscape.

California's landscape today is very much a product of historical events. Most Californians are aware that their modern cities developed in the places around the Spanish missions and along the roads that connected them, but these developments in turn had been based largely on wherever the Spanish missionaries could find suitable land for their settlements and observed existing aggregations of indigenous tribes to recruit into the mission projects. As a result of these procedures, whether intentionally or not, California's historical population centres reflected indigenous views of the most productive and favourable landscapes, although these reflections became increasingly warped over time.

Early European explorers approached California from the south and along the coast, gradually gaining knowledge of the navigable waters, shapes of terrain, types of resources, and distributions of local populations. These oldest written accounts capture a sense of how people explore and learn about a landscape, in this case however with the understanding that the landscape in question already had been inhabited and intimately known by local residents for several millennia. More than 12,000 years previously, the first Californians must have approached the landscape from an opposite direction of the Spanish missionary experience, moving from the Bering Strait in the north and continuing southward along a Pleistocene coastline that no longer is visible today beneath the Pacific Ocean.

The archival records are invaluable for learning about the California landscape at a critical historical juncture, but unfortunately the locations and details often are vague and mysterious. Maps and charts were precious government secrets during the 1500s and 1600s, sometimes destroyed before falling into enemy hands or else containing deliberately misleading information, unique codes, or simply left frustratingly sparse so as to avoid a written record in the first place. The limits of documentary evidence have not stopped bold claims and occasional hoaxes of discovering the places witnessed by Hernán Cortés in the 1530s, Francisco de Ulloa in 1539, Juan Rodríguez Cabrillo in 1542, Francis Drake in 1579, Pedro de Unamuno in 1587, Sebastian Cermeño in 1595, and Sebastien Vizcaíno in 1602. The missions were established over the course of more than 100 years, beginning in Baja California with Misión de Nuestra Señora de Loreta Conchó in 1697 and incrementally expanding as far northward as the north end of San Francisco Bay with Misión San Francisco Solano in 1823. While the missionary expansion still was underway, the Adams-Onis Treaty of 1819 limited Spanish territorial claims at the 42nd Parallel, still in effect today as the boundary between the U.S. States of Oregon and California.

When pealing back the veneer of post-colonial European land-use and indigenous population decline, the native Californian landscape is revealed as among the most densely inhabited regions of North America. Hunter-gatherers lived in sedentary or semi-sedentary communities, collected and stored massive quantities of acorns and other foods, practiced controlled burning and other manipulations of their environment, used standardised shell beads as a form of monetary currency, and maintained complex social and political systems (Arnold 1992, 2001, 2012). These groups occupied every one of the diverse inhabitable regions of California, with equally diverse ethnolinguistic groups accounting for "approximately 20 % of all the languages articulated in North America" (Lightfoot and Parrish 2009:7).

The diversity of the California landscape has been instrumental in supporting the long-term residence of dense populations. Rather than relying too much on a narrow range of foods, people could shift their focus from one resource to another in the event of prolonged drought or other circumstances that occur regularly in California. In addition to the repeated droughts, periodic el Niño events create warm ocean waters with cascading effects in the marine food chain. In this context, both land-based and sea-based subsistence economies developed with the ability and perhaps even a cultural expectation of shifting and reconfiguring according to ever-changing conditions. As long as people were ready to adjust their routines periodically, then ample resources could be available.

Controlled burning evidently comprised another key component of maintaining diversity in the California landscape. Through managing the timing and spatial parameters of burning events, the mosaic of habitats included patches of vegetation of variable ages and compositions as they recovered differentially from the fires. Through weed-pulling and other activities, people could continue to manipulate the development of the fired landscape. Furthermore, fresh young grass may have attracted antelopes and other animals, while the natural habitats of birds, rodents, and other wildlife were variably disrupted or enhanced.

Pyrodiversity through managed burning must be recognised differently from slash-and-burn horticulture. Superficially, both approaches are based on the same principles of allowing the burned organic material to release nutrients into the soil and thus increase plant-growing productivity. The most obvious difference is that horticulturalists would then proceed to plant domesticated crops, whereas the native Californian tribes evidently did not cultivate the domesticated versions of maize, beans, and squash as known in other regions. Moreover, people did not engage in purposeful narrowing of the plant species diversity, instead promoting as much diversity as possible.

The origins and long-term functioning of California's pyrodiversity continue be investigated, but evidence such as charcoal flecking in soil profiles so far suggests managed burning for at least the last 1000 years (Cuthrell et al. 2012). This time range allows fair correlation with ethnohistorically attested Californian tribes and their landscape traditions, although a much older chronological sequence may be expected in this region. The practices of pyrodiversity most likely developed in the context of taking advantage of California's natural biodiversity and adjusting the focus of hunting-gathering regimes in times of droughts and other challenges. Now after some centuries of strict laws against burning, modern California's annual wildfires can be disastrous, and a landscape of fully functional pyrodiversity is difficult to imagine except in the most general terms.

Among the most critical obstacles against researching the long time scale of California's landscape evolution is the fact that the oldest human presence in the region is not yet clarified. Without knowing the context of the first human–environment relations, the available chronological sequence is incomplete and refers only to periods when people already were engaged within a landscape system that necessarily had been inherited and modified over the course of several generations and perhaps for some thousands of years. Based on the limited evidence available, people had migrated into the Americas through the Bering Strait region prior to 12,000 B.C. and perhaps much earlier, then proceeded to occupy ecological niches in a Late Pleistocene landscape that later would transform profoundly with post-glacial warming of the Holocene, increasing resident population levels, and adaptations to the changing environment (Madsen 2015). Despite lingering debates, most researchers accept this general chronological outline, yet the fact remains that the earliest periods of human–environment interactions and dynamics are missing from the Californian archaeological record and indeed missing from most of the Americas.

Throughout the American Continents, the first few millennia of human settlement are notoriously ambiguous in archaeological records, definitely pointing to a human presence prior to 10,000 B.C. but so far lacking a consensus about exactly how much earlier and under what circumstances people migrated through ice-free zones. Fluted stone points of the Clovis cultural tradition are known from more than 1000 sites in North through Central America, dated at least as old as 11,000 B.C. (Miller et al.

2013), but people must have migrated into the Americas earlier in order to create the tent-like structures of a small community campsite at Monte Verde in Chile of South America around 12,500 B.C. (Dillehay et al. 2008). The Monte Verde site now is situated nearly 60 km inland from the Pacific Ocean shoreline of Chile, but it would have been about 90 km inland when the site was occupied prior to the post-glacial flooding of coastal lowland terrain (Dickinson 2011). In those now-flooded coastal lowlands of the Late Pleistocene landscape, ancient archaeological sites of the first Americans likely existed but have not yet been discovered in today's submerged contexts.

The missing millennia of the first Americans very well could be due to the submergence of Pleistocene coastal sites beneath today's oceans on both the Pacific and Atlantic coasts. In particular, the western coast of North America would have been the most accessible migration route for people coming from the Bering Strait and Alaska, although archaeologists so far have not attempted to find submerged sites there. Even without tangible evidence, Jon Erlandson and colleagues formulated a compelling argument that people had migrated along ice-free coastal zones, following the productive habitats of kelp forests and related coastal-marine ecologies of a "kelp highway" along the Pacific coast of North America (Erlandson et al. 2007). More specifically, the first Americans following such a migration route would have been coastal foragers, adapted to the Pleistocene shorelines and estuaries of the far north Pacific Rim, extending their ways of life into the American Continent along the Pacific coast (Madsen 2015). As attractive as this argument may be, it cannot be proven or disproven until archaeologists can develop a realistic way to survey in the submerged continental shelves. Potentially, the post-glacial flooding already has displaced or destroyed the material traces of these ancient sites, so an underwater survey could be pointless. On the other hand, fossils of *Homo erectus* have been recovered from today's seafloor of the Taiwan Strait (Chang et al. 2015), so a number of archaeological sites indeed may be awaiting discovery on the drowned continental shelves of North America. Extremely few archaeologists have attempted to search for these kinds of sites (Faught 2004; Faught and Gusick 2011), and so far no such attempt has been reported for coastal California.

Just within the last few years, researchers have begun to consider seriously how to conceptualise of the submerged Late Pleistocene landscapes of the American coasts (Anderson and Bissett 2015; Clark et al. 2014). Prior to 10,000 B.C., the Late Pleistocene coast of California extended farther westward than can be observed today, comprising a few thousands of sq km of land with several internally variable zones. The approximate shape of the Late Pleistocene coastal terrain can be mapped according to the sea-level history plotted against the depths of the now submerged continental shelf, further refined according to localised effects of California's complicated tectonic movements and other factors. Interdisciplinary studies are beginning to reveal the diversity of ancient landforms now preserved in relict features on the seafloor, consisting of a mosaic of coastal plains, variable sandy and rocky shores, hilly lowlands, and river drainage systems (Masters and Aiello 2007). Beyond identifying the physical shapes of coastal palaeolandforms, supplementary datasets from soil profiles, preserved pollen, and assorted climate indicators provide a fuller understanding of the ancient environment (West et al. 2007).

In the absence of knowing about the potentially submerged offshore sites, California's on-land archaeological record begins more than 2000 years after people already had been living in the Americas. One site has been dated about 10,000 B.C. at Daisy Cave in San Miguel of the Channel Islands (Connolly et al. 1995), and at least a few others may be close to this age. Most of the known archaeological record post-dates approximately 8000 B.C. (Rick et al. 2005), when the Pleistocene–Holocene transitions were underway and creating a significantly transformed landscape. These same later-dated records inherently cannot provide information about initial niche-targeting, adaptations to locally available conditions, or human response to earliest periods of environmental change.

Regardless of what happened during California's archaeologically missing millennia prior to 10,000 B.C., later sites reveal considerable details of how people related with their ecosystems and developed complex landscape systems. Especially in coastal sites with good preservation of faunal remains and other materials in datable contexts, long-term sequences provide invaluable information about human–environment dynamics (Braje 2010). These kinds of records tend to emphasise the impacts of people on the environment, such as over-harvesting of certain shellfish communities, depopulation of birds and other animals, and burning of natural vegetation. Although often overlooked in archaeological records, the natural change in climate, sea level, and coastal ecology certainly affected many of the food resources and food webs, in conjunction with the well documented human-caused impacts and variable cycles of learning how to manage the dynamic conditions.

While the Californian archaeological record offers one of the world's most informative examples of long-term human–environment relations, it nonetheless must be recognised as not yet providing sufficient evidence from the earliest periods of the inhabited landscape. The unbalanced record disallows a complete view of the influences of people and the natural environment on one another, potentially contributing to a number of unfortunate misunderstandings. Most scholars acknowledge that processes of landscape evolution already had been enacted since perhaps 14,000–12,000 B.C. and with significant transformations prior to the available evidence beginning after 10,000 B.C. and more abundantly after 8000 B.C., yet more research will be necessary to clarify those "missing millennia" for a fuller comprehension of California's long-term landscape evolution.

Hawaiian Islands

Among the most intensively studied islands in the Pacific, the Hawaiian Archipelago (Fig. 2.8) presents a worthwhile but in many ways cautionary reference before considering the archaeological landscape of the Mariana Islands. Although well established in the academic literature over several decades of scholarship, for example seen in Kirch's (1985) synthesis and a more recent overview by Bayman and Dye (2013), Hawaiian archaeology should not be mistaken as representative of the Pacific Islands. The Hawaiian and Mariana Islands could hardly be more different in their natural and cultural histories. The Marianas case of this book offers a

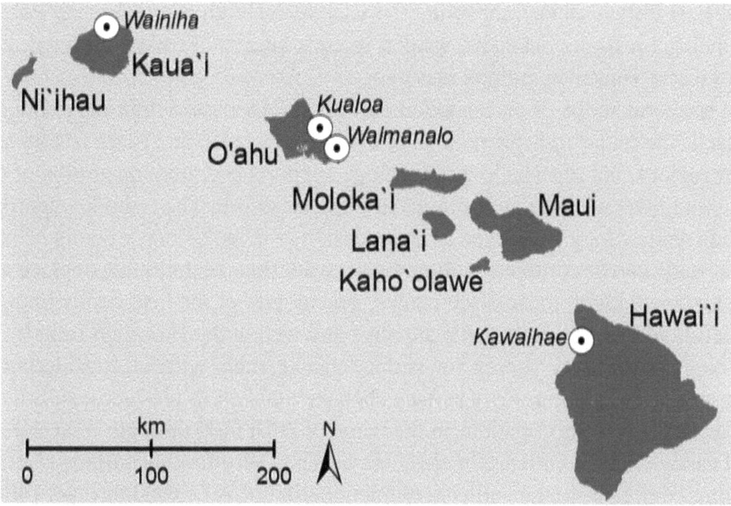

Fig. 2.8 Hawaiian Islands, showing sites mentioned in the text

considerably stronger illustration of long-term landscape evolution than is possible in the Hawaiian Islands, yet the Hawaiian record is instructive in its own right when its limitations are acknowledged.

The Hawaiian environment serves well as model system of natural ecological and social-ecological operations in a functional sense (Vitousek et al. 2010), but Hawaiian archaeology must contend with at least four serious problems. Firstly, the Hawaiian Archipelago comprises the most isolated set of islands in the world, with unique plant and animal species and an accordingly unique culturally inhabited landscape, by its own definition not representative as a model of anything else in the world. Secondly, the brief archaeological chronology since A.D. 1000 disallows accountability for truly long-term change in complex systems. Thirdly, a default research emphasis on the most recent centuries of the ethnohistoric record and surface-visible site ruins has ignored the possibility of chronological change even within the brief frame of 1000 years, while instead the overall picture of Hawaiian archaeology has been synchronic within an unbalanced view of just the few centuries closest to A.D. 1800. Fourth, the contributions of an exquisite Hawaiian ethnohistory are essential for interpreting the later-aged archaeological contexts, but their full potential has been missed when scholars fail to engage in critical review of the sources and multiple layers of interpretation.

Due to the remoteness of these islands, they were among the last in the Pacific to be settled by Polynesian seafaring groups around A.D. 1000 and then again among the last ever known by European explorers with Captain James Cook's arrival at Kaua'i in 1778. By the time of Polynesian settlement about A.D. 1000, nearly every island group of the Pacific already had been occupied, so that the entirety of Hawaiian cultural landscape evolution occurred during a context of an extensively inhabited sea of islands. Additionally, the late settlement date allows firm association of the Hawaiian language, oral traditions, and material culture with a slightly older homeland in

Central East Polynesia encompassing the islands today known as French Polynesia. Of further interest, the extensive written records since 1778 have captured a lively sense of native Hawaiian cultural history.

The temporal scope of an inhabited Hawaiian landscape within only about 1000 years lacks the major transformations of climate, sea level, and other factors as seen in other regions, but this shallow chronology supports potential integration of ethnohistoric traditions with the archaeological material record. The orthodox approach in Hawaiian archaeology has served as an extension of ethnohistory, wherein no archaeological study can be complete without first considering the multitude of place names, stories associated with those place names, genealogies of the historically known ruling chiefs, and other traditions that are alive and well in the Hawaiian Islands. When archaeologists routinely survey for surface-visible ruins without considering older contexts unrelated to the modern surface, then inevitably they produce results that date to the surface-related occupations in the range of A.D. 1400 through 1800 and can be related easily with a familiar ethnohistoric context. Very few subsurface layers have been found that refer to the earlier settlement period of A.D. 1000 through 1400, and oddly enough no investigation yet has attempted expressly to find the oldest sites, leading to repeated examples of later-aged archaeological sites undifferentiated from ethnohistoric contexts.

The shallow time depth of Hawaiian archaeology extends through just a few minor fluctuations in natural environmental transformations, thus creating a false notion that the natural environment did not influence cultural behaviours. Without a record of substantially changing environmental conditions, the major transformations in the Hawaiian landscape logically are attributed to human agency. Nobody would doubt the skills of people in mastering their environment, and indeed the Hawaiian archaeological record reveals profound human-caused impacts on native forests, bird populations, and slope erosion–deposition patterns (Athens et al. 2002). The role of environmental influence on people, however, has not been considered seriously in Hawaiian archaeology except indirectly in terms of noting how people made the best use of their available ecological zones and manipulated them into economically and politically profitable landscapes. These issues tend to be expressed in terms of synchronic ecological functioning as known in the 1700s through 1900s, not accounting for chronological concerns of adapting to ecological niches and changing conditions over extended periods of time.

Perhaps the most significant environmental change within the last 1000 years involved the overall stable, wet, and warm conditions of the Little Climatic Optimum (LCO) at A.D. 1000 through 1300, followed by the unstable, cool, and punctuated stormy conditions of the Little Ice Age (LIA) at A.D. 1300 through 1850 (Nunn et al. 2007). The effects of the LCO–LIA transition are not yet well understood in the Hawaiian record, because the vast majority of the available evidence post-dates A.D. 1400 and therefore is missing the relevant information about how people adapted to the changing conditions prior to this time. Nonetheless, in the few areas with records spanning this time range, at least some difference is detected around A.D. 1300–1400 in the placement of habitation zones, reliance of different kinds of crops, and overall patterns of settlement and land use. Moreover, the emergence of intensive agricultural field complexes, artificial fish ponds, and complex political economies entirely post-dated A.D. 1400 (Carson 2006). Oral traditions and chiefly genealogies point to the A.D. 1400s as the beginning of increased competition of warring chiefs (Cordy 2000), indicative of environmental and social stress.

The trends before and after A.D. 1300 are consistent with the transition from LIA to LCO conditions, but the same outcomes could have been due to population growth following island settlement about A.D. 1000 and reaching a critical threshold by A.D. 1400. This population growth would have been encouraged by the overall favourable conditions of the LCO at A.D. 1000 through 1300, followed by crisis during the extended unfavourable conditions of the LIA that no longer could support the population that had expanded and thrived during prior centuries. The picture is far from clear in the Hawaiian archaeological record, still rather sparse in the range of A.D. 1000 through 1400.

A chronological view of landscape evolution has been under-appreciated in Hawaiian archaeology, while instead a synchronic view of landscape ecology has figured prominently in the last several decades of research. At the time of European records in the late 1700s, the traditional land-use system involved partitioning of each island into a set of pie-slice units known as *ahupua'a*. Each *ahupua'a* contained a series of ecological zones from the interior upland mountain to the sea, often following the natural shapes of stream-cut valleys, so that people living in each ecological zone could maximise the locally specific resources and trade with one another. The zonation was especially clear in steep-sloped terrain with dramatic rain gradients increasing by elevation, wherein each elevation range was most suitable for a different mode of crop growth or other land-use pattern. The *ahupua'a* were more than practical organisations for allocation of resources among the local residents, and in fact the name *ahupua'a* literally means "pig altar" in reference to the tribute of food (and especially pigs) given at the community's altar near the coast for the ruling chiefs to collect.

Each *ahupua'a* is filled with named places and stories of literal and mythological events, and almost every Hawaiian archaeological site can be associated with these traditions. The naming in itself may be viewed as a cultural rendering of the natural world into an inhabited landscape, prompting questions of how and when this rendering occurred. Upon further thought, the place names and traditions as known in the late 1700s conceivably overprinted older traditions, especially when knowing that warring chiefs imposed their own ideals and propaganda when they consumed other lands and even entire islands.

The wonderfully informative Hawaiian ethnohistoric context must have developed over a period of time of changing natural and cultural history of the inhabited landscape, yet this chronological dimension has been under-represented in Hawaiian archaeology. Partly, the focus on the more recent past is due to the sparse knowledge of sites dating to the earliest settlement period of A.D. 1000 through 1400, although external scholars may wonder why Hawaiian archaeologists have not searched more vigorously for the oldest sites in their neatly contained island model systems. The situation has been exacerbated by a status quo methodology of surveying for surface-visible stonework ruins of house foundations, agricultural fields, and religious monuments that invariably date only to the more recent periods of occupation.

Admittedly, the rather brief Hawaiian chronology of approximately 1000 years does not include very much opportunity for preserving subsurface cultural layers hidden from surface survey, but these ancient subsurface layers do of course exist. In coastal plain landforms, depositional units of beach sands and slope-eroded sediments can accumulate 1 m or thicker, covering ancient habitation layers and obscuring their original contexts. These older buried sites logically cannot be identified through the standardised surveys of surface-visible stonework features.

Instead of finding the oldest Hawaiian sites and learning about their original contexts, proxy information has been obtained from palaeoenvironmental archives unrelated to actual archaeological sites. The proxies include preserved botanical remains in lake-bottom and swamp-bottom sediments indicative of forest-clearing by the first human contact in the islands, bones of rats that must have arrived with the first ocean-crossing settlers, and bones of birds that became extinct due to human impacts, all pointing to an age of about A.D. 1000 (Athens et al. 2002). By default, the dating of first human arrival has relied on inferential statistical modelling of the very few available radiocarbon dates from the known early sites in conjunction with the palaeoenvironmental proxies (Dye 2015), effectively refining the most probable dating but surely less desirable than obtaining substantive information directly from the early habitation sites. The reliance on proxies and statistical modelling has made the best of a poor situation with limited evidence, instead of the preferred strategy of finding and examining the actual substantive site records as has been accomplished in the Mariana Islands and elsewhere.

Despite the noted problems in developing a chronologically informed view of Hawaiian landscape evolution, at least three studies have provided long-term records with secure dating and accountability for changing natural and cultural contexts. In Kawaihae of leeward (west) Hawai'i Island, geoarchaeological investigation reveals a sequence of changing cultural use of a transforming coastal environment since A.D. 1200–1400, much different from the ethnohistorically defined cultural landscape of the Kamehameha Dynasty's royal residence during the late 1700s through early 1800s (Carson 2012). In Kualoa of windward (northeast) O'ahu, mythological traditions distinguish at least two periods of different environmental and social contexts, coordinated with geomorphological evidence and an archaeological record extending as old as A.D. 1040–1280 (Carson and Athens 2007). In Wainiha of northern Kaua'i, the sequence of settlement and land use since A.D. 1030–1400 involves changing roles of coastal occupation, as well as low labour-input tree crop arboriculture versus intensive irrigated taro farming, attested in archaeological and oral historical records (Carson 2003, 2004). Other relevant examples exist but are not covered in detail here, such as a reconstruction of the meandering Waimanalo Stream and beach deposits associated with shifting placements of habitations at and around the Bellows Dune Site of O'ahu (Peterson 2005), providing an excellent substantive framework although the dating and context of an apparent early habitation at A.D. 1000–1200 are unclear (Tuggle and Spriggs 2000).

A study at a coastal portion of Kawaihae provides a five-part chronology from A.D. 1200 to –1400 through the present (Fig. 2.9), noting change in the physical landforms, vegetation communities, cultural use of the available setting, and associated

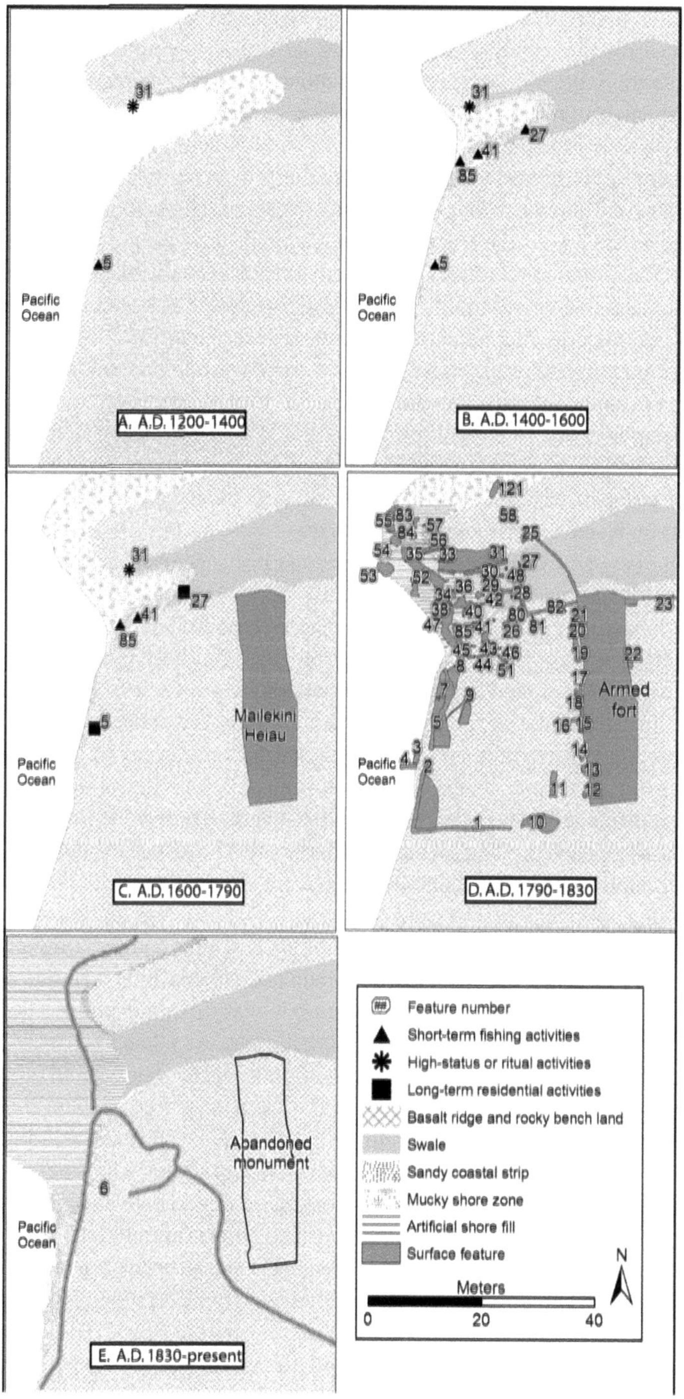

Fig. 2.9 Landscape evolution at Kawaihae, following Carson (2012)

ethnohistoric traditions (Carson 2012). This sequence could not have been known through the orthodox approach of examining surface-visible ruins related with ethnohistories. In this case, the surface ruins and monuments refer to the royal residence of the Kamehameha Dynasty and the birthplace of the historically known Hawaiian Kingdom, where Kamehameha's rival Keoua was killed and sacrificed in the late 1700s, where early European visitors reported dutifully to pay respects to Kamehameha and later his son Liholiho during the formative years of the kingdom through the early 1800s, and where the visible features of the landscape are strongly imbued even today with traditions of the royal residence. The ethnohistoric landscape at Kawaihae offers an effective example of how political elites can formalise public notions of power and authority into material monuments, overprinting older features of the landscape with their own creations and propaganda, further enforced through re-naming of places and installation of new oral traditions at the expense of older memories of the landscape. Prior to being known as the royal residence called Pelekane, this place may have been known as Kikiakoi or another name unclearly remembered today, and the traditions of a religious complex of Mailekini have been lost after the temple was converted into an armed fort by Kamehameha and his entourage.

Looking beyond the surface-visible ruins and historical propaganda at Kawaihae, geoarchaeological study recovered evidence of a chronological sequence of the changing landscape. Initial cultural use of the area at A.D. 1200–1400 entailed two small habitations on opposite sides of a narrow inlet, followed by gradual infilling of the inlet with slope-eroded sediments and enlargement of a low-lying coastal plain landform. In later centuries, people occupied increasingly numerous and larger portions of the available terrain, and they engaged in diverse types of activities such as ordinary habitation, high-status occupation, and religious performance. The royal residence constituted only one portion of the sequence that overlaid and masked much of what had happened previously, and in turn the royal residence eventually was transformed with later developments in the modern era.

Unlike the politically driven ethnohistory at Kawaihae, oral traditions of the landscape at Kualoa are grounded more in mythology and have endured through periodic impositions of different ruling regimes (Carson and Athens 2007). Traditions refer to the slaying of a dragon-like creature whose body was thrown down and thus created the local mountainous terrain of Kualoa, literally meaning "the long back" as in the backbone and body of the dragon. The dragon's tail is said to be a small offshore islet of Mokoli'i, literally meaning "little lizard". These traditions explain the physical landforms while symbolically accounting for a change in social and religious context alluded in other vaguely remembered tales of ancient temples and religious orders that no longer existed in later periods. Before becoming known as Kualoa, the older place name is remembered as Paliku, literally meaning "upright cliff" and referring to the steep mountainside that once bordered the ocean, later transformed by the accumulation of a broad coastal plain within which archaeological deposits have been identified in different stages of the coastal formation (Fig. 2.10). In addition to the mythologically based elements, apparently factual accounts in oral histories mention periodic tidal waves or *tsunami*, with counterparts in the sedimentary profiles showing surge deposits and interruptions of the cultural occupations.

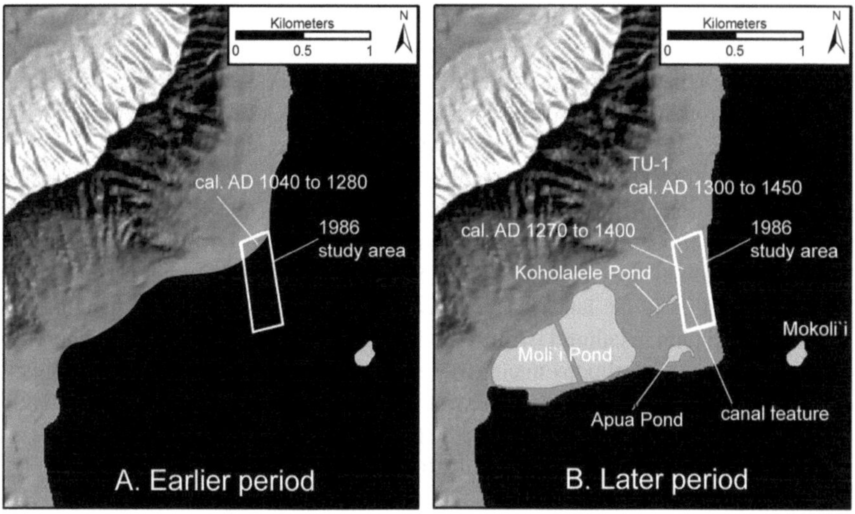

Fig. 2.10 Landscape evolution at Kualoa, modified from Carson and Athens (2007)

One more example of Hawaiian landscape evolution comes from Wainiha in the "separate kingdom" of Kaua'i (Carson 2003, 2004), where cultural traditions proudly underscore the spirit of independence and uniqueness apart from the other islands, in this case with intriguing congruence between the cultural traditions and archaeological evidence. The stream-cut valley of Wainiha, like others in the Halele'a District along the north of Kaua'i, is replete with irrigated terraces for growing taro and exemplary of the economic basis of complex chiefdoms (Earle 1978), yet the extensive transformation of the valley into a taro-producing landscape entirely post-dated A.D. 1400 (Carson 2006). Beneath the constructions of taro fields, remnants of older cultural layers refer to a qualitatively different and less intensive form of land use with scattered charcoal flecking, occasional hearth features, and little or no investment in long-lasting stonework architecture seen in the later periods. The only archaeologically detectable residential occupation prior to A.D. 1400 was close to the beach in the range of A.D. 1030–1400 (Carson 2004), at a time when the interior valley supported low-intensity land use pre-dating the taro fields. The creation of larger-sized and longer-lasting stonework complexes throughout Wainiha post-dated A.D. 1400 and marked the beginning of a significantly different land-use pattern, population demography, and social context.

The archaeological sequence at Wainiha accords well with tales of the banana-eating *Mu* people of this particular part of Kaua'i but not known elsewhere in the Hawaiian Islands, said to be the older inhabitants prior to the more recent ethnohistoric contexts. These traditions reflect memories of "times when (or places where) bananas and other tree crops were more important in a local diet and landscape that otherwise came to be dominated by taro and other root crops" (Carson 2003:100). Furthermore, Kaua'i is known for its unique forms of stone food-pounder artefacts with stirrup-like shapes and other characteristics (McElroy 2004), not seen in other of the Hawaiian Islands and thus suggesting a different use than taro-pounding documented extensively in the Hawaiian record outside Kaua'i. Indeed, throughout the Hawaiian homeland region of Central East Polynesia, food pounders were (and still are) used for making breadfruit paste, and the paste known as *poi* refers to breadfruit in all of these islands except in Hawaii where it refers to taro paste today.

Although still needing more attentive research, Hawaiian landscape evolution highlights the potential contributions of cultural traditions and ethnohistory that can be combined with geomorphology and archaeology. In the few examples reviewed here, substantive evidence informs about chronological change in the physical shape and configuration of the landscape, cultural adaptation and use of the changing settings, and variation in cultural traditions over the last several centuries. This approach contrasts against the orthodox manner of looking at surface stonework ruins and interpreting them through the face values of ethnohistories. Within the limits of a short chronology, the Hawaiian landscape sequence so far includes at least two components before and after A.D. 1400, and new investigations eventually will refine this outline.

Mariana Islands

The next chapters build a detailed account of landscape evolution as seen in the Mariana Islands over a sequence of 3500 years, drawing on diverse lines of evidence in natural and cultural historical records. The time scale in this case is less than in the preceding examples of coastal China and California that span the Pleistocene–Holocene transition, but it is sufficient to account for significant change in climate, sea level, coastal morphology, and several other factors. Moreover, the Marianas chronology is rather refined in periods of a few centuries each, and all of those periods provide substantial physical evidence of the natural and cultural landscape. The Marianas case includes limited input from cultural traditions as compared to the exceptional case of Hawaiian ethnohistory, most directly relevant in later-aged periods.

References

Anderson, D. G., & Bissett, T. G. (2015). The initial colonization of North America: Sea-level change, shoreline movement, and great migrations. In M. D. Franchetti & I. I. I. Robert Spengler (Eds.), *Mobility and ancient society in Asia and the Americas* (pp. 59–88). Switzerland: Springer.

Arnold, J. E. (1992). Complex hunter-gatherer-fishers of prehistoric California: Chiefs, specialists, and maritime adaptations of the Channel Islands. *American Antiquity, 57*, 60–84.

Arnold, J. E. (Ed.). (2001). *The origins of a Pacific coast chiefdom: The Chumash of the Channel Islands*. Salt Lake City: University of Utah Press.

Arnold, J. E. (2012). Detecting apprentices and innovators in the archaeological record: The shell bead-making industry of the Channel Islands. *Journal of Archaeological Method and Theory, 19*, 269–305.

Athens, J. S., Tuggle, H. D., Ward, J. V., & Welch, D. J. (2002). Avifaunal extinctions, vegetation change, and Polynesian impacts in prehistoric Hawai'i. *Archaeology in Oceania, 37*, 57–78.

Bayman, J. M., & Dye, T. S. (2013). *Hawaii's past in s world of Pacific Islands*. Washington, DC: SAA Press, Society for American Archaeology.

Braje, T. J. (2010). *Modern oceans, ancient sites: Archaeology and marine conservation on San Miguel Island, California*. Salt Lake City: University of Utah Press.

Brindley, E. F. (2015). *Ancient China and the Yue: Perceptions and identity on the southern frontier, c. 400 BCE–50 CE*. Cambridge: Cambridge University Press.

Carson, M. T. (2003). Integrating fragments of a settlement pattern and cultural sequence in Wainiha Valley, Kaua'i, Hawaiian Islands. *People and Culture in Oceania, 19*, 83–105.

Carson, M. T. (2004). Resolving the enigma of early coastal settlement in the Hawaiian Islands: The stratigraphic sequence of the Wainiha Beach Site in Kaua'i. *Geoarchaeology, 19*, 99–118.

Carson, M. T. (2006). Chronology in Kaua'i: Colonisation, land use, demography. *Journal of the Polynesian Society, 115*, 173–185.

Carson, M. T. (2012). Ethnohistoric and geoarchaeological landscape chronology at Kawaihae, leeward Hawai'i Island. *Geoarchaeology, 27*, 385–409.

Carson, M. T., & Athens, J. S. (2007). Integration of coastal geomorphology, mythology, and archaeological evidence at Kualoa Beach, windward O'ahu, Hawaiian Islands. *Journal of Island and Coastal Archaeology, 2*, 24–43.

Carson, M.T., & Hung, H.-C. (2014). Semiconductor theory in migration: population receivers, homelands, and gateways in Taiwan and Island Southeast Asia. *World Archaeology, 46*, 502–515.

Chang, C.-h., Kaifu, Y., Takai, M., Kono, R. T., Grun, R., Mastuura, S., Kingsley, L., & Lin, L.-k. (2015). The first archaic *Homo* from Taiwan. *Nature Communications, 6*, 6037.

Chen, C.-Y. (2013). *The Excavation of Liangdao-Daowei Sites Group, Liangdao, Matsu Archipelago and the Reconstruction of Liangdao Man*. Lienjiang County Government, Matsu, Taiwan (in Chinese).

Chen, W.-s., Huang, M.-t., & Lu, T.-k. (1991). Neotectonic significance of the Chimei fault in the coastal range, eastern Taiwan. *Proceedings of the Geological Society of China, 34*, 43–56.

Chen, W.-S., Sung, S.-H., Wu, L.-C., Hsu, H.-de., & Yang, H.-C. (2004). Shoreline changes in the coastal plain of Taiwan since the Last Glacial Epoch. *Chinese Journal of Archaeological Science, 94*, 40–55 (in Chinese).

Clark, J., Mitrovica, J. X., & Alder, J. (2014). Coastal paleogeography of the California-Oregon-Washington and Bering Sea continental shelves during the latest Pleistocene and Holocene: Implications for the archaeological record. *Journal of Archaeological Science, 52*, 12–23.

Connolly, T. J., Erlandson, J. M., & Norris, S. E. (1995). Early Holocene basketry and cordage from Daisy Cave, San Miguel Island, California. *American Antiquity, 60*, 309–318.

Cordy, R. (2000). *Exalted sits the chief: The ancient history of Hawai'i*. Honolulu: Mutual Publishing.

Cuthrell, R. Q., Stripler, C., Hylkema, M., & Lightfoot, K. G. (2012). A land of fire: Anthropogenic burning on the central coast of California. In T. L. Jones & J. E. Perry (Eds.), *Contemporary issues in California archaeology* (pp. 153–172). Walnut Creek, CA: Left Coast Press.

Dickinson, W. R. (2011). Geological perspectives on the Monte Verde archaeological site in Chile and pre-Clovis coastal migration in the Americas. *Quaternary Research, 76*, 201–210.

Dillehay, T. D., Ramírez, C., Pino, M., Rossen, J., & Pino-Navarro, J. D. (2008). Monte Verde: Seaweed, food, medicine, and the peopling of South America. *Science, 320*, 784–786.

Dye, T. S. (2015). Dating human dispersal in Remote Oceania: A Bayesian view from Hawai'i. *World Archaeology, 47*, 661–676.

Earle, T. K. (1978). *Economic and social organization of a complex chiefdom: The Halelea District, Kaua'i, Hawaii*. Anthropological Papers 63. Museum of Anthropology, University of Michigan, Ann Arbor.

Erlandson, J. M., Graham, M. H., Bourque, B. J., Corbett, D., Estes, J. A., & Steneck, R. S. (2007). The kelp highway hypothesis: Marine ecology, the coastal migration theory, and the peopling of the Americas. *Journal of Island and Coastal Archaeology, 2*, 161–174.

Faught, M. K. (2004). The underwater archaeology of paleolandscapes, Apalachee Bay, Florida. *American Antiquity, 69*, 275–289.

Faught, M. K., & Gusick, A. E. (2011). Submerged prehistory in the Americas. In J. Benjamin, C. Bonall, C. Pickard, & A. Fischer (Eds.), *Submerged prehistory* (pp. 145–157). Oxford: Oxbow Books.
Fuller, D. Q., Qin, L., Zheng, Y., Zhao, Z., Chen, X., Hosoya, L. A., & Sun, G.-p. (2009). The domestication process and domestication rate in rice: Spikelet bases from the lower Yangtze. *Science, 323*, 1607–1610.
Hiscock, P. (2008). *Archaeology of ancient Australia*. London: Routledge.
Hung, H.-C., & Carson, M. T. (2014). Foragers, fishers and farmers: Origins of the Taiwan Neolithic. *Antiquity, 88*, 1115–1131.
Kirch, P. V. (1985). *Feathered gods and fishhooks: An introduction to Hawaiian archaeology and prehistory*. Honolulu: University of Hawaii Press.
Ko, A. M., Chen, C. Y., Fu, Q., Delfin, F., Li, M., Chiu, H. L., Stoneking, M., & Ko, Y. C. (2014). Early Austronesians: Into and out of Taiwan. *American Journal of Human Genetics, 94*, 426–436.
Liew, P. M., Pirazzoli, P. A., Hsieh, M. L., Arnold, M., Barusseau, J. P., Fontugne, M., & Giresse, P. (1993). Holocene tectonic uplift deduced from elevated shorelines, eastern coastal range of Taiwan. *Tectonophysics, 222*, 55–68.
Lightfoot, K. G., & Parrish, O. (2009). *California Indians and their environment: An introduction* (California natural history guides). Berkeley: University of California Press.
Liu, L., & Chen, X. (2012). *The archaeology of China: From the late paleolithic to the early Bronze Age*. New York: Cambridge University Press.
Liu, L., Lee, G.-A., Jiang, L., & Zhang, J. (2007). The earliest rice domestication in China. *Antiquity, 81*. online project gallery.
Liu, W., Jin, C.-Z., Zhang, Y.-Q., Cai, Y.-J., Xing, S., Wu, X.-J., et al. (2012). Human remains from Zhirendong, South China, and modern human emergence in East Asia. *Proceedings of the National Academy of Sciences, 107*, 19201–19206.
Madsen, D. B. (2015). A framework for the initial occupation of the Americas. *PaleoAmerica, 1*, 217–250.
Masters, P. M., & Aiello, I. W. (2007). Postglacial evolution of coastal environments. In T. L. Jones & K. A. Klar (Eds.), *California prehistory: Colonization, culture, and complexity* (pp. 35–51). Lanham, MD: Altamira Press.
McElroy, W. K. (2004). Poi pounders of Kaua'i Island, Hawai'i: Variability through time and space. *Hawaiian Archaeology, 9*, 25–49.
Miller, D. S., Holliday, V. T., & Bright, J. (2013). Clovis across the continent. In K. E. Graf, C. V. Ketron, & M. R. Waters (Eds.), *Paleoaerican Odyssey* (pp. 207–220). College Station, TX: Center for the Study of the First Americans, Texas A & M University Press.
Nunn, P. D., Hunter-Anderson, R. L., Carson, M. T., Thomas, F., Ulm, S., & Rowland, M. J. (2007). Times of plenty, times of less: Last-millennium societal disruption in the Pacific Basin. *Human Ecology, 35*, 385–401.
Peterson, J. A. (2005). Holocene landscapes of Waimanalo Bay: Archaeological investigations at Bellows Beach. *Hawaiian Archaeology, 10*, 47–69.
Qin, L. (2013). The Liangzhu culture. In A. P. Underhill (Ed.), *A companion to Chinese archaeology* (pp. 574–596). Malden, MA: Blackwell.
Rick, T. C., Erlandson, J. M., Vellanoweth, R. L., & Braje, T. J. (2005). From Pleistocene mariners to complex hunter-gatherers: The archaeology of the California Channel Islands. *Journal of World Prehistory, 19*, 169–228.
Rolett, B. V., Zheng, Z., & Yue, Y. (2011). Holocene sea-level change and emergence of Neolithic seafaring in the Fuzhou Basin, Fujian, China. *Quaternary Science Reviews, 30*, 788–797.
Shen, G., Gao, X., Gao, B., & Granger, D. E. (2009). Age of Zhoukoutian *Homo erectus* determined with $^{26}Al/^{10}Be$ burial dating. *Nature, 458*, 198–200.
Soares, P., Alshamali, F., Pereira, J. B., Ferdandes, V., Silva, N. M., Alfonso, C., Costo, M. D., Musilova, E., Macaulay, V., Richards, M. B., Cerny, V., & Pereira, L. (2012). The expansion of mtDNA haplogroup L3 within and out of Africa. *Molecular Biology and Evolution, 29*, 915–927.
Stanley, D. J., Chen, Z., & Song, J. (1999). Inundation, sea-level rise and transition from Neolithic to Bronze Age cultures, Yangtze Delta, China. *Geoarchaeology, 14*, 15–26.

Sung, W.-H. (1969). Changbin culture: The first discovery of Palaeolithic culture in Taiwan. *Newsletter of Chinese Ethnology, 9*, 1–27 (in Chinese).

Tsang, C.-H. (2005). Recent discoveries at the Takenkeng culture sites in Taiwan: Implications for the problem of Austronesian origins. In L. Sagart, R. Blench, & A. Sanchez-Mazas (Eds.), *The peopling of East Asia: Putting together archaeology, linguistics and genetics* (pp. 63–73). London: Routledge Curzon.

Tsang, C.-H., Chen, W. S., Li, K. T., & Zeng, Y. X. (2009). *Report of the Baxiandong cave sites, Changbin, Taidong County, the first year*. Academia Sinica, Taipei (in Chinese)

Tsang, C.-H., Chen, W. S., Li, K. T., & Zeng, Y. X. (2011). *Report of the Baxiandong cave sites, Changbin, Taidong County, the second year*. Academia Sinica, Taipei (in Chinese).

Tuggle, H. D., & Spriggs, M. (2000). The age of the Bellows Dune Site O18, O'ahu, Hawai'i, and the antiquity of Hawaiian colonization. *Asian Perspectives, 39*, 165–188.

Vitousek, P. M., Chadwick, O. A., Hartshorn, A. S., & Hotchkiss, S. C. (2010). Intensive agriculture in Hawai'i: The model system approach. In P. V. Kirch (Ed.), *Roots of conflict: Soils, agriculture, and sociopolitical complexity in ancient Hawai'i* (pp. 31–44). Santa Fe: School for Advanced Research Press.

Wang, S.-Y., Lu, H.-Y., Liu, J.-Q., & Negendank, J. F. W. (2007). The early Holocene optimum inferred from a high-resolution pollen record of Huguangyan Maar Lake in southern China. *Chinese Science Bulletin, 52*, 2829–2836.

West, G. J., Woolfenden, W., Wanket, J. A., & Scott Anderson, R. (2007). Late Pleistocene and Holocene environments. In T. L. Jones & K. A. Klar (Eds.), *California prehistory: Colonization, culture, and complexity* (pp. 11–34). Lanham, MD: Altamira Press.

Xu, S., Xiao, J., Xiao, X., Gao, Y., Han, Y., & Qi, G. (2011). Impact of environmental evolution and coastline change on the Neolithic cultures in the eastern part of Jiang-Huai area. *Marine Geology and Quaternary Geology, 31*, 127–134 (in Chinese).

Yancheva, G., Nowaczyk, N. R., Mingram, J., Dulski, P., Schettler, G., Negendank, J. F. W., et al. (2007). Influence of the intertropical convergence zone on the East Asian monsoon. *Nature, 445*, 74–77.

Yang, X., Barton, H. J., Wan, Z., Li, Q., Ma, Z., Li, M., Zhang, D., & Wei, J. (2013). Sago-type palms were important plant foods prior to rice in southern subtropical China. *PloS One, 8*, e63148. doi:10.1371/journal.pone.0063148.

Yasuda, Y. (2008). Climate change and the origin and development of rice cultivation in the Yangtze River Basin, China. *Ambio: A Journal of the Human Environment, 37*, 502–506.

Zhang, Q., Cheng, Z., Liu, C., & Jiang, T. (2005). Environmental change and its impacts on human settlement in the Yangtze Delta, P. R. China. *Catena, 60*, 267 277.

Zhang, C., & Hung, H.-C. (2008). The Neolithic of southern China: Origin, development, and dispersal. *Asian Perspectives, 47*, 299–329.

Zhang, C., & Hung, H.-C. (2010). The emergence of agriculture in southern China. *Antiquity, 84*, 11–25.

Zhang, C., & Hung, H.-C. (2012). Later hunter-gatherers in southern China, 18,000–3000 B.C. *Antiquity, 86*, 11–29.

Zhang, C., & Hung, H.-C. (2013). Jiahu 1: Earliest farmers beyond the Yangtze River. *Antiquity, 87*, 46–63.

Zhang, Q., Jiang, T., Shi, Y., King, L., Liu, C., & Metzler, M. (2004). Paleo-environmental changes in the Yangtze Delta during past 8000 years. *Journal of Geographical Sciences, 14*, 105–112.

Zong, Y. (2004). Mid-Holocene sea-level highstand along the southeast coast of China. *Quaternary International, 117*, 55–67.

Zong, Y., Huang, G., Switzer, A. D., Yu, F., & Yim, W. W. S. (2009). An evolutionary model for the Holocene formation of the Pearl River Delta, China. *The Holocene, 19*, 129–142.

Chapter 3
Environmental Setting and Dynamics

Before attempting to examine the long-term human–environment interactions in any region, the basic environmental conditions and their changing dynamics first must be acknowledged. In this way, the material findings at any single site can be understood in relation to a securely contextualised time and place. This book illustrates how ecological conditions have varied through time and across geographic space, in many cases correlated with trends and patterns in archaeological evidence. The study must begin, though, with a solid grasp of the natural environment, within which human actions took place and archaeological records were formed.

If the natural environment and cultural setting always have been inherently interconnected, then logically any distinction between natural and cultural history is untenable. Nonetheless, the Marianas archaeological record provides a clear example of the first contact between human beings and an isolated natural environment. This first contact can be appreciated as the meeting of natural and cultural factors, but the following centuries and millennia can be understood as a unified natural-cultural history.

In later chapters of this book, archaeological sites are interpreted as cultural manifestations within larger landscapes that consist of multiple inter-related attributes, some of which are more easily defined than others. As preparation, the present chapter reviews the essential landscape attributes that in principle could exist without human presence, for instance in terms of geological structure, sea-level history, coastal geomorphology, slope erosion–deposition patterns, soil formation, plant and animal communities, climate and weather, and water sources. These attributes are considered in the case of the Mariana Islands in terms of how they may have influenced or been influenced by cultural activities.

M.T. Carson, *Archaeological Landscape Evolution*,
DOI 10.1007/978-3-319-31400-6_3

Geological Structure

The Mariana Islands first began to form more than 40 million years ago (Cloud et al. 1956; Tracey et al. 1964) at the interface of two of the world's major tectonic plates (Fig. 3.1). The enormous Pacific Plate collided with (and subducted partly beneath) the Philippines Plate (Hussong and Uyeda 1981). The plate collision created a long trough, oriented roughly north–south. This trough now is known as the Mariana Trench, which includes the deepest measured spot in the world's seabed hydrosphere at 10.9 km, called Challenger Deep.

Along the colliding tectonic plates, magma periodically erupted into volcanic masses, eventually forming two great arcs of more than one dozen islands (see Fig. 1.2). The oldest of these formations are in the south, and the youngest are in the north. The primary volcanic land-mass formations were in place by 3–5 million years ago, although several transformations have continued since then.

Most extensively in the older southern arc, large colonies of corals grew around the volcanic masses, originally just below sea level (Fig. 3.2). Due to tectonic uplift and periods of changing sea level, the coral colonies died and became fossilised as

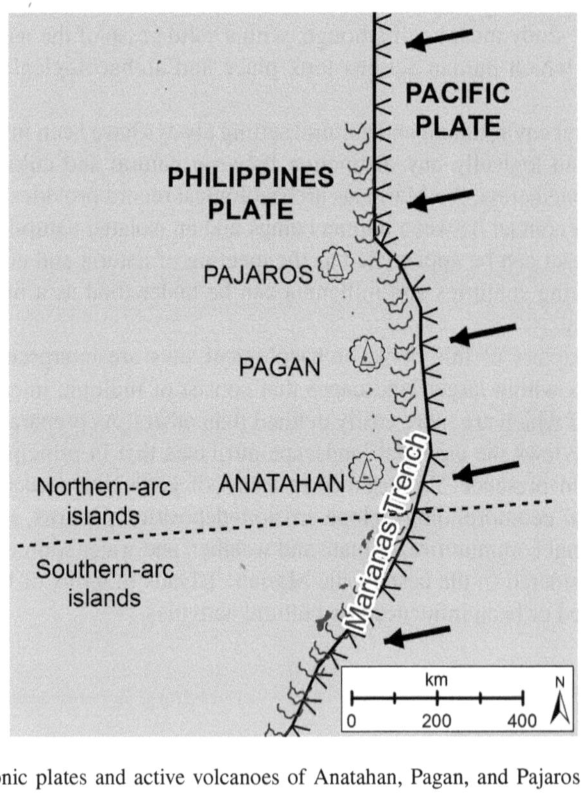

Fig. 3.1 Tectonic plates and active volcanoes of Anatahan, Pagan, and Pajaros in the Mariana Islands

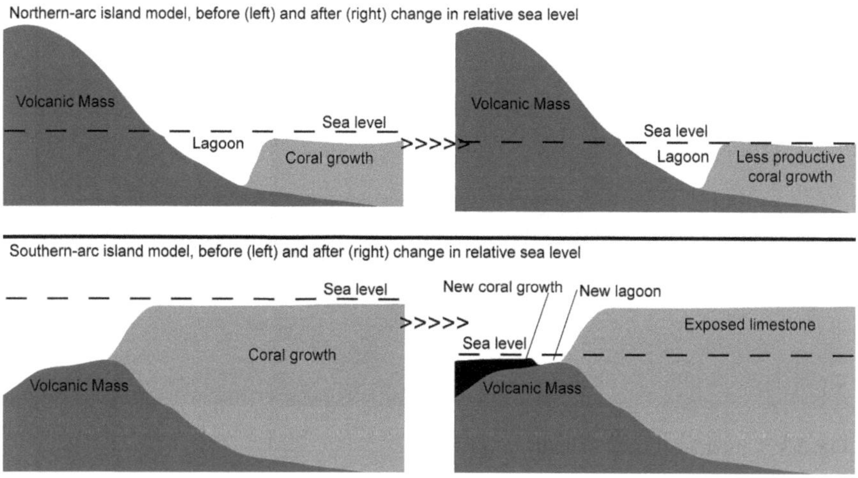

Fig. 3.2 Major stages of island formation, before and after tectonic uplift and other change in relative elevation of land versus sea

limestone terraces or plateaus, built over one another in stages from about 5 million through 100,000 years ago. These landforms now constitute broad parts of the southern-arc islands that generally are larger than the northern-arc islands lacking these formations. An extensive limestone plateau comprises northern Guam today (Fig. 3.3), while volcanic hills are exposed in southern Guam (Fig. 3.4).

In the geologically younger northern arc, volcanic activity continues even now, notably not only at Anatahan and Pagan (Figs. 3.5 and 3.6) but also at the farthest north island of Pajaros. The northern-arc Mariana Islands are characterised by rough volcanic terrain. The shapes of volcanic cones are easily detectable, not yet eroded into stream-cut valleys.

Certain geological features have become recognised as focal points or landmarks, and some have been linked with mythic traditions. Mountain peaks and other distinctive formations can be identified from various locations around an island and even from a distance at sea. Different portions of the island of Guam have been interpreted as representing the body parts of the first living man, named Puntan in mythic traditions (Cunningham 1992:3). Similarly in the island of Saipan, landforms can be recognised as the body parts of mythic ancestors (McKinnon et al. 2014), and in some cases a mountain peak can be understood alternatively as male or female when observed from different points of view.

The basic geological structure of the Marianas existed long before human arrival, but several of the terrain features have continued to transform through ongoing processes. In fact, many ancient site layers now are buried beneath more recent modifications of the terrain. By understanding these processes, we can discover ancient sites and learn about their original contexts.

Fig. 3.3 Limestone plateau of northern Guam

Fig. 3.4 Volcanic mountains of southern Guam

Sea-Level History

The Mariana Islands originated entirely in an Oceanic setting, never connected to a larger land mass or continent (Fig. 3.7). These islands are so distantly removed from other land that they are considered part of the Remote Oceanic region, where islands developed completely outside contact with larger land masses. Even during periods of

Fig. 3.5 Anatahan Volcano

Fig. 3.6 Pagan Volcano

much lower sea level, these islands were isolated in the remote Pacific. The lowest sea levels (as much as 100 m lower than present) occurred during major ice ages, when much of the world's ocean water was trapped inside ice sheets. For approximately the last 10,000 years, the islands have been overall stable relative to sea level within a few metres, but those small fluctuations have been significant for the island environment.

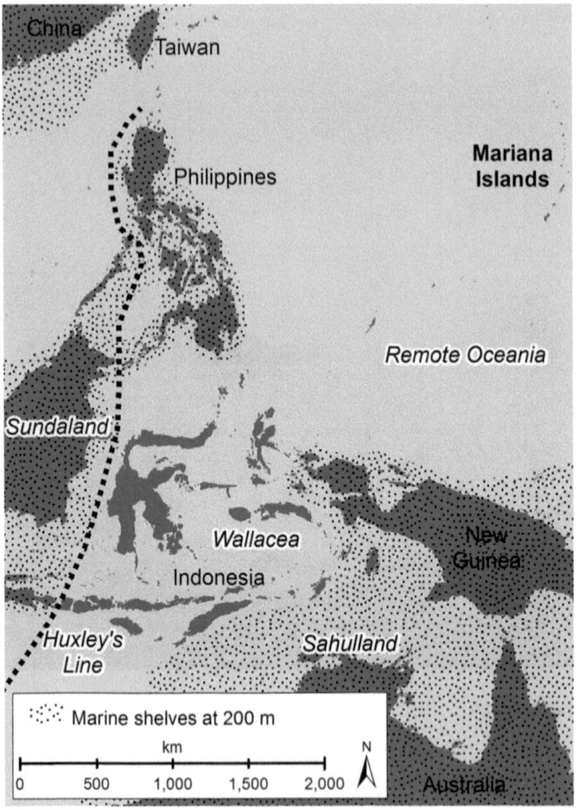

Fig. 3.7 Asia-Pacific region with approximate areas of exposed land during the last major Ice Age, noting the related zones of natural biodiversity

The local sea-level history of the Marianas now is well documented for the last few thousand years (Dickinson 2000, 2001, 2003), including the entire time range of human habitation (Fig. 3.8). When people first settled in these islands, the sea level was about 1.8 m higher than at present, and accordingly the coastal landforms, coral reefs, and related ecologies all differed remarkably from modern conditions (Carson 2011, 2014). Today's broad sandy beaches, especially in zones beneath 2 m elevation (Fig. 3.9), are quite recent formations that did not exist during most of the span of human habitation in the Mariana Islands.

Driven primarily by sea-level change, the coastal environment was transformed considerably throughout the time range of human presence in the Mariana Islands. This information is essential for understanding original contexts of ancient sites. An overall lowering sea level created conditions for larger coastal plain landforms, but coastal ecologies needed some time to adjust. Additionally, coastal ecosystems were more stable during some centuries and less stable during others.

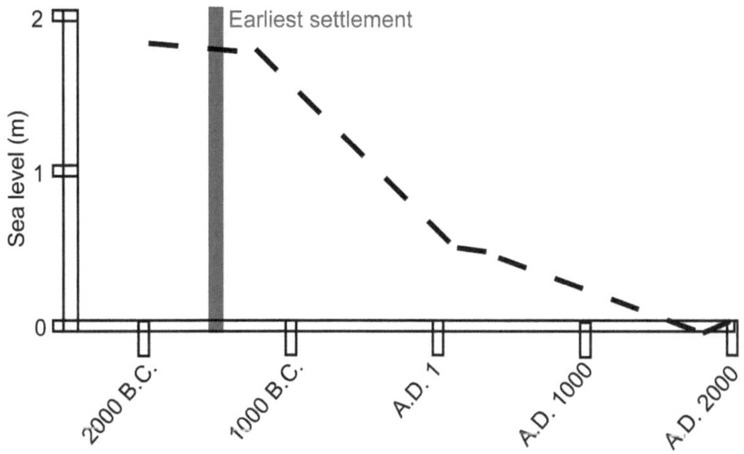

Fig. 3.8 Marianas sea level history. Based on data from Dickinson (2000, 2001, 2003)

Fig. 3.9 Beach at Ritidian, Guam, entirely beneath 2 m elevation today, composed of recent storm-surge sands

Coastal Geomorphology

Ancient sites can be found buried deeply beneath today's coastal landforms (Fig. 3.10), but they reflect palaeohabitats much different from today's conditions. Abundant evidence now leaves no doubt that many ancient sites originally were coastal-oriented, yet the physical compositions of coastal zones have transformed substantially over time. Although a generic "coastal setting" may have been consistent, the specific characteristics underwent major and minor change. The original contexts can be clarified by understanding the local sea-level history, in conjunction with the formation of different layers of beach sands.

Two major types of beach sands may be classified as storm-surge debris and as lagoon facies deposits. Modern beach surfaces have accumulated as layers of storm-surge debris, overlaying older lagoon facies deposits from a period of slightly higher sea level more than 3000 years ago. The lagoon facies deposits are found in places where the sea level lowered and left behind patches of large-bodied and mostly non-eroded algal bioclasts and other materials in stranded beds. The storm-surge deposits are found in places where loads of pulverised calcareous material accumulated during successive events of high tides and low-pressure storms.

Fig. 3.10 Deeply buried cultural layer at Ritidian ancient shoreline, sealed beneath hardened calcrete. The author is standing on the palaeo-reef of *Heliopora* sp. coral dated 2455–2068 B.C. Photograph by Diego Camacho

Deep excavations have exposed the different layers of beach sands, also noting periods of rapid deposition versus other periods of temporary stability (Fig. 3.11). More research has confirmed the depths, spatial distributions, and direct radiocarbon dating of buried coral reefs (Fig. 3.12), as well as locations of former reef margins (Fig. 3.13). Further studies have achieved direct radiocarbon dating of *Halimeda* sp. algal bioclasts (Carson and Peterson 2012) that originally were deposited in thick beds on palaeolagoon floors or in thin traces at high-tide marks (Fig. 3.14).

Fig. 3.11 Beach profile of Charterhouse Condominiums project area in Tumon, Guam, showing a temporarily stable surface about A.D. 1–200. The author is cleaning the construction trench profile, with scale bar in 20-cm increments near the *left side* of the image. Photograph by John A. Peterson

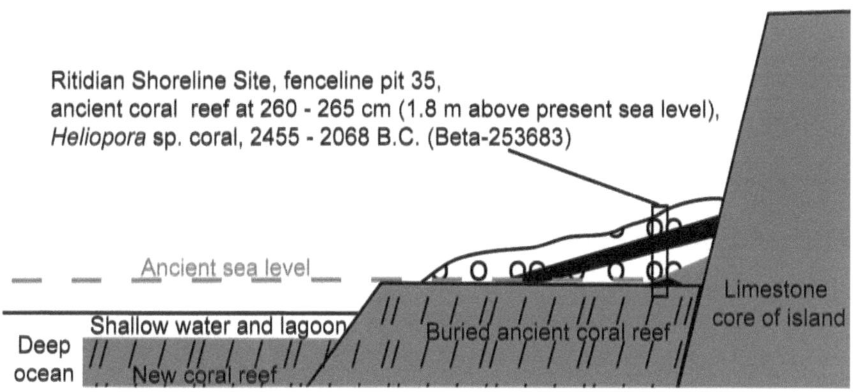

Fig. 3.12 Profile of buried coral reef at Ritidian. Schematic diagram is not to scale

Fig. 3.13 Ancient and modern coral reef margins, near eastern end of Ritidian

Fig. 3.14 *Halimeda* sp. algal bioclasts at Ritidian beach, following storm-surge event. The individual bioclasts each are less than 1 cm in length

Within the long sequence of transforming coastal habitats, fluctuations are detectable in the level of the ocean itself, the configuration of sandy beaches, the structures of reefs and lagoons, types of shellfish and other organisms, and of course the activities of people. Layers of sands provide obvious chronological markers. In addition to the sands, a fuller range of evidence reveals how the coastal zones evolved over time in relation to archaeological site settings, coastal and marine resource-use, and cultural impacts on the ecosystem.

Slope Erosion–Deposition Patterns

Continually over time, hill slopes have eroded, so that volumes of weathered clay, silt, and rock have been re-deposited in lower slopes and cliff-base areas. These sedimentary layers potentially seal over older archaeological layers, while they provide a new ground surface for other cultural activities, sometimes in a succession of repeated episodes (Fig. 3.15). This pattern is most actively visible along sloping surfaces, especially in stream-cut valleys where upland sediments can be carried downstream. The results also are seen in the piles of debris that accumulate along the foot-base of most limestone cliffs and other steep slopes.

Where slope-eroded sediments are found in lower elevations, their date of re-deposition can be informative about the past environmental conditions. Greater erosion tends to occur during punctuated storminess or times of deliberate forest-clearing, due to increased water flow, less vegetation ground cover, or both factors combined. The root cause of any single erosional event may not be clear, but individual dating results can be checked against other knowledge to seek possible correlations.

In some of the larger Pacific Islands, thick and broad sheets of slope-eroded sediments have contributed to buildup of vast coastal plains. The slope-erosional clays and silts in some cases were more than 1 m thick and covered several dozens of sq km, ideal for supporting lush vegetation and new kinds of crop growth. Matthew Spriggs (1997) argued that the coastal plain buildup in Vanuatu and New Caledonia resulted from human-caused endeavours of inland forest-clearing, undertaken for agricultural production by the early settlers in these islands about 1100–900 B.C. Moreover, agricultural land use could expand into the newly formed or vastly augmented coastal plains, rich with terrigenous nutrients.

Large coastal plain landforms in the Mariana Islands in most cases do not share the slope-erosional origins as described above for certain other Pacific Islands. Instead, they formed either as uplifted limestone terraces or else as accumulations of beach sands during and after a period of lowering sea level. In most cases, only thin upper drapes of slope-eroded materials are noted over most of these landforms, generally less than 20 cm and not nearly as spectacular as the cases described in Vanuatu and elsewhere (Spriggs 1997). Thicker layers are found in rare cases of the mouths of stream valleys and river basins of southern Guam, but these do not compare with the extensive buildup of coastal-plain landforms in other Pacific Islands.

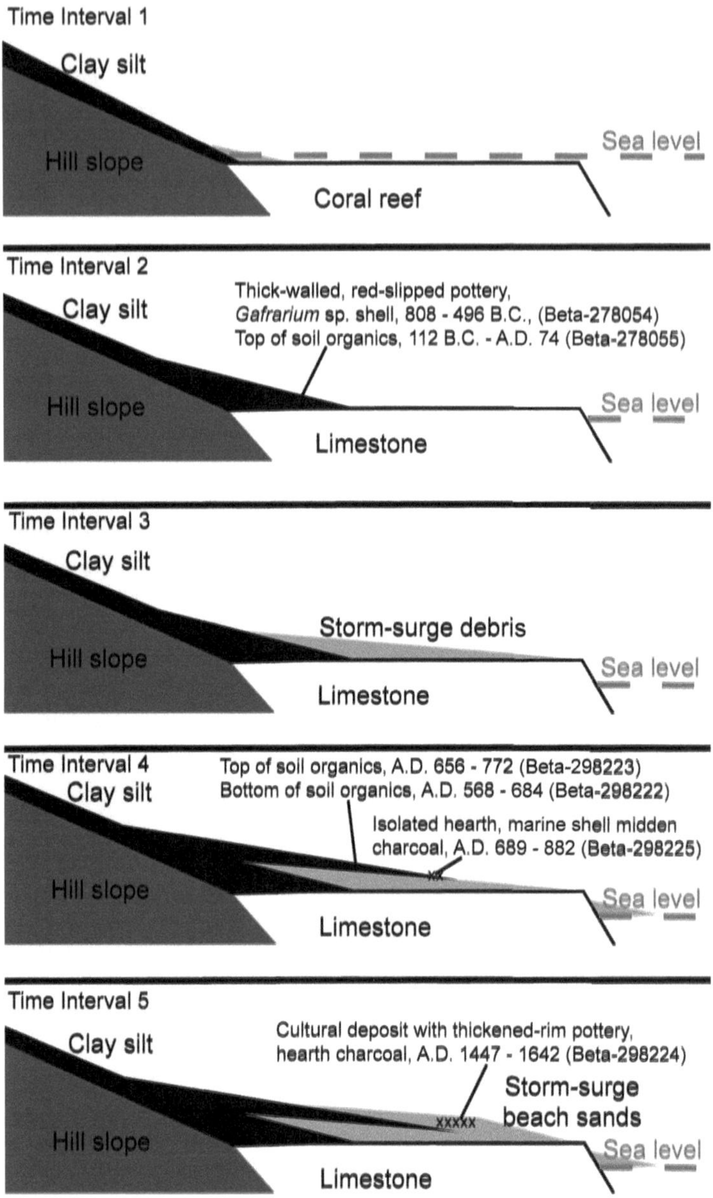

Fig. 3.15 Hill slope erosion and coastal plain development in five major time intervals at Ipan, Guam. Schematic diagram is not to scale

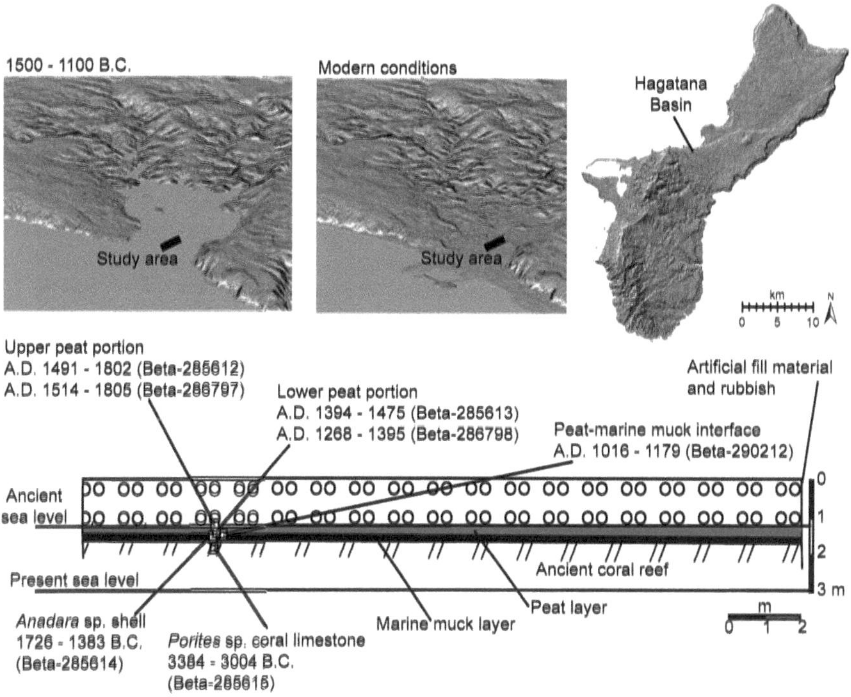

Fig. 3.16 Chronological development of Hagatna Basin, Guam. Based on data from Carson (2011, 2014)

The large river basins of Hagatna and Pago (both in Guam) offer two of the best opportunities to document slope-eroded sediments as re-deposited accumulations in lowland zones. Hagatna comprised a shallow lagoon during the time of first human arrival about 1500 B.C. (Fig. 3.16), transitioned into a freshwater swamp after a declining sea level, and slowly accumulated slope-eroded sediments while forming a peat deposit in the base of the swamp. The mouth of Pago River similarly was a submerged environment during the first 2500 years of human settlement in the region (Fig. 3.17), and slope-eroded sedimentary buildup began above sea level only within the last 1000 years.

Soil Formation

Soils and sediments are essential for any archaeological field study, because archaeological materials can be found overlaying, within, or beneath them. These stratigraphic relationships help to define the dating and original contexts of archaeological sites. Admirably useful towards these studies, standard soil type guides are available for the Mariana Islands (Young 1988, 1989). At a global scale, considerable scholarship has been devoted to archaeological studies of soils and sediments (e.g. Holliday 1992).

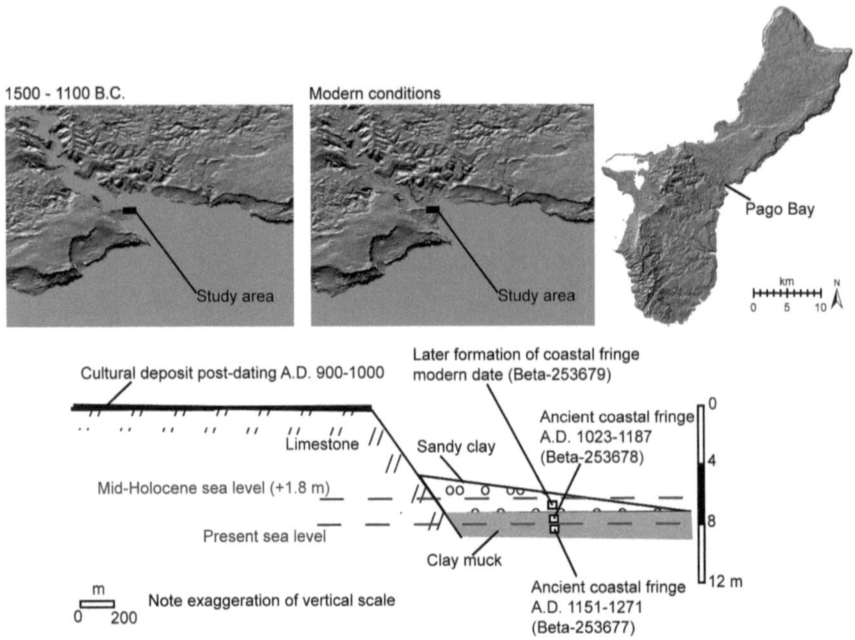

Fig. 3.17 Chronological development of Pago Bay, Guam. Based on data from Carson (2011, 2014)

Classic soil profiles, with A–B–C horizons reflecting formal pedogenesis, are not always obvious (Fig. 3.18). Overall in the Marianas, warm temperatures and plentiful rainfall contribute to vigorous vegetation growth, in principle resulting in accelerated soil formation. Nonetheless, qualities of the parent material in many cases diminish the rate of soil-forming processes, with only thin organic zones of A horizons overlaying weakly formed B and C horizons.

Soil types in the Mariana Islands relate to their origins in three major terrain landforms of limestone plateau, volcanic hills, and beach sand deposits. The formal development of soil horizons can differ greatly from one setting to another. The processes are affected by several factors, including the physical properties of the parent material, degree of slope, temperature, water flow and content, and vegetation growth.

In the limestone plateau terrain of the southern-arc Mariana Islands, many large areas are exposed as rough surfaces, but some pockets contain shallow rocky clays and silts, generally 25 cm or less. This material is derived from air-blown dust from the Asian continent, in situ decomposition of vegetation, and local weathering of the limestone itself. Inside irregular depressions and cavities in the limestone, thicker layers can accumulate, as re-deposited buildup washed or eroded from the adjacent higher ground.

In the volcanic hilly terrain of southern Guam and of the northern-arc Mariana Islands, rocky clays and silts occur in varying depths. The material is derived originally from local weathering of the volcanic base rock, but it has been re-distributed

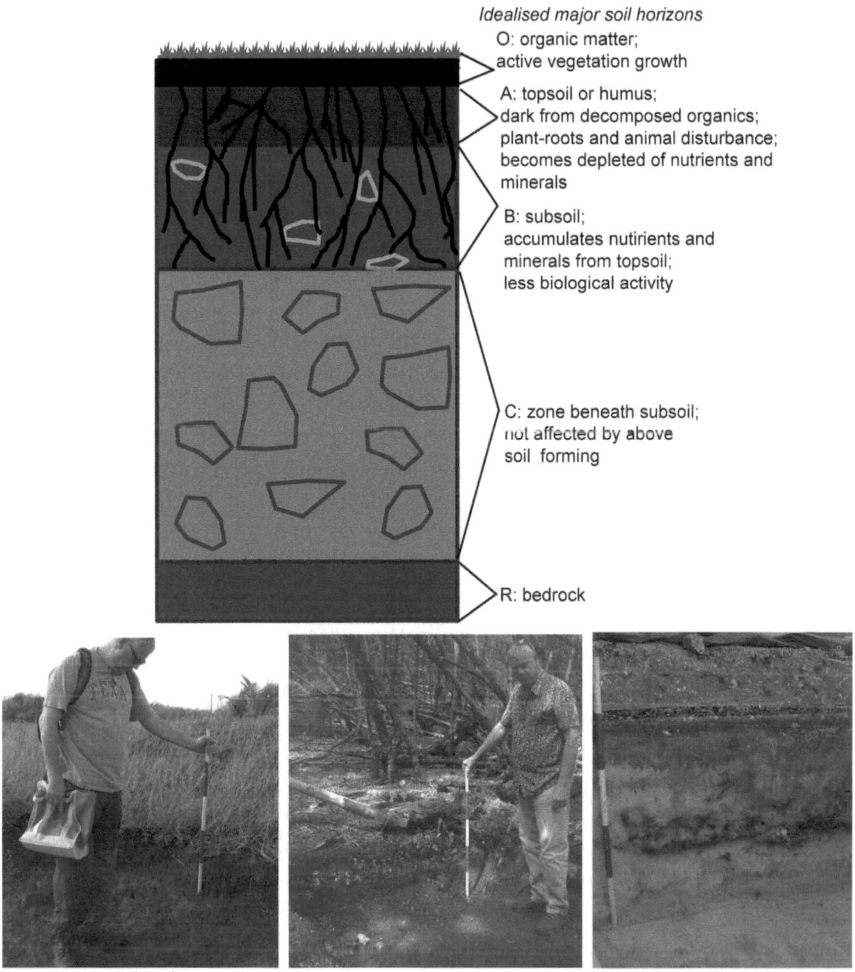

Very thin A and B horizons over C horizon and bedrock at Upland Talofofo, Guam

More distinct A and B horizons over C horizon and bedrock at Pago, Guam

Ancient cultural layers within weakly formed A-B horizons of beach sands at Tumon, Guam

Fig. 3.18 Idealised soil horizons, with Inset real-life examples of variations in Guam. John A. Peterson examines terrigenous soil profiles at Upland Talofofo (*left*) and Pago Bay (centre), while a calcareous beach sand profile is shown at Tumon (*right*). All scale bars are in 20-cm increments

by patterns of erosion and deposition along slopes (see Fig. 3.15). In most places, these rocky clays and silts are 35 cm or less in thickness, but they can be deeper in zones bearing accumulations of slope-erosional buildup.

In both the limestone and volcanic hilly terrain settings, archaeological materials often are visible on or near the surface, with little or no opportunity for buried deposits. The surviving materials almost always date within the last 1000 years or less, due to the poor preservation conditions in exposed settings or inside highly acidic matrix. Pottery and stone (but usually not shells and bones) potentially can survive much longer, but so far none have been found in these settings except very rarely up to 2000 or possibly 2500 years old.

In beach sand deposits, soil formation tends to be quite slow, because of the extreme alkaline conditions, fast water drainage through the porous matrix, little opportunity for plant growth, and tendency for rapid movement of the sand grains (both built up or swept away) that can interrupt soil-forming processes. Despite these factors, organic horizons do form in beach sand deposits, typically in the stable backbeach settings removed from active shores, further augmented by cultural activities that introduce more organic content. In the same zones, especially near the bases of inland hill slopes or limestone cliffs, thin layers of slope-eroded rocky clays and silts can support stronger soil formation.

Beach sands offer excellent preservation conditions of a broad range of archaeological materials, and they frequently occur in distinguishable layers associated with measurable time periods. Beach sand deposits contain some of the most informative archaeological sites in the Mariana Islands, sometimes in stratified layers 2 m or deeper. Accordingly, coastal geomorphology has proven indispensable for understanding where to search for ancient sites and how to interpret their original contexts.

In rare cases in the Marianas, ancient soil surfaces (palaeosols) are found in buried contexts, covered by more recent sediments that in turn may have supported a new set of soil-formation horizons. In the Mariana Islands, rapid burial and preservation of an old soil surface typically involves a storm-surge deposit, but it also could occur with slope-eroded sedimentation or another event causing a sharp change in the depositional environment. A good example of a palaeosol is found in the coastal terrace of Ipan (also "Ypan") in southeast Guam (see Fig. 3.15), where a layer of clay and silt at one time had developed into a formal A horizon and supported a habitation site about 2500 years ago, later covered by storm-surge debris. Another type of rather ephemeral buried soil surface can be found in many of the sandy beaches of the Marianas, for example at Tumon in Guam, where a weakly formed organic horizon (typical of a sandy backbeach) had been buried rapidly beneath storm-surge sands about 2000–1800 years ago (see Fig. 3.11).

Plant and Animal Communities

The only living organisms (including human beings) in the Mariana Islands originally must have arrived from elsewhere. Marine organisms naturally had the most opportunities to reach these isolated islands. Birds and bats also could reach the Marianas, albeit with some difficulty, and they depended on pre-existing natural vegetation for their dietary survival. Diverse plant species thrived in the humid tropical environment for more than 10,000 years (Ward 1994), but they were limited by the kinds of plants that could float over the ocean, translocate in air-currents, or "hitch-hike" via marine or avian wildlife. Large land-dwelling mammals could not make the ocean-crossing voyage without human mediation, and so far only monitor lizards and small reptiles have been documented in the Marianas palaeontological records of terrestrial fauna (Pregill 1998; Pregill and Steadman 2009).

Prior to Spanish colonial occupation, the biotic communities (both plants and animals) in the Marianas mostly can be understood as isolated developments at the outer extremity of a Southeast Asian environment, with an overlay of human-introduced taxa. For thousands of years before any people came to these shores, the native plants and animals evolved in isolation from a restricted "bottleneck" fraction of the more diverse populations known in the larger land masses of Southeast Asia. Distinctive plant and animal communities had evolved in the regions that formerly were connected by land bridges and continental shelves, resulting in the biogeographic zones now known as Sundaland, Wallacea, and Sahulland (see Fig. 3.7). The Mariana Islands simply were too far away from these regions to have inherited the same biological populations. An overall impoverished biota holds true across the islands of Remote Oceania, as compared to the greater diversity of Island Southeast Asia and Near Oceania that derived from continental origins.

The Remote Oceanic islands are by no means desolate, but rather their natural biodiversity lessens with distance from continental landmass. In these settings, the native plants and animals evolved in varying degrees of isolation, in some cases with peculiar results of species found only in one island or in one group of islands. The plant taxa in particular tend to include only very few items useful for human subsistence, in contrast to the situation of abundant edible plants and animals in Near Oceania.

When people first resided in the Mariana Islands, they found plentiful natural resources in many respects, yet the environment was deprived of the plant foods and other economically useful plants known elsewhere. Protein foods were ample and perhaps even lavish in coastal, marine, and forest zones. Essential starches and other dietary nutrients from plants, however, were scarce, and accordingly several plant taxa were imported by people across the ocean.

The Mariana Islands, like other islands of Remote Oceania, lacked most of the plants and animals that supported human survival in Island Southeast Asia and Near Oceania. People successfully translocated several foreign plants and animals into the remote island world, effectively re-producing familiar landscapes. This process illustrates what Edgar Anderson (1952) described as the tendency of people to transform new territories into "transported landscapes". Extreme cases may be considered as examples of "ecological imperialism" noted by Alfred Crosby (1986).

Transported landscapes came to dominate much of Remote Oceania (Kirch 2000:109), yet the effects were perhaps less pronounced in the Mariana Islands. Plant foods (namely taro, yams, banana, and assorted tuber and tree crops) certainly were imported by the first settlers (Athens et al. 2004; Athens and Ward 2004), but formalised agricultural field systems such as terraces, mounds, and other constructions did not develop here as they did elsewhere in the Pacific Islands. Additionally, viable stocks of domesticated animals did not accompany the founding generations of people in the Marianas, although pigs, dogs, and chickens were key parts of the transported landscapes in other islands (Wickler 2004). Native birds and other wildlife suffered dramatic population loss with first human arrivals almost everywhere in Remote Oceania (Steadman 1995), but this pattern has not been documented in the Mariana Islands.

Rather than the siege of ecological imperialism unleashed elsewhere in Remote Oceania, notions of a "managed forest" or more broadly a managed ecosystem (of land and sea habitats) may be more realistic. Tree crops and root-tuber crops supplied a nutritional base without investing in terrain-changing earthworks, irrigation systems, and other overt manipulations of the ecological system. Coastal and marine zones likewise provided key resources without demanding large-scale indelible transformations of the natural environment.

When considering the imported plant species in the Marianas, the role of rice cultivation has been a curious anomaly here, not known in any other Remote Oceanic island prior to European Contact. Rice presumably was pre-Spanish in origin, yet so far it has not been dated archaeologically (Hunter-Anderson et al. 1995). The first written record of rice in the Mariana Islands was in Andres de Urdaneta's report to imperial authorities in Spain in 1537, referring to rice as a traded commodity in September 1526 (translated in Barratt 2003:41–42). The Mariana Islands lacked the large-scale formalised field systems that otherwise re-shaped entire landscapes in other regions where rice-farming was prevalent (Bellwood 2011). Rice may have been grown in small household plots of just a few sq m each, as was the case in China prior to the development of formal irrigation systems (Hung 2014). Rice may have been reserved for special occasions, as hinted in historical records of Juan Pobre's visit in the island of Rota in 1602 (translated in Driver 1993:30).

Another curious anomaly in the Marianas is the absence of evidence for rats until after A.D. 900–1000, as compared to the arrival of rats invariably with the first people elsewhere throughout the Pacific Islands (Storey et al. 2013; Wickler 2004). This delayed timing in the Marianas coincided with the oldest dates of sites with stone pillar-raised houses known as *latte*, as well as the first large decline in native bird populations (Pregill and Steadman 2009). Further correlations may yet be expected with increasing human population size and overseas contacts.

The modern setting is characterised by unusually high numbers of foreign-introduced plants and animals to some extent forcing de-population of native taxa. Most forests today are dominated by modern and historical introductions of *tantangtangan* trees, *limondechina* thorny bushes, and tough vines of false rattan. The brown tree snake (introduced from the New Guinea region in the middle twentieth Century) has been responsible for massive decline of native birds. Water buffalo (*carabao*), deer, pigs, dogs, cats, cattle, horses, and other animals all were introduced since the Spanish colonial era, with profound effects on the fragile island ecosystem that had evolved for thousands of years without these invasive components.

Climate and Weather

At least a few traits of the local climate have influenced cultural history and practice in the Mariana Islands. These same traits also have influenced the basic facts of life in regards to what kinds of plants and animals can survive in this environment. In simple terms, three key factors are (1) hot and humid conditions year-round; (2) 6 months of sustained heavy rains, juxtaposed with another 6 months of lesser rains each year; and (3) periodic exposure to strong typhoon systems.

The local climate indisputably has shaped daily activities and long-term planning in the Marianas. Prior to Spanish arrivals, clothing was minimal and made for allowing bodies to cool without constrictive covering, and houses likewise were designed for facilitating cooling effects of cross-breeze. Daily hard labour tasks always required preparation of drinkable water or availability of coconuts, and only limited actions could be planned for the hottest times of mid-day or afternoon. During extended heavy raining periods, many activities could not be managed at all, and major projects typically must be completed within the more favourable seasons. Given the rapid and thick vegetation growth in this climate, any endeavours of construction, landscaping, and gardening required intensive and ongoing investment in maintenance, and a brief period of neglect quickly could make a site easier to abandon than to resurrect. Moreover, periodic typhoons inevitably would cause irreparable destruction, so large-scale capital investments were in a sense impractical.

Due to their location on the earth's globe, the Mariana Islands are situated in a wet and warm environment, characterised generally as a humid tropical climate. Annual rainfall typically exceeds 2000 mm (78 in.), and daily temperatures are stable year-round at 29–31 °C (84.2–87.8 °F). Heavy rains occur regularly, more so in the southern islands than in the northern islands. Humidity noticeably decreases in farther north latitude.

The Mariana Islands are affected by the same monsoon weather patterns that affect East Asia, so that warm and moist air creates a "rainy season" for 6 months of each year, approximately May through October. During these 6 months, rainfall is expected almost every day. During the opposing 6 months of a "dry season" November through April, several consecutive days can pass without rain, but overall these months bring 40 % of the annual total rainfall.

The region's strong weather systems periodically can develop into typhoons, and indeed the Mariana Islands are located within "typhoon alley" of the northwest Pacific (Fig. 3.19). Every day of the year, the residents of Guam are on notice for the possibility of a typhoon developing within 72 h. Typhoons minimally sustain winds of 118 km per hour (64 knots), and they bring exceptionally heavy rainfall. These storms create intense low pressure and strong storm surge along coasts and low-elevation zones.

Winds and weather systems come mostly (but not entirely) from the east, so that each island includes a "windward" east side and a "leeward" west side. The land mass of each island creates slight difference in rainfall from windward to leeward,

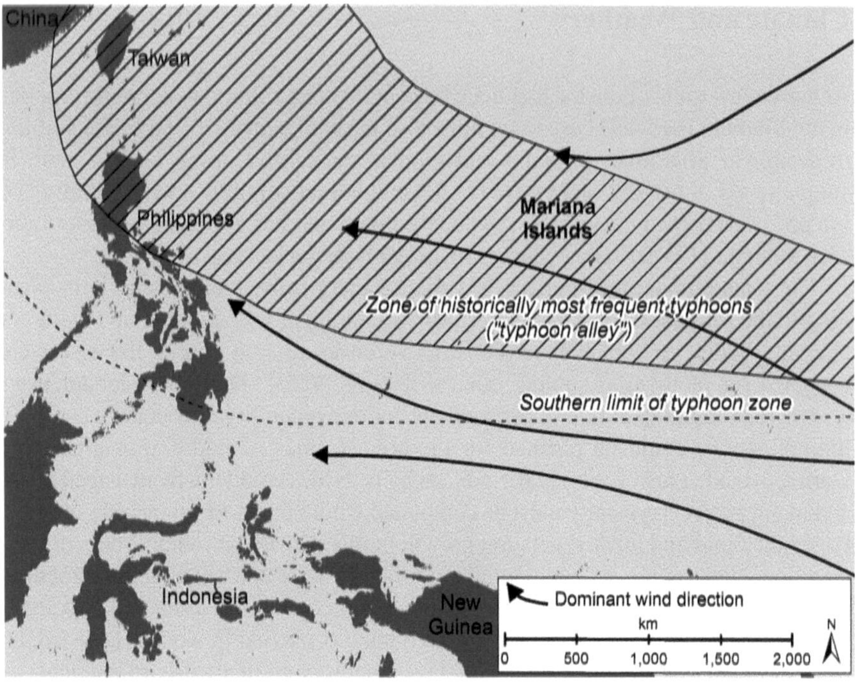

Fig. 3.19 General patterns of winds and typhoons in the Asia-Pacific region

although the rain shadow effect is minimal compared to other regions with taller mountain masses. The windward/leeward distinction certainly exists in the Marianas, but it does not carry the life-controlling implications of wet/dry contrast as known in larger and more mountainous Pacific Islands with prominent effects of orographically produced rainfall in windward and upland zones, steep rain gradients along mountain slopes, and leeward rain shadows. Many of the islands of Polynesia incorporate strongly differentiated wet windward versus dry leeward sides, with profound effects on social-ecological systems, yet the effects in the Marianas are greatly subdued by comparison.

The windward/leeward distinction in the Marianas has been formalised in recognising the windward side as the "back" (*tatte*) of an island as compared to the leeward side as the "front" (*me'nan*) of an island. This description may seem contrary to the direction of weather systems that approach an island from the windward side, arguably the front face in such a perspective. The "front" in fact refers to the leeward (west) side of an island in the Marianas, because canoes traditionally approach an island for landing on this side. Otherwise, an approach from the "back" or windward (east) side very well could result in a crash landing against the coast.

Notions of "wet and dry" in the Marianas refer to seasons of the year, but they do not apply to the separation of windward versus leeward environmental zones as in

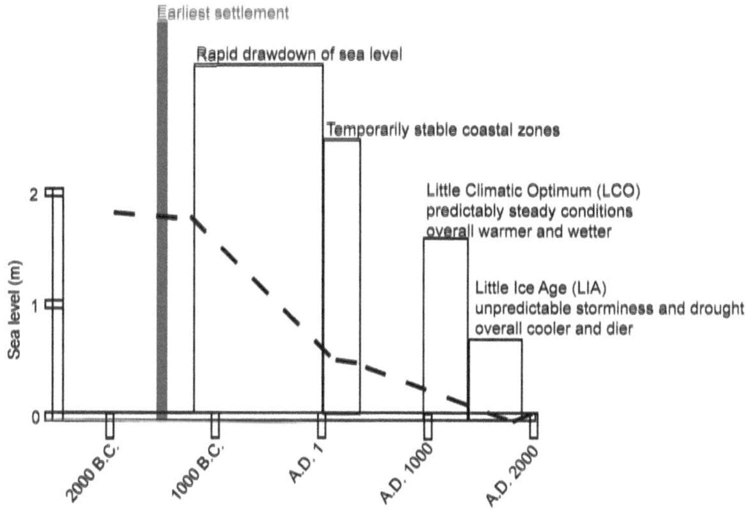

Fig. 3.20 General trends in climate and sea level in the Marianas region

other Pacific Islands. At any given time, the climate is reasonably similar all across the Mariana Islands. Nonetheless, local residents are very much aware of subtle differences in rainfall, temperature, wind exposure, humidity, and other factors that vary from one locality to another.

Conditions of year-round warmth and humidity, seasonal rainfall pattern, and periodic typhoons generally persisted throughout the time range of human presence in the Marianas, as well as for most of the last 10,000 years of the world's Holocene climate. When people first arrived in these islands about 3500 years ago, they found a natural ecosystem that already had adapted to the local conditions for at least a few thousand years. Likewise, cultural traditions adapted to the realities of the local climate and weather patterns, although substantial change in climate at different times may have prompted people to adjust their behaviours. Periodic seasonal shifts in winds would not have altered the overall trends and patterns of life and landscape, but they would have allowed for variable activities of sailing directions and other opportunities.

Fluctuations in climate have followed global trends (Fig. 3.20), within which two factors are most concerning in the Marianas. Firstly, even low-magnitude change in sea level (often driven by global climate) could create important transformations of coastal zones and ecologies, as apparently occurred after 1100–1000 B.C. Secondly, any change in the regularity of rainfall and storminess could create overall environmental instability and potential crisis of plant and animal resources, especially pronounced when a long period of stability was followed by a change to less stable conditions, for instance as occurred after A.D. 1300 with a change from the Little Climatic Optimum to the Little Ice Age (Nunn 2000).

Water Sources

Human life depends on availability of fresh water, possible in the Mariana Islands through collection and storage of rain water, as well as access to streams, seeps, and aquifers (Fig. 3.21). Rainfall is plentiful overall, but people have needed to develop techniques for storage and purification of water. Pottery vessels likely were important for capturing and holding water, as well as for purifying through boiling. Rain-fed streams occur only in a few places, notably in southern Guam and southeastern Saipan, but fresh lenses of water potentially can be found in all islands in underground aquifers and seeps. Freshwater wetlands or ponds are found in a number of places of Guam, Tinian, and Saipan.

Water may have been a primary factor in supporting greater populations in the southern islands and lesser numbers of people in the northern islands. The southern-arc islands with larger mass provide the most opportunities for accessing fresh water in streams (Fig. 3.22), in drips from cave ceilings (Fig. 3.23), in standing pools or portions of aquifers exposed in caves (Figs. 3.24 and 3.25), and in seeps emerging in low-elevation coastal plain landforms (Fig. 3.26). Additionally, these same southern-arc islands tend to receive slightly more annual rainfall, so they offer more confidence in sustaining a water supply between raining episodes.

A change in base sea level can affect the accessibility of ground water sources (see Fig. 3.21). When the sea level lowered after about 1100–1000 B.C., the floating lens of fresh water accordingly lowered with it. As a result, pools of water inside

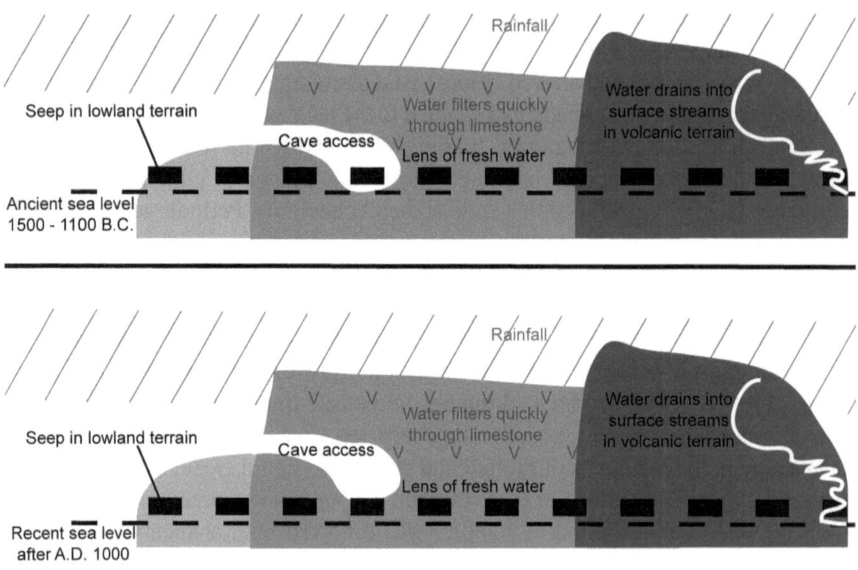

Fig. 3.21 Potential water access in aquifer-tapping caves, natural seeps, and rain-fed streams with potential effects of a change in sea level. Schematic diagram is not to scale

Fig. 3.22 Stream in southern Guam, with dam constructed during the Japanese occupation of the 1940s

Fig. 3.23 Water drips constantly but at variable rates from ceilings of limestone caves, with more volume of water according to greater surface area

Fig. 3.24 Standing pools of water can accumulate in caves where the floors do not allow drainage into a lower water table

Fig. 3.25 Portions of an island's aquifer can be exposed inside some deep caves

Fig. 3.26 Drinkable water can be accessed in coastal seeps. Photograph by Hiro Kurashina

caves became less accessible, and seeps in coastal lowlands shifted to new locations. Meanwhile, stream drainages were altered very slightly, so that the water could drain into the lower sea level, but these alterations became more substantial in combination with other factors of slope erosion and re-deposition. Caves provided predictable supplies of naturally filtered water dripping from ceilings, although the rate of dripping varied considerably from one cave to another.

References

Anderson, E. (1952). *Plants, man and life*. Boston: Little, Brown and Company.

Athens, J. S., Dega, M. F., & Ward, J. V. (2004). Austronesian colonisation of the Mariana Islands: The palaeoenvironmental evidence. *Bulletin of the Indo-Pacific Prehistory Association, 24*, 21–30.

Athens, J. S., & Ward, J. V. (2004). Holocene vegetation, savannah origins and human settlement of Guam. In V. Attenbrow & R. Fullagar (Eds.), *A Pacific odyssey: Archaeology and anthropology in the western Pacific: Papers in honour of Jim Specht* (pp. 15–30). Sydney: Australian Museum. Records of the Australian Museum, Supplement 29.

Barratt, G., (2003). *The Chamorros of the Mariana Islands: Early European records, 1521–1721*. Occasional Historical Papers Series, Number 10. Commonwealth of the Northern Mariana Islands Division of Historic Preservation, Saipan.

Bellwood, P. (2011). The checkered prehistory of rice movement southwards as a domesticated cereal: From the Yangzi to the Equator. *Rice, 4*, 93–103.

Carson, M. T. (2011). Palaeohabitat of first settlement sites 1500–1000 B.C. in Guam, Mariana Islands, western Pacific. *Journal of Archaeological Science, 38*, 2207–2221.

Carson, M. T. (2014). Paleoterrain research: Finding the first settlement sites of Remote Oceania. *Geoarchaeology, 29*, 268–275.

Carson, M. T., & Peterson, J. A. (2012). Radiocarbon dating of algal bioclasts in beach sites of Guam. *Journal of Island and Coastal Archaeology, 7*, 64–75.

Cloud, P. E., Jr., Schmidt, R. G., & Burke H. W. (1956). *Geology of Saipan, part 1: General geology*. United States Geological Survey Professional Paper 280-A. Washington, DC: Government Printing Office.

Crosby, A. W. (1986). *Ecological imperialism: The biological expansion of Europe, 900–1900*. Cambridge: Cambridge University Press.

Cunningham, L. J. (1992). *Ancient Chamorro society*. Honolulu: Bess Press.

Dickinson, W. R. (2000). Hydro-isostatic and tectonic influences on emergent Holocene paleo-shorelines in the Mariana Islands, western Pacific Ocean. *Journal of Coastal Research, 16*, 735–746.

Dickinson, W. R. (2001). Paleoshoreline record of relative Holocene sea levels on Pacific islands. *Earth-Science Reviews, 55*, 191–234.

Dickinson, W. R. (2003). Impact of Mid-Holocene hydro-isostatic highstand in regional sea level on habitability of islands in Pacific Oceania. *Journal of Coastal Research, 19*, 489–502.

Driver, M. G. (1993). *Fray Juan Pobre in the Marianas, 1602*. Miscellaneous Series Number 8. Micronesian Area Research Center, University of Guam, Mangilao.

Holliday, V. T. (Ed.). (1992). *Soils in archaeology: Landscape evolution and human occupation*. Washington, DC: Smithsonian Institution Press.

Hung, H. C. (2014). Rice in China. In H. Selin (Ed.), *Encyclopedia of the history of science, technology, and medicine in non-western cultures*. Dordrecht: Springer Science and Business Media. doi:10.1007/978-94-007-3934-5_10026-1.

Hunter-Anderson, R. L., Thompson, G. B., & Moore, D. R. (1995). Rice as a prehistoric valuable in the Mariana Islands, Micronesia. *Asian Perspectives, 34*, 69–89.

Hussong, D. M., & Uyeda, S. (1981). Tectonic processes and the history of the Mariana Arc: A synthesis of the results of deep sea drilling project Leg 60. In M. Lee & R. Powell (Eds.), *Initial reports of the deep sea drilling project* (Vol. 60, pp. 909–929). Washington, DC: Government Printing Office.

Kirch, P. V. (2000). *On the road of the winds: An archaeological history of the Pacific Islands before European contact*. Berkeley: University of California Press.

McKinnon, J., Mushynsky, J., & Cabrera, G. (2014). A fluid sea in the Mariana Islands: Community archaeology and mapping the seascape of Saipan. *Journal of Maritime Archaeology, 9*, 59–79.

Nunn, P. D. (2000). Environmental catastrophe in the Pacific Islands around A.D. 1300. *Geoarchaeology, 15*, 715–740.

Pregill, G. K. (1998). Squamate reptiles from prehistoric sites in the Mariana Islands. *Copeia, 1998*, 64–75.

Pregill, G. K., & Steadman, D. W. (2009). The prehistory and biogeography of terrestrial vertebrates in Guam, Mariana Islands. *Diversity and Distributions, 15*, 983–996.

Spriggs, M. (1997). Landscape catastrophe and landscape enhancement: Are either or both true in the Pacific? In P. V. Kirch & T. L. Hunt (Eds.), *Historical ecology in the Pacific Islands: Prehistoric environmental and landscape change* (pp. 80–104). New Haven, CT: Yale University Press.

Steadman, D. W. (1995). Prehistoric extinctions of Pacific Island birds: Biodiversity meets zooarchaeology. *Science, 267*, 1123–1131.

Storey, A. A., Clarke, A. C., Ladefoged, T., Robins, J., & Matisoo-Smith, E. (2013). DNA and Pacific commensal models: Applications, construction, limitations, and future prospects. *Journal of Island and Coastal Archaeology, 8*, 37–65.

Tracey, J. I., Jr., Schlanger, S. O., Stark, J. T., Doan, D. B., & May, H. G. (1964). *General Geology of Guam*. Geological Survey Professional Paper 403-A. Washington, DC: United States Government Printing Office.

Ward, J. V. (1994). A Holocene pollen record from the Pago River Valley, Guam. In R. L. Hunter-Anderson (Eds.) *Archaeology in Manenggon Hills, Yona, Guam* (Vol. II, pp. 9.34–9.51). Report prepared for MDI Guam Corporation. Micronesian Archaeological Research Services, Mangilao, Guam.

Wickler, S. K. (2004). Modelling colonisation and migration in Micronesia from a zooarchaeological perspective. In M. Mondini, S. Munoz, & S. K. Wickler (Eds.), *Colonisation, Migration and marginal areas: A zooarchaeological approach* (pp. 28–40). Oxford: Oxbow Books.

Young, F. J. (1988). *Soil survey of territory of Guam*. Washington, DC: United States Department of Agriculture, Soil Conservation Service.

Young, F. J. (1989). *Soil survey of the Islands of Aguijan, Rota, Saipan, and Tinian, Commonwealth of the Northern Mariana Islands*. Washington, DC: United States Department of Agriculture, Soil Conservation Service.

Chapter 4
Marianas Archaeology in Local and Regional Perspectives

As a context for exploring the changing landscape contexts in later chapters of this book, this chapter considers the role of Marianas archaeology for learning about the surrounding Asia-Pacific region and for learning about the past in more general terms. Much of the significance of the Marianas case study here depends on how this particular case relates with the changing landscapes of a larger part of the world beyond the shores of these small islands. In the international arena, the Mariana Islands are known for unexpectedly old habitation dates, remarkable isolation within the Remote Oceanic region, and megalithic ruins of the later *latte* period. Meanwhile in the perspective of many local residents, archaeology offers one of several ways to learn about the past and about native Chamorro cultural heritage and identity, so an account of Marianas landscape evolution may be regarded as necessarily coordinating archaeology with other studies.

Marianas Settlement in Asia-Pacific Context

The Mariana Islands occupy a critical position in the Asia-Pacific region, as important for modern global military strategies as for archaeological studies of the distant past. The Marianas constitute a frontier joining Asia and the Pacific. In some ways, the Marianas may be viewed as a remote island extension of Southeast Asia, but in other ways they resemble an isolated Pacific Oceanic setting with strong retention of Southeast Asian roots. The frontier zone characteristics apply to the geology, plant and animal life, and natural environment as much as to cultural heritage.

The geographic location of the Marianas lies in the northwest of Remote Oceania (sees Fig. 1.1 and 3.7). Deep ocean effectively isolates the Remote Oceanic region from the rest of the world, with remote-distance water gaps of at least 350 km (Green 1991), and the Mariana Islands are situated much farther than this minimal qualification of 350 km from the next nearest other islands. This distance requires a

M.T. Carson, *Archaeological Landscape Evolution*,
DOI 10.1007/978-3-319-31400-6_4

voyage outside the range of inter-visible landfalls, so that sailors necessarily would lose sight of land for some time (Irwin 1998). This same barrier greatly inhibited transfer of plants and animals from larger land masses and biogeographic zones of Island Southeast Asia and Near Oceania. People crossed short water gaps and lived as far as Australia and New Guinea prior to 40,000 years ago (Bellwood and Hiscock 2013; Mulvaney and Kamminga 1999), but they did not undertake water-crossing migrations more than 350 km until much later.

Over many thousands of years, hundreds of generations of people lived as hunter-gatherers in Island Southeast Asia and Near Oceania, living in non-sedentary camps and relying mostly (but not always entirely) on "wild" foods that could be found there naturally in great abundance. The campsites apparently were occupied seasonally or for short periods, while people moved from one place to another repeatedly but not continuously. This pattern of broad-ranging mobility differed significantly from a long-term sedentary lifestyle invested in one place. Arboriculture and limited horticulture were practiced in some areas, such as in the New Guinea Highlands (Denham 2011) and probably some parts of the Philippines and Indonesia (Denham 2013; Paz 2002, 2005), but people in these areas did not engage in formalised agriculture or develop structured residential settlement systems until much later.

In the Island Southeast Asian and Near Oceanic environment of abundant and diverse biomass, life-sustaining foods were accessible in numerous forms year-round. The natural ecologies were highly resilient against periodic small-scale shifts in climate. By comparison, the prospects in Remote Oceania were considerably less attractive in their natural state of lower biodiversity, so any attempts to live there long term would require the radical concepts of importing plants and perhaps animals across the ocean and of deliberately modifying the island ecologies in support of human populations.

The environments of hunter-gatherers for thousands of years provided reliably ample foods and resources for small groups of people, most comfortable if these groups could maintain low-density impacts. Hunters, foragers, and fishers lived mostly in small campsites, apparently with a degree of mobility from one place to another. Stone tools were the most durable artefacts in high-domed cave shelters, rocky cliff overhangs, and other ancient sites, while pottery was notably absent. Also missing were the durable footprints of sedentary villages and agricultural fields.

Throughout Island Southeast Asia and Near Oceania, no pottery at all is evident until the sudden appearance of red-slipped earthenware (Fig. 4.1). Pottery-bearing deposits created obvious horizons in stratigraphic layers, amenable for archaeological dating across the Asia-Pacific region (Bellwood et al. 2011). Pottery-making emerged in Taiwan by 4000 B.C. or possibly earlier (Hung 2008; Hung and Carson 2014), derived from much older traditions across the Taiwan Strait in coastal southeast China. Thereafter, horizons of red-slipped pottery have been verified in the northern Philippines by 2000 B.C. (Bellwood and Dizon 2013; Hung 2008; Hung et al. 2011), next in the Mariana Islands about 1500 B.C. but perhaps earlier (Carson and Kurashina 2012), in various parts of Indonesia after 1500 B.C. (Bellwood 1997:219–241; Simanjuntak 2008), and finally about 1350 B.C. in the Bismarck Archipelago east of New Guinea (Summerhayes 2007). The pottery-making

Fig. 4.1 Examples of early decorated red-slipped pottery, from 2011 to 2013 excavations landward of House of Taga, Tinian

traditions in the Bismarck Archipelago later were transferred with people migrating into other parts of Island Melanesia and West Polynesia after 1200 B.C. and continuing through 800 B.C. (Kirch 1997).

The "pottery trail" can be traced across the Asia-Pacific region, as the hard evidence of the first people who settled in Remote Oceania (Carson et al. 2013). This material archaeological horizon first emerged as a new tradition overlaying several millennia of hunter-gatherer lifestyle in Island Southeast Asia. As outlined above, its slightly later-dated manifestations represented the first human occupations ever to occur in the small and isolated islands of Remote Oceania (Carson 2014b).

The first settlement of Remote Oceania most often has been studied from a Polynesian perspective, tracing the distant ancestral origins of Polynesians in Hawai'i and New Zealand back westwards to a heartland region of Fiji, Tonga, and Samoa and then reaching farther back to an older homeland in the Bismarck Archipelago in Near Oceania (Kirch 2000; Kirch and Green 2001). In this view, Remote Oceanic settlement emerged as the legacy of people who made dentate-stamped Lapita-style pottery and who crossed from Near to Remote Oceania (Kirch 1997; Spriggs 1997). The oldest known Near Oceanic Lapita sites are found in the Bismarck Archipelago about 1350 B.C. (Denham et al. 2012; Kirch 2001; Summerhayes 2007). Eventually, the Lapita Cultural Complex spread into the Remote Oceanic islands of Southern Melanesia and West Polynesia after 1200 B.C. (Green et al. 2008) but mostly in the range of 1100–800 B.C. (Bedford et al. 2006; Burley and Dickinson 2001; Burley et al. 2012; Nunn and Petchey 2013; Sand 1997).

Near Oceanic Lapita origins for Remote Oceanic Polynesians have become abundantly clear since the 1950s (Kirch 2000), but in fact the first settlement of Remote Oceania occurred earlier and in an entirely different location, about 1500 B.C. in the Mariana Islands of far western Micronesia (Carson 2014b; Carson and Kurashina 2012). This possibility was known ever since Alexander Spoehr's (1957) discovery of early red-slipped and decorated pottery in the Marianas (see also Pellett and Spoehr 1961). After finding similar red-slipped pottery in the Philippines, Spoehr (1973:274) proposed an explicit research challenge: "Although the archaeological evidence is limited, work has now progressed to the point where a systematic attack can be made on the question of the prehistoric relations of Micronesia with Island Southeast Asia. The time level involved in these relations remains uncertain, but I believe will be found to antedate 1000 B.C. and may well have commenced at a much earlier date. In addition to the Philippines, the northern Celebes, Moluccas, and Halmahera must be brought within the sphere of archaeological investigation of the problem".

Appreciable advance has been made toward the goal of a "systematic attack" of finding the origins of Marianas settlers somewhere in Island Southeast Asia, but the results often are not integrated into larger Asia-Pacific archaeology narratives. The major regional research focus has involved tracing Polynesian origins as far back as Lapita sites in the Bismarck Archipelago, with little or no update about archaeological discoveries in the Mariana Islands or in the vast region of Island Southeast Asia. Despite several encouraging discoveries about early Marianas settlement (Athens et al. 2004; Athens and Ward 2004; Bonhomme and Craib 1987; Butler 1994; Carson 2008, 2012a; Craib 1993, 1999; Dilli et al. 1998; Haun et al. 1999; Kurashina and Clayshulte 1983a, b; Kurashina et al. 1981; Moore et al. 1992), the evidence was not accepted into mainstream views of Asia-Pacific archaeology until recently (Carson 2014b). Meanwhile, archaeological knowledge of Island Southeast Asia has grown considerably more extensive than had been the case some decades ago (Bellwood 1997), but only very few attempts have been made for cross-regional comparison of those findings with the records of the Mariana Islands or anywhere else in Pacific Oceania (Carson et al. 2013; Hung et al. 2011).

The earlier and different direction of Remote Oceanic settlement in the Marianas required the longest ocean-crossing migration of its time in human history, exceeding 2000 km (Craib 1999; Hung et al. 2011). This impressive feat can be compared to the longest Lapita ocean-crossing of 900 km in Melanesia-Polynesia (Fig. 4.2). Longer distance of seaborne migration did not occur until much later in the Pacific, about A.D. 1000 for settlement of the farthest margins of East Polynesia in Hawai'i and Rapa Nui (Easter Island) and as late as A.D. 1300 in New Zealand (Kirch 2010).

If early-period Marianas settlement could be proven, then it would necessitate a re-writing or at least a serious re-evaluation of Asia-Pacific archaeology. First settlement of Remote Oceania would need to be acknowledged as having occurred 1500 B.C. in the Marianas, older than the appearance of Lapita pottery in Melanesia-Polynesia a few centuries later. Moreover, the origins of earliest Marianas settlement necessarily would need to be traced to an ancestral homeland that pre-dated Lapita in Near Oceania and instead pointed to older traditions in Island Southeast Asia. These implications diverged rather disturbingly from the orthodox view of Lapita as the founding ancestry of the Remote Oceanic world.

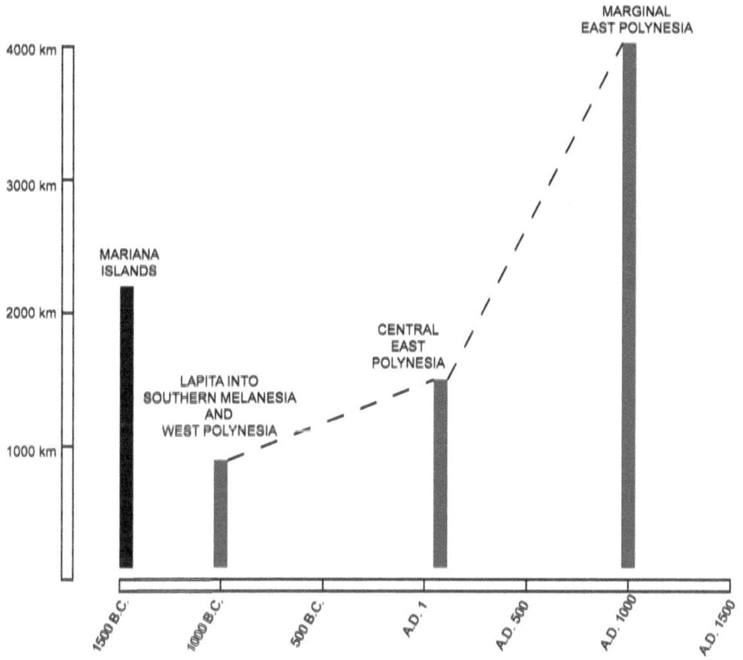

Fig. 4.2 Pacific voyaging distances and chronology

Despite the potentially controversial implications, Spriggs (1999:20) identified early-period Marianas settlement as the "smoking gun" of an Island Southeast Asian source for the populations who later entered Remote Oceania (see also Spriggs 2007:113–114). Otherwise, Remote Oceanic origins have been hotly debated from a Lapita-focused perspective. Cultural roots of Lapita could be attributed in variable ways to an indigenous in situ development in the Bismarck Archipelago (Allen 1996; Allen and Gosden 1996; Terrell and Welsch 1997), to intrusive populations from Island Southeast Asia (Bellwood 1991; Bellwood 2007; Bellwood et al. 1995), or to a combination of both factors (Green 2000; Spriggs 1997).

For a stronger grasp of Marianas archaeology in a broader Asia-Pacific context, external observers rightfully have called for more intensive research of the earliest sites (Bellwood 2007; Shutler 1999), but this topic received almost no direct fieldwork attention internally within the Marianas until just recently. Throughout more than 100 years of archaeological practice in the Marianas (reviewed by Carson 2012b), beginning with Antoine-Alfred Marche's expedition in 1887 (Marche 1889, 1982), surprisingly little work has related to the earliest settlement period. Instead, nearly all efforts focused on the period of megalithic house-pillar ruins called *latte*, dating after A.D. 1000 (Carson 2012c; Laguana et al. 2012). Others have focused on historical studies of Spanish-era events of the 1600s through 1800s and even more recent events such as World War II.

Studies of early-period Marianas sites can address several important questions about the precise dating of first settlement, contexts of ancient sites, initial encounters between human groups and the Remote Oceanic environment, and long-distance links with possible homeland areas. These and other issues are discussed in Chaps. 8 and 15, within the general scope of ancient life and landscape. An emergent synthesis of first Marianas settlement has been possible within the last few years (Carson 2014b), building on investigations of ancient terrain landforms and habitats (Carson 2011, 2014a), critical review of radiocarbon dating (Carson and Kurashina 2012), and comparison of the early-period pottery and other artefacts with findings in Island Southeast Asia and Near Oceania (Carson et al. 2013; Hung et al. 2011).

Foundations of Chamorro Heritage

Marianas archaeology of course involves much more than the earliest sites, and undeniably the vast majority of research has concentrated on much later periods, especially concerning the centuries associated with *latte* house-pillar ruins, dated approximately A.D. 1000 through 1700. The native Chamorro culture during this period has become a familiar and proximal reference point of indigenous identity and heritage, as the last time when a native culture operated prior to the devastation brought by the Spanish-Chamorro wars of the late 1600s. *Latte* serve as icons of Chamorro culture today (Fig. 4.3), with special concerns linking archaeological research and cultural heritage (Kurashina et al. 1999).

Latte sites and complexes are found in each of the Mariana Islands (Russell 1998; Yawata 1945), easily accessible in surface-visible and near-surface contexts that captivate modern perceptions of Marianas cultural history and archaeology (Figs. 4.4 and 4.5). These impressive sites attracted the curiosity of European visitors such as Lord Anson in 1742 (Barratt 1988), as well as the first archaeologists in the region such as Antoine-Alfred Marche in 1887 (Marche 1889, 1982) and Hans Hornbostel in the 1920s (Hornbostel 1925; see also Thompson 1932). Substantial research is possible

Fig. 4.3 The Latte of Freedom in Adelup, Guam. The concrete structure stands about 24–25 m tall, currently housing the Guam Hall of Governors

Fig. 4.4 Drawing of *latte* ruins at House of Taga in Tinian, drawn by Peircy Brett during Lord Anson's visit in 1742. Copy of image at the Micronesian Area Research Center, University of Guam

Fig. 4.5 Preserved *latte* set in the US Navy Royal Palms housing complex, Guam. Scale bar is in 20-cm increments

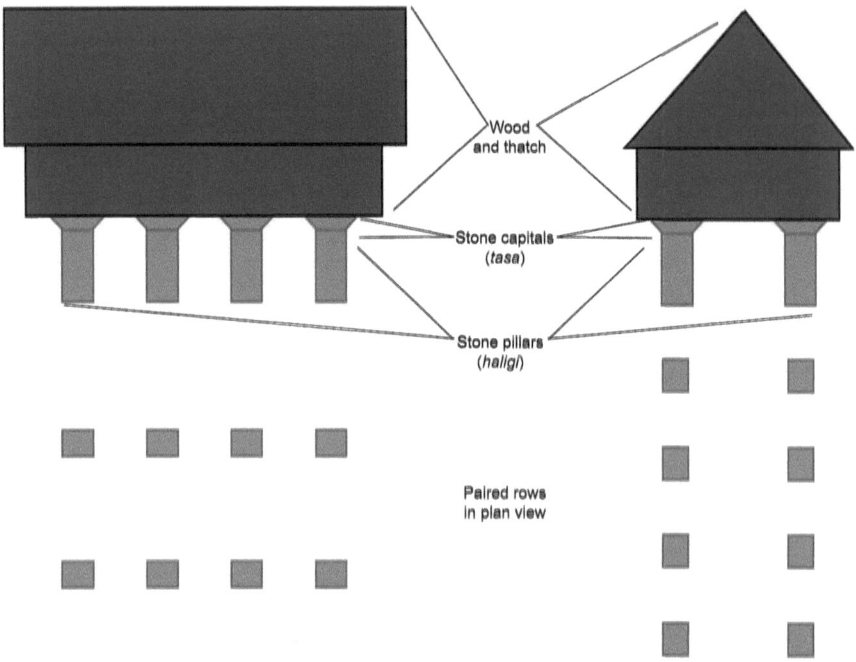

Fig. 4.6 Major structural components of *latte*, showing pillars (*haligi*), capitals (*tasa*), and superstructure

with very little or even zero subsurface excavation work, for example relying on visual observation, mapping, and recording in combination with possibly some controlled collections of artefacts and midden on the surface of abandoned sites.

By the time archaeologists were interested in studying *latte* or any other vestiges of Marianas archaeology, the *latte* houses had been abandoned for more than a century, and only the stone pillars (*haligi*) and capital stones (*tasa*) remained without any surviving wooden or thatch super-structure (Fig. 4.6). Nonetheless, the apparent pile-raised house forms and available historical notes communicate a strong continuity with houses known in Island Southeast Asia (Laguana et al. 2012). The rudimentary technological function of supporting a house is unquestionable (Thompson 1940) although any house conceivably may have involved various activities of a dwelling, storage facility, workshop, high or low status residence, place of family lineage, ancestral burial, and other associations.

Latte represent an ancient tradition of pile-raised Austronesian houses, found throughout Island Southeast Asia, Near Oceania, and parts of Micronesia and Melanesia (Fox 1993; Waterson 1997). By comparison, house forms differed in later-occupied regions such as Polynesia, where instead houses typically were built at ground level on stonework foundations, without pile-raiser supports (Green 1970, 1986, 1993). The persistence of a classic Western Austronesian house form in the Marianas therefore suggests a different cultural history than was the case for the descendants of Lapita pottery-makers in Polynesia.

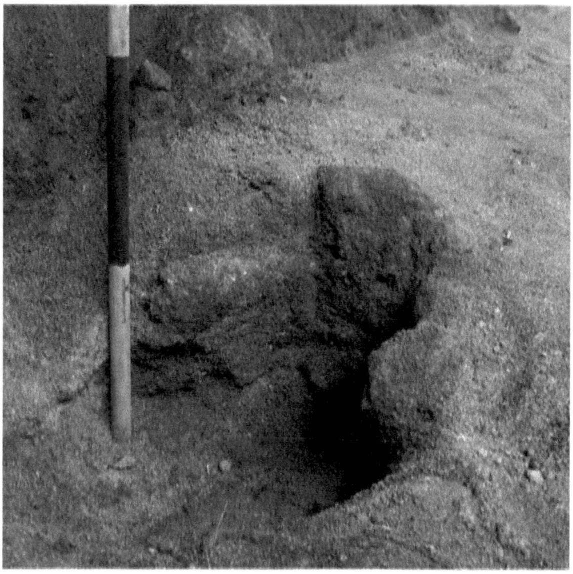

Fig. 4.7 Example of post mould with stone bracing, landward of House of Taga, Tinian. Scale bar is in 20-cm increments

In a large-scale cross-regional view, the formalisation of *latte* pillars and capitals into stone was unique to the Marianas (Peterson 2012). In Island Southeast Asia, these components were made of wood. In a few parts of Micronesia, stone pillars were used without overtopping capitals. While varied hypotheses have been proposed, the reasons for this anomaly in the Marianas are not yet understood.

The use of stone for *latte* began around A.D. 1000, whereas older houses were supported by wooden posts. In older archaeological layers, wooden posts have disintegrated, leaving behind cylindrical moulds filled with dark staining from the decomposed wood and in-filled sediment (Fig. 4.7). In many cases, bracing stones are found surrounding the post moulds.

Stone elements of *latte* signalled a new cultural development in the Marianas, manifest in the archaeological record as the "*latte* period" (Carson 2012c). *Latte* sites are coincident with a suite of other material markers such as thickened-rim large earthenware pots, slingstones, and stone grinding basins or mortars (*lusong*). Given the abundance of this material throughout the islands, Marianas archaeology has been virtually synonymous with studies of the *latte* period, even now more than 50 years after Alexander Spoehr's (1957) discovery of deeply stratified and older cultural layers by far pre-dating the *latte* period.

Latte are recognised as a tangible link to the native Chamorro past. The abandoned *latte* sites today are respected as homes of ancestral spirits (Kurashina et al. 1999). The ancestors continue to make active contributions in daily life today, and they may be encountered in or around *latte* ruins where their spiritual connections are strongest, as described by Farrer and Sellmann (2014).

Knowledge of the *latte* period is well attested archaeologically, yet living heritage relations with older time periods have not yet been pursued in their fullest potential. Partly due to the comparatively few studies of any sites older than the *latte* period, not much has become popularised or integrated into general education about these most ancient centuries. A time depth of thousands of years has been important in framing Chamorro identity, but this aspect has been unclear in its detail and familiarity as compared to the *latte* period. In an archaeological perspective, the *latte* period witnessed significant change during the centuries of A.D. 1000 through 1700, and certainly the 2500 years prior to the oldest *latte* sites involved much more than a singular "pre-*latte*" period.

The long-standing problem of cultural chronology is tackled in this book, as detailed in Chaps. 7 through 14. After the disclosure of primary datasets, larger research issues will be addressed in the concluding chapters. Over the last several decades, in the absence of relevant studies reported here, questions logically have arisen about the age of Marianas settlement, potential influences of cross-regional overseas connections, a possible population replacement at the beginning of the *latte* period, the development of social and political systems, and more. These questions have been addressed from multiple viewpoints, not always involving archaeology.

References

Allen, J. (1996). The pre-Austronesian settlement of Island Melanesia: Implications for Lapita archaeology. In W. H. Goodenough (Ed.), *Prehistoric Settlement of the pacific* (pp. 11–27). Philadelphia: American Philosophical Society. Transactions 86, Part 5.

Allen, J., & Gosden, C. (1996). Spheres of interaction and integration: Modelling the culture history of the Bismarck Archipelago. In J. M. Davidson, I. Geoffrey, B. Foss Leach, A. Pawley, & D. Brown (Eds.), *Oceanic culture history: Essays in honour of Roger Green* (pp. 183–197). Dunedin: New Zealand Journal of Archaeology Special Publication. New Zealand Archaeological Association.

Athens, J. S., Dega, M. F., & Ward, J. V. (2004). Austronesian colonisation of the Mariana Islands: The palaeoenvironmental evidence. *Bulletin of the Indo-Pacific Prehistory Association, 24*, 21–30.

Athens, J. S., & Ward, J. V. (2004). Holocene vegetation, savannah origins and human settlement of Guam. In V. Attenbrow & R. Fullagar (Eds.), *A Pacific Odyssey: Archaeology and anthropology in the Western Pacific—Papers in honour of Jim Specht* (pp. 15–30). Sydney: Australian Museum. Records of the Australian Museum, Supplement 29.

Barratt, G. (1988). *H.M.S. centurion at Tinian, 1742: The ethnographic and historic records.* Micronesian Archaeological Survey Report Number 26. Commonwealth of the Northern Mariana Islands Division of Historic Preservation, Saipan.

Bedford, S., Spriggs, M., & Regenvanu, R. (2006). The Teouma Lapita Site and the early human settlement of the Pacific Islands. *Antiquity, 80*, 812–828.

Bellwood, P. (1991). The Austronesian dispersal and the origin of languages. *Scientific American, 265*, 88–93.

Bellwood, P. (1997). *Prehistory of the Indo-Malaysian Archipelago* (Rev. ed.). University of Hawai'i Press, Honolulu.

Bellwood, P. (2007). Southeast China and the prehistory of Austronesians. In L.-T. Jiao (Ed.), *Lost maritime cultures: China and the Pacific* (pp. 36–53). Honolulu: Bishop Museum Press.

Bellwood, P., Chambers, G., Ross, M., & Hung, H.-c. (2011). Are 'cultures' inherited? Multidisciplinary perspectives on the origins and migrations of Austronesian-speaking peoples prior to 1000 BC. In B. Roberts & M. V. Linden (Eds.), *Investigating archaeological cultures: Material culture, variability, and transmission* (pp. 321–354). New York: Springer.

Bellwood, P., & Dizon, E. (Eds.). (2013). *4000 years of migration and cultural exchange: The archaeology of the Batanes Islands, Northern Philippines.* Terra Australis 40. Canberra: Australian National University E Press.

Bellwood, P., Fox, J. J., & Tryon, D. (Eds.). (1995). *The Austronesians: Historical and comparative perspectives*. Canberra: Department of Anthropology, Australian National University.
Bellwood, P., & Hiscock, P. (2013). Holocene Australia and the Pacific Basin. In C. Scarre (Ed.), *The human past* (pp. 264–305). London: Thames and Hudson.
Bonhomme, T., & Craib, J. L. (1987). Radiocarbon dates from Unai Bapot, Saipan: Implications for the prehistory of the Mariana Islands. *Journal of the Polynesian Society, 96*, 95–106.
Burley, D. V., & Dickinson, W. R. (2001). Origin and significance of a founding settlement in Polynesia. *Proceedings of the National Academy of Sciences, 98*, 11829–11831.
Burley, D. V., Weisler, M. I., & Zhao, J.-x. (2012). High precision U/Th dating of first Polynesian settlement. *PLoS One, 7*, e48769. doi:10.1371/journal.pone.0048769.
Butler, B. M. (1994). Early prehistoric settlement in the Mariana Islands: New evidence from Saipan. *Man and Culture in Oceania, 10*, 15–38.
Carson, M. T. (2008). Refining earliest settlement in Remote Oceania: Renewed archaeological investigations at Unai Bapot, Saipan. *Journal of Island and Coastal Archaeology, 3*, 115–139.
Carson, M. T. (2011). Palaeohabitat of first settlement sites 1500–1000 B.C. in Guam, Mariana Islands, western Pacific. *Journal of Archaeological Science, 38*, 2207–2221.
Carson, M. T. (2012a). Evolution of an Austronesian landscape: The Ritidian Site in Guam. *Journal of Austronesian Studies, 3*, 55–86.
Carson, M. T. (2012b). History of archaeological study in the Mariana Islands. *Micronesica, 42*, 312–371.
Carson, M. T. (2012c). An overview of latte period archaeology. *Micronesica, 42*, 1–79.
Carson, M. T. (2014a). Paleoterrain research: Finding the first settlement sites of Remote Oceania. *Geoarchaeology, 29*, 268–275.
Carson, M. T. (2014b). *First settlement of Remote Oceania: Earliest sites in the Mariana Islands*. New York: Springer.
Carson, M. T., Hung, H.-c., Summerhayes, G., & Bellwood, P. (2013). The pottery trail from Southeast Asia to Remote Oceania. *Journal of Island and Coastal Archaeology, 8*, 17–36.
Carson, M. T., & Kurashina, H. (2012). Re-envisioning long-distance Oceanic migration: Early dates in the Mariana Islands. *World Archaeology, 44*, 409–435.
Craib, J. L. (1993). Early occupation at Unai Chulu, Tinian, Commonwealth of the Northern Mariana Islands. *Bulletin of the Indo-Pacific Prehistory Association, 13*, 116–134.
Craib, J. L. (1999). Colonisation of the Mariana Islands: New evidence and implications for human settlement in the western Pacific. In J.-C. Galipaud & I. Lilley (Eds.), *Le Pacifique de 5000 à 2000 avant le Présent: Suppléments à l'Histoire d'une Colonisation* (pp. 477–485). Paris: Institut de Recherche pour la Développement.
Denham, T. P. (2011). Early agriculture and plant domestication in New Guinea and Island Southeast Asia. *Current Anthropology, 52*(S4), S379–S395.
Denham, T. P. (2013). Early farming in Island Southeast Asia: An alternative hypothesis. *Antiquity, 87*, 250–257.
Denham, T. P., Christopher, B. R., & Jim, S. (2012). Dating the appearance of Lapita pottery in the Bismarck Archipelago and its dispersal to Remote Oceania. *Archaeology in Oceania, 47*, 39–46.
Dilli, B. J., Alan, E. H., Susan, T. G., & Brian, D. (1998). *Archaeological mitigation program, Mangilao Golf Course project area, Mangilao Municipality, Territory of Guam. Volume I: Introduction, research design, and data recovery results*. Report prepared for Mr. Jetan Sahni. Paul H. Rosendahl, Ph.D., Inc., Hilo, HI.
Farrer, D. S., & Sellmann, J. D. (2014). Chants of re-enchantment: Chamorro spiritual resistance to colonial dominion. *Social Analysis, 58*, 127–148.
Fox, J. J. (Ed.). (1993). *Inside Austronesian houses: Perspectives on domestic designs for living*. Canberra: Department of Anthropology, Australian National University.
Green, R. C. (1970). Settlement pattern archaeology in Polynesia. In R. C. Green, M. Kelly (Eds.) *Studies in oceanic culture history* (Vol. 1, pp. 13–32). Pacific Anthropological Records Number 11. Department of Anthropology, Bishop Museum, Honolulu.
Green, R. C. (1986). Some basic components of the Ancestral Polynesian settlement system: Building blocks for more complex Polynesian societies. In P. V. Kirch (Ed.), *Island societies:*

Archaeological approaches to evolution and transformation (pp. 50–54). Cambridge: Cambridge University Press. New Directions in Archaeology.

Green, R. C. (1991). Near and Remote Oceania: Disestablishing "Melanesia" in culture history. In A. Pawley (Ed.), *Man and a half: Essays in Pacific Anthropology and Ethnobiology in honour of Raplh Bulmer* (pp. 491–502). Auckland: The Polynesian Society.

Green, R. C. (1993). Community-level organisation, power and elites in Polynesian settlement pattern studies. In M. W. Graves, & R. C. Green (Eds.) *The evolution and organisation of prehistoric society in Polynesia* (pp. 9–12). Monograph Number 19. Auckland: New Zealand Archaeological Association.

Green, R. C. (2000). Lapita and the cultural model for intrusion, integration and innovation. In A. Anderson & T. Murray (Eds.), *Australian archaeologist: Collected papers in honour of Jim Allen* (pp. 372–392). Canberra: Coombs Academic Publishing, Australian National University.

Green, R. C., Jones, M., & Sheppard, P. (2008). The reconstructed environment and absolute dating of SE-SZ-8 Lapita site on Nendo, Santa Cruz, Solomon Islands. *Archaeology in Oceania, 43*, 49–61.

Haun, A. E., Joseph, A. J., Melissa, A. K., & Susan, T. G. (1999). *Archaeological investigations at Unai Chulu, Island of Tinian, Commonwealth of the Northern Mariana Islands*. Report prepared for Naval Facilities Engineering Command, Pacific Division. Paul H. Rosendahl, Ph.D., Inc., Hilo, HI.

Hornbostel, H. (1925). *Unpublished field notes, 1921–1924*. Record on file at Library of Bishop Museum, Honolulu.

Hung, H.-C. (2008). *Migration and cultural interaction in southern coastal China, Taiwan and the northern Philippines, 3000 BC to AD 100: The early history of Austronesian-speaking populations*. Ph.D. dissertation. Australian National University, Canberra.

Hung, H.-c., & Carson, M. T. (2014). Foragers, fishers and farmers: Origins of the Taiwan Neolithic. *Antiquity, 88*, 1115–1131.

Hung, H.-C., Carson, M. T., Bellwood, P., Campos, F., Piper, P. J., Dizon, E., et al. (2011). The first settlement of Remote Oceania: The Philippines to the Marianas. *Antiquity, 85*, 909–926.

Irwin, G. J. (1998). The colonisation of the Pacific Plate: Chronological, navigational and social issues. *Journal of the Polynesian Society, 107*, 111–143.

Kirch, P. V. (1997). *The Lapita peoples: Ancestors of the oceanic world*. Cambridge: Blackwell.

Kirch, P. V. (2000). *On the road of the winds: An archaeological history of the Pacific Islands before European contact*. Berkeley: University of California Press.

Kirch, P. V. (Ed.). (2001). *Lapita and Its transformations in near Oceania: Archaeological investigations in the Mussau Islands, Papua New Guinea, 1985–88* (Introduction, stratigraphy, chronology, Vol. 1). Berkeley: University of California. Archaeological Research Facility Contributions Number 59.

Kirch, P. V. (2010). Peopling of the Pacific: A holistic anthropological perspective. *Annual Review of Anthropology, 39*, 131–148.

Kirch, P. V., & Green, R. C. (2001). *Hawaiki, Ancestral Polynesia: An essay in historical anthropology*. Cambridge: Cambridge University Press.

Kurashina, H., & Clayshulte, R. N. (1983a). Site formation processes and cultural sequence at Tarague, Guam. *Bulletin of the Indo-Pacific Prehistory Association, 4*, 114–122.

Kurashina, H., & Clayshulte, R. N. (1983b). *Site formation processes and cultural sequence at Tarague, Guam*. Miscellaneous Publications Number 6. Micronesian Area Research Center, University of Guam, Mangilao.

Kurashina, H., Moore, D., Kataoka, O., Clayshulte, R., & Ray, E. (1981). Prehistoric and protohistoric cultural occurrences at Tarague, Guam. *Asian Perspectives, 24*, 57–68.

Kurashina, H., Stephenson, R. A., Iverson, T. J., & Laguana, A. (1999). The megalithic heritage sites of the Marianas: Latte stones in past, present and future contexts. In W. Nuryanti (Ed.), *Heritage tourism and local communities* (pp. 259–282). Yogyakarta, Indonesia: Gadjah Mada University.

Laguana, A., Kurashina, H., Carson, M. T., Peterson, J. A., Bayman, J. M., Ames, T., Stephenson, R. A., Aguon, J., & Putra, H. (2012). Estorian i latte: A story of latte. *Micronesica, 42*, 80–120.

Marche, A.-A. (1889). Rapport general sur une mission aux Îles Mariannes. *Nouvelles Archives des Missions Scientifiques et Littéraires, Nouvelle Série, 1*, 241–280.

Marche, A.-A. (1982). *The Mariana Islands*. (R. D. Craig, Ed. & S. E. Cheng, Trans.). Publication Series Number 8. Micronesian Area Research Center, University of Guam, Mangilao.

Moore, D. R., Rosalind, L. H.-A., Judith R. A., & Eleanor F. W. (1992). *Archaeology at Chalan Piao, Saipan*. Report prepared for Jose Cabrera. Micronesian Archaeological Research Services, Mangilao, Guam.

Mulvaney, D. J., & Kamminga, J. (1999). *Prehistory of Australia* (Rev. ed.). Washington, DC: Smithsonian Institution Press.

Nunn, P. D., & Petchey, F. (2013). Bayesian re-evaluation of Lapita settlement in Fiji: Radiocarbon analysis of the Lapita occupation at Bourewa and nearby sites on the Rove Peninsula, Viti Levu Island. *Journal of Pacific Archaeology, 4*, 21–34.

Paz, V. (2002). Island Southeast Asia: Spread or friction zone? In P. Bellwood & C. Renfrew (Eds.), *Examining the farming-language dispersal hypothesis* (pp. 275–285). Cambridge: McDonald Institute for Archaeological Research.

Paz, V. (2005). Rock shelters, caves, and archaeobotany in Island Southeast Asia. *Asian Perspectives, 44*, 107–118.

Pellett, M., & Spoehr, A. (1961). Marianas archaeology: Report on an excavation on Tinian. *Journal of the Polynesian Society, 70*, 321–325.

Peterson, J. A. (2012). Latte villages in Guam and the Marianas: Monumentality or monumenterity? *Micronesica, 42*, 183–208.

Russell, S. (1998). *Tiempon I Manmofo'na: Ancient Chamorro Culture and History of the Northern Mariana Islands*. Micronesian Archaeological Survey Report Number 32. Commonwealth of the Northern Mariana Islands Division of Historic Preservation, Saipan.

Sand, C. (1997). The chronology of Lapita ware in New Caledonia. *Antiquity, 71*, 539–547.

Shutler, R., Jr. (1999). The relationship of red-slipped and lime-impressed pottery of the southern Philippines to that of Micronesia and the Lapita of Oceania. In J.-C. Galipaud & I. Lilley (Eds.), *Le Pacifique de 5000 a 2000 avant le Present: Supplements a l'Histoire d'une Colonisation* (pp. 521–529). Paris: Institut de Recherche pour le Developpement.

Simanjuntak, T. (Ed.). (2008). *Austronesian in Sulawesi*. Yogyakarta, Indonesia: Center for Prehistoric and Austronesian Studies.

Spoehr, A. (1957). *Marianas prehistory: Archaeological survey and excavations on Saipan, Tinian and Rota* (Fieldiana: Anthropology, Vol. 48). Chicago: Chicago Natural History Museum.

Spoehr, A., (1973). *Zamboanga and Sulu: An archaeological approach to ethnic diversity*. Ethnology Monographs, Number 1. Department of Anthropology, University of Pittsburgh, Pittsburgh.

Spriggs, M. (1997). *The Island melanesians*. Cambridge: Blackwell.

Spriggs, M. (1999). Archaeological dates and linguistic sub-groups in the settlement of the Island Southeast Asian-Pacific region. *Bulletin of the Indo-Pacific Prehistory Association, 18*, 17–24.

Spriggs, M. (2007). The Neolithic and Austronesian expansion within Island Southeast Asia and into the Pacific. In S. Chiu & C. Sand (Eds.), *From Southeast Asia to the Pacific: Archaeological perspectives on the Austronesian expansion and the Lapita cultural complex* (pp. 104–140). Taipei: Center for Archaeological Studies, Research Center for Humanities and Social Sciences, Academia Sinica.

Summerhayes, G. R. (2007). The rise and transformations of Lapita in the Bismarck Archipelago. In S. Chiu & C. Sand (Eds.), *From Southeast Asia to the Pacific: Archaeological perspectives on the Austronesian expansion and the Lapita cultural complex* (pp. 141–184). Taipei: Center for Archaeological Studies, Research Center for Humanities and Social Sciences, Academia Sinica.

Terrell, J. E., & Welsch, R. L. (1997). Lapita and the temporal geography of prehistory. *Antiquity, 71*, 548–572.

Thompson, L. (1932), *Archaeology of the Mariana Islands*. Bishop Museum Bulletin 100. Honolulu: Bishop Museum Press.

Thompson, L. (1940). The function of latte in the Marianas. *Journal of the Polynesian Society, 49*, 447–465.

Waterson, R. (1997). *The living house: An anthropology of architecture in Southeast Asia* (3rd ed.). London: Thames and Hudson.

Yawata, I. (1945). Peculiar forms of the stone-piles on the Mariana Islands: Mariana sekichuretsu iko no tokushu keishiki. *Zinruigaku Zasshi, 59*, 418–424 (in Japanese).

Chapter 5
Coordinating Perspectives of the Past

This book examines ancient life and landscape in a 3500-year chronology that can be difficult but not impossible to reconcile with other perspectives. For instance, geological time scales tend to be on the order of thousands or tens of thousands of years that do not readily match with archaeological time periods of just a few centuries. At another extreme, many historical studies are concerned with how traditional cultures are represented today or within memory of some generations, mostly referring to the historically reported last few centuries and with arguable relevance to the more distant archaeological past. Often, the archaeological past is homogenised as a singular vague period of "prehistory" prior to the availability of written records.

Archaeologists and others conceptualise the past in a variety of ways, for example, as constantly changing, as more or less stable over time, or as some combination of variably stable and unstable factors. In this regard, the French scholar Fernand Braudel (1949) provided a useful analogy of the ocean as representative of human history, within which some trends and rhythms are noticeable as the waves and tides rapidly fluctuating on the surface, while others endure in profound depths with only very slow change over long periods of time. These insights sometimes are reduced to a simplistic dichotomy of the *longue durée* as long-term durable and stable structures of history, in contrast to *histoire événementielle* as rote listing of superficial historical events. Indeed, Braudel's work followed the core ideas of distinguishing between *longue durée* and *histoire événementielle* as originally formulated by François Simiand (1903), but a rigid dichotomy may not ever have been intended.

In this book, multiple time scales and rhythms of change are understood as concurrent processes, and each is equally worthy of study. In this liberal view, chronology is of paramount importance for identifying what aspects of a landscape have changed or remained the same, at what magnitude, when, and over what period of time. Accordingly, basic descriptive datasets serve as the necessary foundation to discuss how and why things changed in the ways that they did, thus accounting for all of the variable currents and rhythms that were involved in shaping natural and cultural history.

M.T. Carson, *Archaeological Landscape Evolution*,
DOI 10.1007/978-3-319-31400-6_5

Archaeology at its core examines the material remains of past human behaviour, so it differs fundamentally from other approaches that aim to learn about the past through memory, experience, text, or other means. The substance of archaeology is limited to the partial remnants of artefacts and by-products that happened to survive in abandoned condition for centuries or millennia. The surviving materials are interpreted as sub-sample reflections of a larger cultural system, keeping in mind that the original cultural context must have been diverse and vibrant beyond the constraints of the durable material record.

Archaeology is strongest when working with material-based observations in datable contexts. These strengths are enhanced when the material findings are compared across multiple geographic areas or across different time periods. Accordingly, this book explores the chronological sequence of Marianas archaeology over the last 3500 years, as a basis for addressing questions of cross-regional relations and long-term change.

Chronology is fundamental in archaeological study, as a means to examine change through time (Fig. 5.1). Archaeological site deposits naturally contain older materials in deeply buried layers, as compared to younger materials in near-surface layers. These stratigraphic positions allow relative dating of older versus younger contexts, while radiocarbon and other techniques can assign measurable years to those relative dates. Archaeologists further can characterise the artefacts, cultural behaviours, and environments of each identifiable time period, noting how they may or may not have changed from one time period to another.

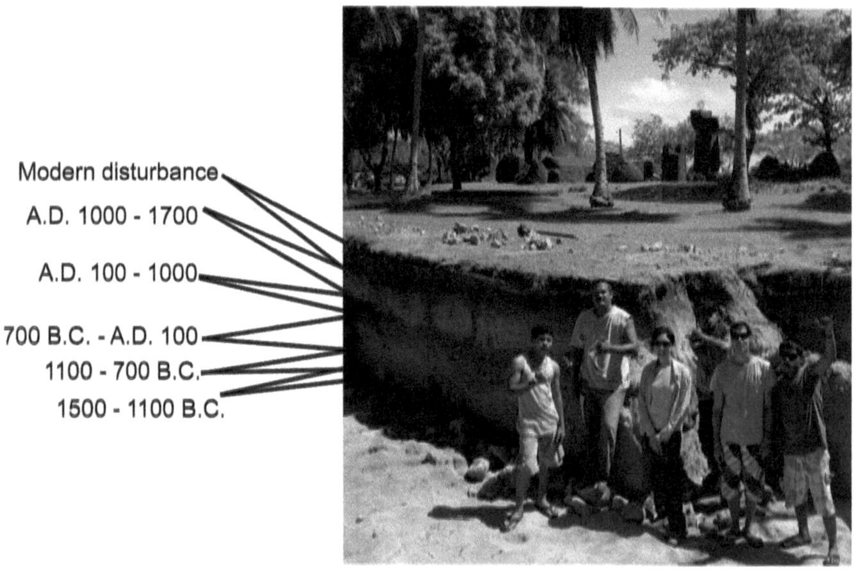

Fig. 5.1 Stratigraphic layers, inland of House of Taga, Tinian. The *latte* monument is visible in the far background toward the shoreline

This book concentrates on archaeology, yet certainly written and oral histories, linguistics, human biology, and ancient faunal and botanical records all make valid contributions for learning about the past. These fields of study naturally deal with their own datasets and ask quite different questions. They often point to different aspects of the past, separate time periods, and variable portions of what Simiand (1903) and Braudel (1949) described as the multiple concurrent time scales and rhythms of historical change.

Despite their dissimilarities, selected approaches to the past can overlap in a limited fashion, or sometimes they challenge each other toward entirely new interpretations. Some of the most intriguing discoveries have arisen from interdisciplinary research. Univocal synthesis may not be possible or even desirable, but rather a multi-perspective approach can be beneficial toward harmonising an otherwise discordant mass of information.

A multi-perspective approach to the past has been described as akin to "triangulation" of spatial points (Kirch and Green 2001), using two (or more) points to calculate the position of another point. In theory, multiple known vantage points (in this case, the datasets from archaeology and other disciplines) allow better precision about the target of interest (in this case, the subject of the human past). This approach can be effective only when the multiple lines of evidence point to the same cultural group and time period of interest, or else serious criticisms arise about incompatibility of source data. Within limits of each dataset, however, fascinating pictures can emerge that otherwise might not be possible from any single perspective.

Historical Perspectives

Marianas historical studies are strong overall, based on both written records and oral traditions that have accumulated over the last few centuries. The written sources offer veritable treasure troves of information about early Spanish encounters since the 1500s and continuing through modern times. Oral traditions refer primarily to this same time span but also extend deeper into the pre-Spanish past. As with all historical studies, the sources need to be evaluated and interpreted, keeping in mind about potential bias in the records.

The written history of the Marianas mostly was made from the observations and opinions of non-Chamorro people since the 1500s, and the greatest amounts of material were written following the profound transformative effects of Spanish imperialism. The Mariana Islands received little outside attention for more than a century after Ferdinand Magellan's landing in 1521. The major contact-induced transformations unfolded after the 1660s, when Jesuit missionaries and Spanish military forces enacted a plan of long-term and large-scale occupation, reaching a crescendo in the Spanish-Chamorro Wars. By 1700, the Spanish-imposed program of *reducción* exterminated large sectors of the native Chamorro population and relocated the survivors into just a few easily controlled villages in Guam, Rota, and Saipan.

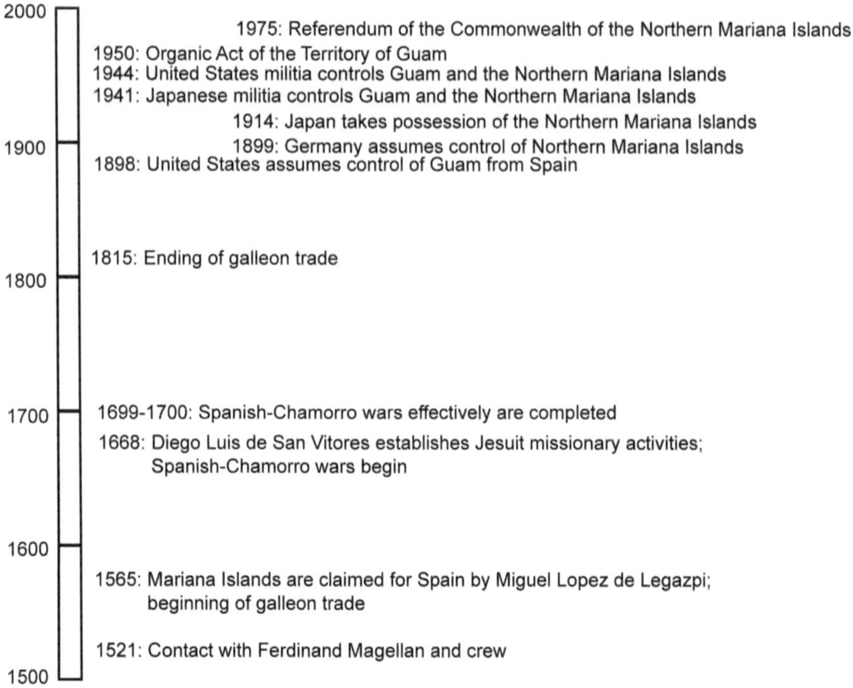

Fig. 5.2 Historical episodes of foreign rule in the Mariana Islands

Demographic, economic, social, religious, political, and ecological impacts were compounded over successive centuries of foreign rule under Spanish, German, Japanese, and US affiliations (Fig. 5.2). Spain claimed ownership of the Marianas through Miguel Lopez de Legazpi's proclamation in 1565, maintained for more than 300 years until the end of the Spanish-American War. At first, Spanish interests were limited to a way station along the sealane of galleons voyaging between the Philippines and the Americas. Eventually, Spanish interests grew to invest in agriculture and other economic production in this strategic spot in the Asia-Pacific region. Spain surrendered Guam to the USA in 1898 and then sold the islands north of Guam to Germany in 1899. At the outbreak of World War I, the German administration was removed from the Marianas, and Japan took possession of these islands in 1914. During World War II, on the same day of the attack at Pearl Harbor in Hawai'i, Japan captured Guam from the USA and held control of all of the Mariana Islands. In 1944, the US military recaptured Guam and also seized control of all of the Mariana Islands. Guam and the Northern Mariana Islands continue to operate as two separate entities although both have been under US allegiance with heavy overtones of US military tactical presence.

Foreign imperialism has been devastating, yet the Chamorro people were not dispossessed of their heritage. Nobody can doubt that many traditions in oral history, language, and daily practice have remained strong or have adapted to evolving circumstances. Meanwhile, historical records have been embraced as a means to learn about Chamorro ancestral identities.

Profuse amounts of historical references reveal how life changed in the Marianas throughout nearly 500 years of imperial colonial contexts. Indispensable first-hand accounts come from Magellan's visit in 1521 (in Stanley 1874) and Juan Pobre de Zamora's adventure in 1602 (in Driver 1993). The total amount of documentary material since the 1500s is overwhelming, as summarised by Barratt (2003a) and compiled in thousands of pages of original and translated source documents by Lévesque (1992–1999). In 1673, Peter Coomans reported a history of the Jesuit mission in the Marianas, but it was largely forgotten until much later (Coomans 1997). Early Jesuit missionary history was more popularly memorialised in the records of the life of the famed Jesuit martyr, Diego Luis de San Vitores, reported by Francisco Garcia in 1683 (2004). The first officially published "history of the Mariana Islands" was written by Charles le Gobien (1700), with much attention to the Spanish-Chamorro wars, *reducción*, and attendant transformations of native Chamorro culture still taking place while this history was being written (Fig. 5.3). This work was annotated and updated more

Fig. 5.3 Map of Guam, prepared by Alonso Lopez in 1671. Photograph of original printing published in le Gobien (1700). The sheet of parchment is slightly curved inside the original binding of the book's pages, curated at University of Hawai'i Library, Pacific Rare Books Collections

than a century later by Louis Claude de Freycinet in 1819 (in Barratt 2003b). These and other records have enabled a history of the transition from Spanish conquest to imperial occupation, about 1690 through 1740 (Hezel 2000). During his time as District Officer of the German Mariana Islands, Georg Fritz (1904) produced a history and ethnography of the Chamorro people, greatly interested in recovering native Chamorro culture from the imprint of foreign rule (see also Fritz 2001). These and countless other references support endless historical studies in the Marianas.

In addition to the source documents written in foreign languages and perspectives, oral traditions convey more of the native Chamorro voices about Marianas history and heritage. Chamorro legends, folklore, and songs provide insights into cultural notions and values, generally about how the world came to be the way it is and about how people should behave in this world. Examples relate to mythic origins of the first people, formation of the islands, and encounters between Chamorro and Spanish cultures. Gertrude Hornbostel recorded some of these traditions in the 1920s, included with the notes of archaeological studies by Hans Hornbostel (Hornbostel 1925; see also Thompson 1932). Others were compiled by Mavis Warner van Peenen in the 1940s (Van Peenen 2008). Many more traditions survive without written documentation, slowly entering the academic literature only very recently (Farrer and Sellmann 2014; McKinnon et al. 2014).

Numerous oral traditions and cultural practices continue to be sustained, for instance related to the meanings of traditional place names, events that occurred in specific places, and traditional ways of viewing the world (Fig. 5.4). Despite broad

Fig. 5.4 Chamorro blessing at Ritidian, prior to archaeological field training session. Cultural practitioner Leonard Iriarte (*right*) initiates archaeologist James M. Bayman (*left*)

agreement of the importance of these studies, commitment into written form has been much slower than elsewhere in the Asia-Pacific region. Oral traditions never were intended to be transmitted through printed words, so significant elements become lost or incomprehensible without their essential context of performance, sharing, and integration into real-life activities. Other issues arise from multiple viewpoints and interpretations of the same or similar traditions, wherein the diverse opinions are not always reconcilable.

Both written and oral histories can address several social issues in details that are not possible in most archaeological investigations, but they must be understood in relation to their given time frames. Historical studies offer their clearest information about sites of the 1500s or later, and of course certain aspects of Chamorro culture and society may have endured over much longer periods. New research yet needs to address what parts of historically known Chamorro culture can be extended farther into the past, exactly how far back in time, and why such may or may not have been the case.

Historical research has become absolutely essential for learning about Chamorro cultural heritage and identity, as exemplified in the works of Lawrence Cunningham (1992), Scott Russell (1998), and Carlos Madrid (2006). In some cases, archaeological information has been incorporated into historical perspectives of Chamorro culture. In other cases, historical references have been brought to bear on interpreting specific archaeological sites. In no instance should archaeology and history be conflated, but occasionally they can be complementary.

Linguistics

Language strongly signifies the identity and history of any individual, group, or society. In almost every instance of cultural action or interaction, whenever a person speaks a certain language or uses a specific speech pattern, he or she signals membership in a community of speakers who share aspects of cultural practice, values, and heritage. Anyone speaking the Chamorro language can be recognised as a resident of the Mariana Islands or as someone with close ties to one or another Chamorro community. Within this general category of Chamorro language-speakers, variations are noted from one community to another in pronunciation, vocabulary, and expression. In modern English-speaking groups, great diversity is noted regionally as a signifier of hometown geography, while additional subgrouping can be discerned according to social buzzwords, technical jargon, and fashionable phrases. Much the same can be said historically of almost every language community worldwide.

Every language inevitably changes through time, from one generation of speakers to the next and compounded over multiple generations. Among the world's best examples are the historically documented transformations of the English language. A Modern English-speaker would encounter some frustration when communicating with William Shakespeare or other Early Modern English-speakers in the late 1500s through early 1600s. The difficulties would be nearly insurmountable if trying to communicate more than just superficially with Geoffrey Chaucer or other Middle English-speakers of the 1300s. Going farther back in time, the English language becomes virtually unrecognisable when compared to how it is spoken by present-day communities, yet today's English-speakers can trace their linguistic heritage (although often not their biological heritage) to distant ancestors who spoke an archaic form of the English language.

Today's Chamorro language has been greatly affected by centuries of Spanish influence, with overlays of more recent Japanese and American English (Topping et al. 1975), but the core language remains confidently identifiable as a member of the Austronesian family (Blust 2013). Most modern Chamorro-speakers would face a great challenge if trying to communicate with a pre-Spanish inhabitant of a *latte* village in the 1300s or 1400s, much like the above-noted examples of comparing Modern English with its ancestral forms of past centuries. The challenge would be magnified when comparing the modern Chamorro language with the language of the first settlers of the Mariana Islands about 3500 years ago.

Although the Chamorro language has changed remarkably over the last 3500 years, the language itself must be recognised as having undergone much the same long-term transformations of any language in the world. The Chamorro language changed, but it did not become extinct, in many ways reflecting how its speakers adapted to new conditions and challenges with each generation. The events of language history reflect larger patterns and trends in the community of language-speakers, in what linguist Malcolm Ross (1997, 1998) described as the relation between "language events" and "speech-community events".

A series of changes can be identified in a community over time, each with possible manifestations in the language history as well as potentially in archaeological and other records. Examples of events in the larger community may include (but certainly are not limited to) the arrival of the first people in the Marianas, adaptation to a changing island environment, internal population increase, encounters with communities from overseas, exposure to foreign imperialism, and now a new phenomenon of revitalising the native Chamorro language. Many scholars would be quick to stress that these "events" are understood better as processes that unfolded over several years, decades, or even centuries.

Linguistic studies necessarily are conducted among living speakers of a language, so these studies cannot measure the time when a specific event (or process) occurred in a language history. Relative chronologies can be developed when comparing features of related languages, for example, ascertaining the origins of Chamorro language as older than some but younger than other Austronesian languages in the Asia-Pacific region (Blust 2000; Reid 2002; Zobel 2002). Importantly, these comparative studies account not only for vocabulary but also for phonology and syntax, thereby providing a stronger sense of each language as a whole.

When seeking Chamorro linguistic origins, two points about linguistic change gain central importance. First, all languages are viewed as experiencing ongoing change. Second, past relationships may be ascertained by comparing the contemporaneous states of two or more ancestrally related languages. For instance, the Chamorro language appears related to other Austronesian languages of the Asia-Pacific region, but its current form differs significantly from other Austronesian

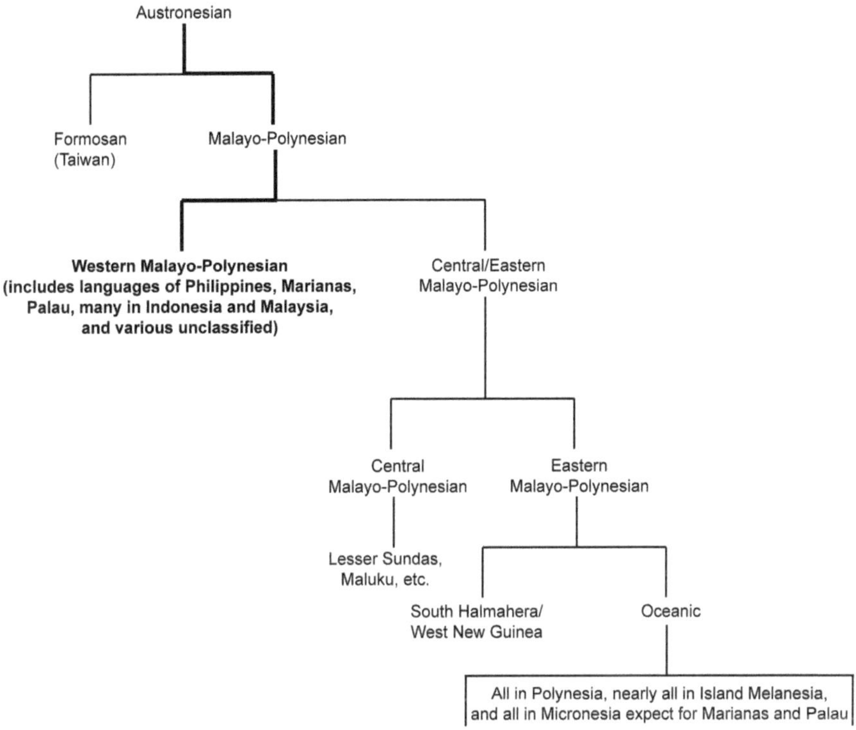

Fig. 5.5 Position of the Chamorro language in the Austronesian family. Based on data from Blust (2013)

languages today. When discounting Spanish and other modern influences, the position of Chamorro may be identified in relation to other Austronesian languages. These studies involve intensive examination of languages for identifying the features of long-term language change, and they should not be mistaken as results of modern cross-contact of communities. Modern contact-induced overlays bring new vocabulary and pronunciation of words, whereas ancestral language retentions are revealed in archaic pronunciation and phrasing.

The Chamorro language undeniably is unique when compared to any other Austronesian language in the Pacific Islands (Fig. 5.5). The language retains older features that have not survived elsewhere in the Pacific, while it similarly does not include many of the innovations that occurred elsewhere more recently. Chamorro most certainly is *not* a member of the Oceanic (Oc) subgroup spoken broadly throughout Micronesia, Melanesia, and Polynesia (Blust 2013). Instead, the Chamorro language shows its strongest affinities with the languages of the northern and central Philippines (Blust 2000, 2009; Reid 2002). The Chamorro language diverged from this ancestry about the same time when other Malayo-Polynesian languages were developing in the Philippines and apparently for the first time in Indonesia, so Chamorro in this sense conceivably can be grouped with a period of diversification from Philippines origins (Zobel 2002).

Chamorro can be grouped primarily with languages of the Philippines and secondarily in a more generalised sense with languages spoken more broadly in Island Southeast Asia. Blust (2000, 2009) and Reid (2002) proposed a linguistic homeland in the northern or central Philippines, based on their understanding of past transformations of the Malayo-Polynesian languages of the Philippines, as well as some of the unique features of Chamorro. In particular, Blust (2009) noted the importance of the regional continuity of words for "typhoon", retained in Chamorro. Otherwise, an origin of Chamorro from outside the typhoon zone would have resulted in a different word form and a different set of related vocabulary when the Chamorro-speakers came to live in "typhoon alley" (see Fig. 3.19).

The evidence most parsimoniously suggests that the ancient speakers of Chamorro language never lived outside the typhoon zone and monsoonal weather system throughout their traceable language history. In this view, prior to settling in the Mariana Islands, the ancestral language-speakers must have lived no farther south than the central Philippines. Given the absence of linguistic markers that would group Chamorro directly with Austronesian languages of Taiwan, the only remaining option for a Chamorro language homeland is within the northern or central Philippines.

An intriguing picture emerges when coupling the linguistic analysis with archaeological dates of first settlement in different parts of the Asia-Pacific region (Fig. 5.6). The results have been highlighted as one of the world's most convincing examples of linking language groups with human migrations (Blust 1995; see also Bellwood et al. 1995). In this view, Austronesian culture history can be traced in a large-scale sequence of three steps: (1) from Taiwan (and presumably from coastal southeast China) at least as early as 4000 B.C.; (2) across to the Philippines by 2000 B.C. and expanding into other parts of Island Southeast Asia by 1500 B.C.; and (3) into the Bismarck Archipelago after 1500 B.C. and eventually into the remote Pacific Islands around 1000 B.C. Within this sequence, the Chamorro language originated within "step 2", generally associated with today's West Malayo-Polynesian (WMP) languages of Island Southeast Asia.

The Chamorro language cannot be grouped with older Austronesian language origins in Taiwan, but rather it groups closely with the languages that developed outside Taiwan when Austronesian-speakers were living in the Philippines and perhaps other parts of Island Southeast Asia (Hung et al. 2011). In this view, Chamorro linguistic origins diverged from a hypothetical Proto Malayo-Polynesian (PMP) language at some point after 2000 B.C., and the language must have been spoken by 1500 B.C. when the oldest sites were inhabited in the Marianas (Carson and Kurashina 2012). These origins can be linked to a context of Malayo-Polynesian expansion into various parts of Island Southeast Asia about 1500 B.C. (if not earlier), generally into places where other people already had been living for thousands of years, but one of these population expansions brought people for the first time into the remote islands of the Pacific, specifically in the Mariana Islands (Bellwood et al. 2011). These events necessarily pre-dated the development of Oceanic (Oc) languages elsewhere in the Pacific Islands after 1500–1350 B.C. (Denham et al. 2012; Summerhayes 2007).

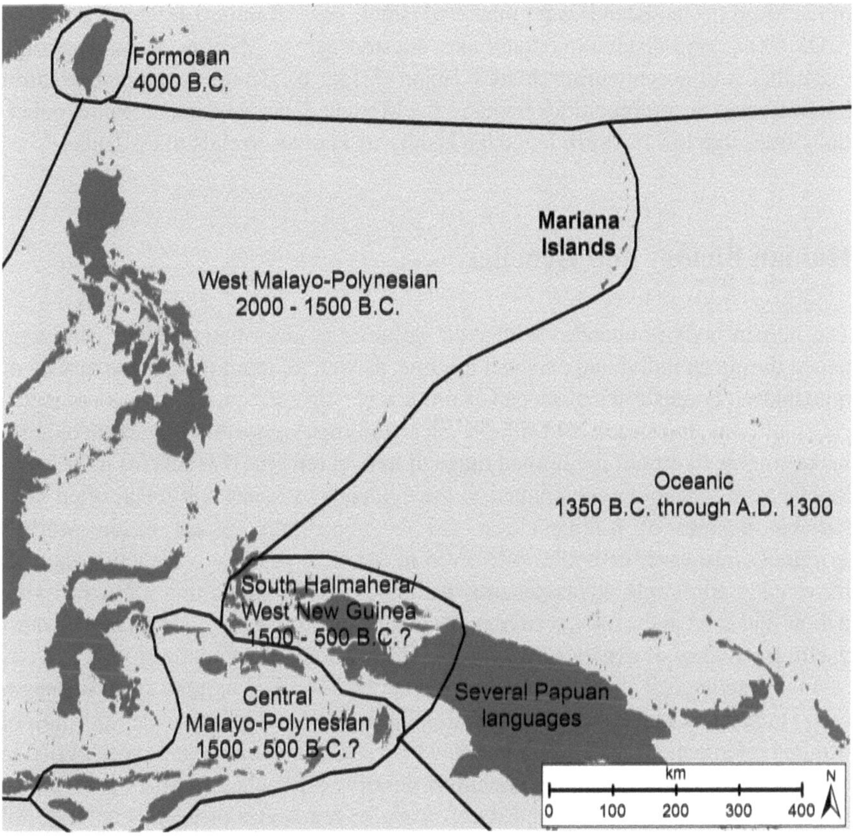

Fig. 5.6 Major language groupings in the Asia-Pacific region, coordinated with patterns of settlement chronology. Based on data from Blust (2013)

Within the limits of linguistic analysis, a specific village locus of a homeland for the Chamorro language cannot be identified, but most likely it included at least one community in the central or northern Philippines as the primary contributor of the language. Realistically, the founding population of Chamorro-speakers may have been drawn from multiple communities with different languages or dialects. After a few generations of separation from the homeland context, the Chamorro language already was developing its own distinguishing characteristics. Given the long distance of more than 2000 km between the Mariana Islands and any conceivable Austronesian-speaking homeland, the Chamorro language must have developed largely in isolation. The effects of isolation are further stressed by archaeological knowledge that no other islands in Remote Oceania were inhabited until some centuries later.

Later in Chamorro language history, opportunities were increasingly abundant for cross-cultural language contacts. Inter-island exchange networks created broad-ranging connectivity, especially in Micronesia (Peterson 2012). By A.D. 200, most of the Micronesian Islands were inhabited (Intoh 1997; Rainbird 1994, 2004). After A.D. 1000, long-distance contacts are attested across Micronesia and possibly extending into neighbouring areas (Carson 2013a, b). These developments must have affected the communities living in the Mariana Islands to some degree, potentially traceable in Chamorro language history as in other strands of evidence.

Human Biology and Genetics

The human body is encoded with great volumes of information about what happened during an individual person's lifetime, as well as about the genetic history of humankind. Forensic scientists can examine a person's accumulated years of medical conditions, injuries, and health. Physical anthropologists and bio-archaeologists make similar studies of the limited range of human remains that survive in archaeological sites, generally constrained to the evidence in bones and teeth, often after variable degrees of fragmentation and decomposition, except in the world's extremely rare cases of bodies preserved in ice or in peat bogs. Genetics analysts have made admirable advancements in the science of examining DNA in living people and even in the preserved remains of ancient archaeological specimens, most useful for studies of genetics origins and interrelations of populations.

In archaeological sites, human remains can be described in three general categories: (1) primary burial, where skeletal remains are found in articulated position of original interment; (2) secondary burial, where skeletal elements have been buried in disarticulated or other modified condition, possibly after other mortuary procedures; or (3) isolated or displaced contexts of single or few bones or teeth (Fig. 5.7). All three of these occurrences have been reported at several sites in the Mariana Islands.

Burial features are known abundantly at sites of the *latte* period about A.D. 1000 through 1700, largely because these sites are the best preserved and most numerous in the Mariana Islands. Older burials have been found in a few cases, and so far the oldest burial population in the Marianas has been reported at Naton Beach of Tumon Bay in Guam, dated about 700–500 B.C. (DeFant 2008). The most ancient sites so

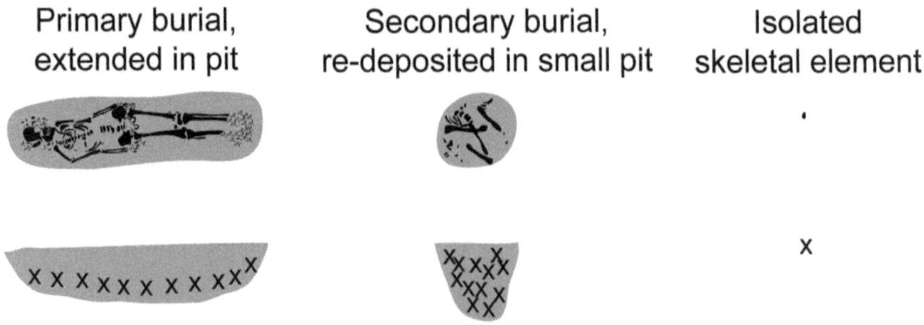

Fig. 5.7 Three major types of human burial features of primary, secondary, and isolated elements in plan and section views

far have not yielded human burials, most likely due to burial practice in places that were spatially separated from the residential habitation sites that happen to have been excavated. Archaeological sites cannot be expected to contain evidence of a deceased person's remains that were sent out to sea, burned, intentionally hidden elsewhere, or exposed to decay in natural weathering elements.

From analysis of ancient burial populations, bio-archaeologists and physical anthropologists have been able to characterise a number of traits of Chamorro people, mostly relevant to the *latte* period as noted (Hanson and Butler 1997; Hanson and Pietrusewsky 1997). Overall health is estimated as good, with low prevalence of disease and pathology, although individual people certainly endured stressful lives of physical labour hardship (Pietrusewsky et al. 1997). Teeth often are found to have been stained red, likely from chewing betel nut, used as a narcotic stimulant (Zumbroich 2008), but the lifelong effects on health are unclear in the archaeological remains. In some cases, teeth were modified intentionally by patterns of incision or abrasion (Ikehara-Quebral and Douglas 1997), sometimes in conjunction with teeth blackening (Zumbroich 2011), perhaps signifying rites of passage.

Knowledge from physical anthropology does not yet match the chronological depth of archaeological information in the Marianas. Given the imbalance of numerous burials from the *latte* period and comparatively few from any earlier time range, little can be stated confidently about chronological change in the Chamorro population. Possible temporal trends are presently unknown in terms of burial practice, demographic composition, or health conditions. The general findings of the *latte* period arguably can be extended into the more distant past, but definite conclusions will need to await future data. Caution must be advised when working with the larger represented population size and thus an inherently different biological diversity of the *latte* period as compared to any earlier periods.

A large-scale view of Chamorro genetics history has become possible through analysis of DNA in modern living populations, and the results are overall congruent with those from archaeology and linguistics. Initial studies identified Chamorro people as belonging to unique genetic lineages as compared to other groups in the Pacific Islands (Lum and Cann 1998, 2000). More detailed studies of mitochondrial DNA (mtDNA, inherited only through maternal lines) confirmed that Chamorro people have retained genetic markers that are not shared with any other Pacific Islands populations, but instead they can be linked with populations now living in the Philippines and Indonesia (Vilar et al. 2013). By estimating the number of generations needed for creating the amount of genetic mutations and variance seen today, Vilar et al. (2013) proposed that the founding Chamorro population diverged from a source population in Island Southeast Asia about 120 generations ago, in other words about 2000 B.C. if calculating 33.3 years per maternal child-bearing generation.

Following the initial founding settlement in the Marianas, Vilar et al. (2013) further postulated an overlay of much later change in the local population size and diversity. According to Vilar et al. (2013), this change probably occurred approximately at A.D. 1000, around the beginning of the *latte* period when archaeological evidence suggests a substantial increase in resident population size. Also at this time, long-distance inter-island contacts were strong throughout Micronesia and probably connected into neighbouring regions as well, so the change in genetic structure may have been due to arrivals of new immigrants. In one point of view, the evidence seen in the DNA record could be attributed to an internal population growth that brought increased diversity and an overall change in the genetic pool.

Faunal Records

Bones, teeth, and shells of non-human animals are common in archaeological deposits, but they also can be found outside archaeological sites and potentially predating human occupation of any area, including the Mariana Islands. Faunal records most often reveal information about the kinds of animals that people ate or that could be found in a given environment. They additionally can be analysed to learn about effects of hunting or collecting strategies, long-distance import of animals from overseas, or response to environmental change.

In Marianas archaeological sites, faunal records reflect the limited natural occurrence of animals in this Remote Oceanic setting, lacking the diversity found in Island Southeast Asia and Near Oceania. The most abundant faunal records refer to shellfish and fish, with variable contributions from birds, fruit bats, turtles, and snails. Monitor lizards and other reptiles are represented only extremely rarely in archaeological sites (Fig. 5.8), but they are more common in non-cultural cave deposits (Pregill and Steadman 2009). Rat bones are evident as early as A.D. 1000 (Pregill and Steadman 2009) although rats appeared with the first human inhabitants in nearly all other Pacific Islands (Matisoo-Smith 1994). No domesticated animals

Fig. 5.8 Monitor lizard (*hilitai* or *Varanus* sp.) outside a cave near the east end of Ritidian, Guam

are evident at all in the Mariana Islands until after Spanish influence, as compared to the prevalence of domesticated pigs, dogs, and chickens that were imported to almost all other islands in Remote Oceania (Wickler 2004).

The lack of pre-Spanish domesticated animals and long-delayed arrival of rats in the Marianas, although peculiar when compared to the faunal records in other Pacific Islands, may in fact be a reasonable outcome of the exceptionally long voyaging distance between the Mariana Islands and any other source area of these animals. As has been mentioned, the initial colonising voyage to the Marianas exceeded 2000 km, at its time the longest ocean-crossing migration in human history at 1500 B.C. This migration was the first ever accomplished into the Remote Oceanic world, and conceivably the settlers were not prepared to bring animals with them. Perhaps any animals aboard the canoes may have been eaten during the journey of some weeks at sea.

Marianas faunal records further appear peculiar for a rather mild impact on local bird populations, as compared to the massive avifaunal extinctions in other Remote Oceanic islands immediately after first human settlement (Steadman 1995). Bird bones are indeed found in small numbers in the oldest cultural layers in the Marianas, but so far these bones do not include the large numbers of extinct species as seen elsewhere in Pacific Oceania. One explanation may relate to the large sections of uninhabited territory, including whole islands, in the earliest centuries of Marianas settlement 1500–1100 B.C., in contrast to the numerous populated areas throughout Island Melanesia and West Polynesia during Lapita settlement 1100–800 B.C. (Kirch 1997). An additional factor may have been the lack of domesticated animals that could affect birds or their habitats. Marianas bird populations declined slightly after A.D. 1000 (Pregill and Steadman 2009), likely related to expansion of human settlement throughout the islands during the *latte* period, curiously coincident with the arrival of rats that potentially could harm the local avifauna. In practical terms, however, the low numbers of bird bones in the oldest layers may be affecting the ability to identify the real trends and patterns, and future research of larger collections of bird bones may yet solve this mystery.

Many faunal analysts focus on vertebrate remains, but in fact the Marianas faunal records are dominated by the remains of invertebrate shellfish that are greatly informative about the changing natural environment and impacts of cultural harvesting. According to analysis of archaeo-shellfish records, a "nearshore resource depression" has been described shortly after first settlement, due to the combined factors of lowering sea level, adjusting coastal ecology, and stress from predation by people acting as an invasive species (Fig. 5.9). The earliest sites reveal an emphasis on ark clam *Anadara* sp. shellfish that thrived in swampy environs or shallow-water sea-grass beds (Amesbury 1999, 2007), but these habitats no longer existed just a few centuries later (Carson 2008, 2012). As the coastal zones underwent major transformations after 1100–1000 B.C., people adapted by shifting their harvesting to target middle and outer-reef dwellers such as *Turbo* spp. and *Trochus* spp. shellfish. Meanwhile, taxa of chitons, sea urchins, and limpets quickly were depleted in the zones nearest to the habitations although they continued to survive in healthy numbers elsewhere. Much later, when new coral reef ecologies were established, nearshore *Strombus* sp. shellfish overpoweringly dominated site middens after A.D. 1000 (Fig. 5.10).

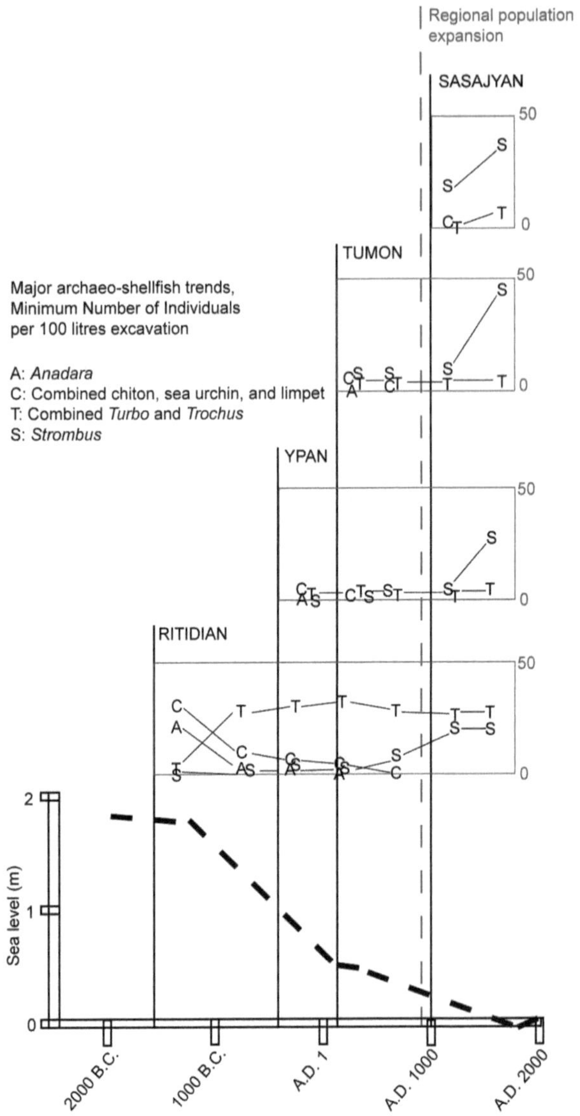

Fig. 5.9 Nearshore resource depression, showing combined effects of natural and cultural impacts of shellfish taxa. Modified from Carson (2014: 93–96)

In addition to the shellfish remains, Marianas sites often contain numerous shells of terrestrial invertebrate snails that tend to be overlooked. Most common are *Phythia* sp. and *Partula* sp. shells. *Pythia* sp. could exist in a variety of habitats, but the tree snail *Partula* sp. needs to live directly in a forested setting. The oldest site deposits of Marianas palaeo-coastlines tend to occur in unstable sandy shores that could not possibly have supported forest growth directly although forests were

Fig. 5.10 Abundant *Strombus gibberulus* shells from the Charterhouse Condominiums project excavations in Tumon, Guam. These specimens comprise 100 % of the shellfish midden from a single 10-cm level in a 1 by 1 m excavation square

available at a short distance inland in elevated landforms (Carson 2014). In stratified coastal site deposits, the in situ development of forest growth is confirmed by a sudden appearance of both *Pythia* sp. and *Partula* sp. shells, evidenced by 700–500 B.C. and then continuing in intensity thereafter (Carson 2012).

Botanical Records

Given the constraints of the Remote Oceanic setting with limited natural occurrence of plants and animals, several food items must have been transported from overseas. A few key plant foods, such as coconut and a local seeded breadfruit (*Artocarpus mariannensis*), existed naturally in the Marianas (Zerega et al. 2004, 2006), but nearly all others (including today's more popular non-seeded breadfruit, *Artocarpus altilis*) must have been imported by human agency. In terms of protein-rich animal foods, fish and shellfish were plentiful in the coasts and oceans, while birds and fruit bats were available in the forests, but plant foods were necessary for basic human nutrition. Seaweeds contributed to dietary survival, but people nonetheless needed at least some amount of starchy plant foods, such as taro, yam, banana, breadfruit, sago palms, and cycads. For instance, the intensively marine-oriented sea nomads of Island Southeast Asia can collect seemingly unlimited supplies of protein from the sea, but still they need to maintain symbiotic relations with land-based agriculturalists for obtaining essential plant foods (Sather 1995).

Traditional farming in the Marianas did not involve large-scale impacts of artificial fields, terraces, irrigation canals, water-tapping pits, mulching mounds, or other constructions that have come to characterise the landscapes of other Pacific Islands. Rather, informal management of forests, orchards, and gardens seems to have been practised for growing steady supplies of banana, breadfruit, coconut, betel nut, cycads, sugarcane, taro, yam, and other tree and tuber-root crops (Safford 2009). Elsewhere in the Pacific, ethnographic studies of cultivation and land-use strategies have helped to clarify the otherwise ambiguous and ephemeral material signatures of non-intensive low labour-input household gardening (Carson 2006), and these lessons may yet be applied in the archaeological landscapes of the Mariana Islands.

The role of rice in the ancient Marianas landscape will require new research. As with the pillar-raised house form (eventually formalised in stone as *latte*), the presence of rice in the Marianas suggests links with an older Malayo-Polynesian cultural background in Island Southeast Asia, not reproduced in other parts of Remote Oceania. Rice is attested in the Marianas in historical records as early as the 1500s, but so far its initial date of arrival in the Marianas has not been discovered in datable pre-Spanish archaeological context. According to early historical records, such as Juan Pobre de Zamora's account from 1602 (in Driver 1993), rice contributed only very little to the regular food base in the Marianas, reserved only for special occasions. No formal rice fields have been documented in the Marianas, so local rice cultivation most likely occurred in small rain-fed ("dryland") plots instead of extensive wetland paddies.

Clarification of ancient subsistence crops may yet be possible from studies of palaeo-botanical remains, both from archaeological sites and "off-site" natural deposits. Swamp-bottom and lake-bottom sediments effectively trap ancient pollen and other preserved plant materials from the surrounding environment, often in datable contexts (Athens et al. 2004; Athens and Ward 2004). Investigations of soil chemistry, starch residues, and other cultivation indicators still are being developed for analysis of archaeological site sediments in the Marianas (Dixon et al. 2012; Horrocks et al. 2015). For these approaches, suitable reference collections from the Marianas have been the most difficult factor for enabling confident taxonomic identifications.

Coring samples from lake-bottom and swamp-bottom sediments have provided unique opportunities for direct dating of the first anthropogenic (human-generated) impacts on the fragile island ecosystem, but the results are sometimes difficult to comprehend (Fig. 5.11). A definite horizon is noted in each coring profile, where a sudden influx of charcoal particles reflects human-caused forest-clearing at one point in time that never happened during the prior thousands of years. This same horizon includes preserved pollen spores that illustrate a decline in native forest and increase in overseas imported taxa.

Despite the clear and consistent anthropogenic impact horizon, the dating has been problematic. Most of the studies point to a date about 1500 B.C., but some ambiguity has allowed for interpretation of dates as old as 2200–2000 B.C. (Athens et al. 2004). The older end of the possible date range could be correct, but it would suggest a curious human impact a few centuries older than any known archaeological site in the region and rivalling the oldest Malayo-Polynesian settlements in Island Southeast Asia.

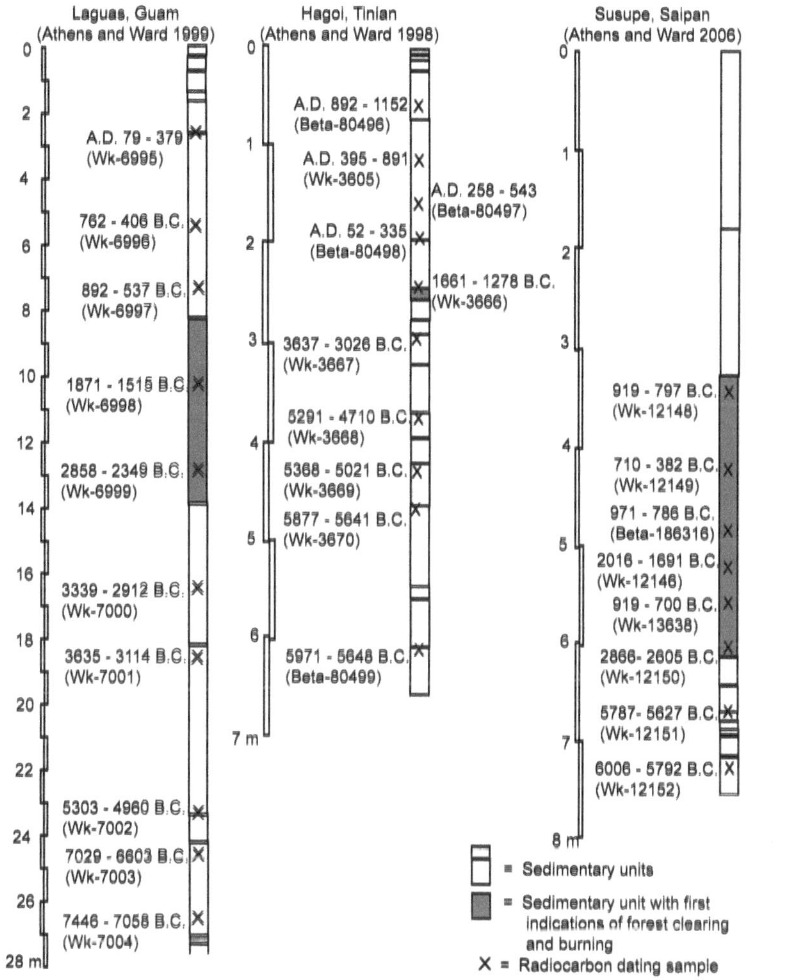

Fig. 5.11 Horizon of initial anthropogenic impact shown in palaeoenvironmental coring studies

Questions revolve around mismatch between the palaeo-botanical remains (charcoal and pollen) and the actual dated material (organic content of sediment and occasionally small shells). The charcoal, pollen, shells, and sediment particles naturally derived from separate sources and eventually settled at the bottoms of the lakes and swamps where they were preserved, so their individual radiocarbon dating results could refer to somewhat different ages. These issues become more complicated when recognising that the initial evidence of human impact involved sudden forest clearing and burning, contributing directly to greater amounts of slope-eroded sediment that potentially included a range of materials of varied sources and associated radiocarbon ages.

The effects of slope erosion and re-deposition are most visible in the coring records from Laguas in Guam and Susupe in Saipan, as compared to the rather small-scale or even negligible slope erosional deposition at Hagoi in Tinian (see Fig. 5.11). Notably, Laguas and Susupe both are situated in low-lying elevations at the base of sloped terrain, where they act as productive traps of slope-eroded sediments, especially accelerated after the initial human-caused forest clearing. Meanwhile, Hagoi is situated in a nearly flat limestone plateau, where slope erosion is barely noticeable at all.

Pending further studies, the most reasonable solution for the coring records is to accept that the sedimentary units accumulated over some centuries, leading to intermixing of material within each discrete layer, so that each preserved sedimentary unit today contains vertically mixed dates. Comparing one complete sedimentary unit against another, the findings generally appear in predictable stratigraphic order, with the oldest at the bottom and the youngest at the top of the sequence of superimposed layers. Dating can be quite precise in cases where the sedimentary units appear as thin laminated occurrences of a few mm each, one superimposed over the other in rapid succession. In other cases, dating can be frustratingly imprecise, where thicker sedimentary units contain a long range of dating results, especially problematic when those results reflect vertical inter-mixing.

The palaeoenvironmental coring studies have produced consistent results of an anthropogenic impact horizon, but the dating has been variable (see Fig. 5.11). The most precise dating so far has been at Hagoi in Tinian (Athens and Ward 1998), where the impact horizon occurred within a thin sedimentary unit of just a few cm dated 1661–1278 B.C., closely matching the archaeological dating of first settlement in the region about 1500 B.C. A study at Laguas in Guam again defined a clear impact horizon (Athens and Ward 1999), but the associated sedimentary unit was more than 5 m thick with dates spanning several centuries. The most recent work at Susupe in Saipan verified the same impact horizon (Athens and Ward 2006), but dating results were inter-mixed and contradictory to their vertical positions in the associated sedimentary unit of 3 m thickness.

References

Amesbury, J. R. (1999). Changes in species composition of archaeological marine shell assemblages in Guam. *Micronesica, 31*, 347–366.

Amesbury, J. R. (2007). Mollusk collecting and environmental change during the prehistoric period in the Mariana Islands. *Coral Reefs, 26*, 947–958.

Athens, J. S., Dega, M. F., & Ward, J. V. (2004). Austronesian colonisation of the Mariana Islands: The palaeoenvironmental evidence. *Bulletin of the Indo-Pacific Prehistory Association, 24*, 21–30.

Athens, J. S., & Ward, J. V. (1998). *Paleoenvironment and prehistoric landscape change: A sediment core record from Lake Hagoi, Tinian, CNMI*. Report prepared for United States Naval Facilities Engineering Command. Honolulu: International Archaeological Research Institute.

Athens, J. S., & Ward J. V. (1999). *Paleoclimate, vegetation, and landscape change on Guam: The Laguas core*. Report prepared for United States Naval Facilities Engineering Command. Honolulu: International Archaeological Research Institute.

Athens, J. S., & Ward, J. V. (2004). Holocene vegetation, savannah origins and human settlement of Guam. In V. Attenbrow & R. Fullagar (Eds.), *A Pacific Odyssey: Archaeology and anthropology in the Western Pacific—Papers in honour of Jim Specht* (pp. 15–30). Sydney: Australian Museum. Records of the Australian Museum, Supplement 29.
Athens, J.S., & Ward, J. (2006). *Holocene paleoenvironment of Saipan: Analysis of a core from Lake Susupe*. Micronesian Archaeological Survey Report Number 35. Saipan: Commonwealth of the Northern Mariana Islands Division of Historic Preservation.
Barratt, G. (2003a). *The Chamorros of the Mariana Islands: Early European Records, 1521–1721*. Occasional Historical Papers Series, Number 10. Commonwealth of the Northern Mariana Islands Division of Historic Preservation, Saipan.
Barratt, G. (Trans.). (2003b). *An Account of the Corvette l'Uranie's Sojourn at the Mariana Islands, 1819 by Louis Claude de Freycinet, Supplemented with the Journal of Rose de Freycinet*. Commonwealth of the Northern Mariana Islands Division of Historic Preservation, Saipan and Micronesian Area Research Center, University of Guam, Mangilao.
Bellwood, P., Chambers, G., Ross, M., & Hung, H.-c. (2011). Are 'cultures' inherited? Multidisciplinary perspectives on the origins and migrations of Austronesian-speaking peoples prior to 1000 BC. In B. Roberts & M. V. Linden (Eds.), *Investigating archaeological cultures: Material culture, variability, and transmission* (pp. 321–354). New York: Springer.
Bellwood, P., Fox, J. J., & Tryon, D. (Eds.). (1995). *The Austronesians: Historical and comparative perspectives*. Canberra: Department of Anthropology, Australian National University.
Blust, R. (1995). The prehistory of the Austronesian-speaking peoples: A view from language. *Journal of World Prehistory, 9*, 453–510.
Blust, R. (2000). Chamorro historical phonology. *Oceanic Linguistics, 39*, 83–122.
Blust, R. (2009). The historical value of single words. In E. Bethwyn (Ed.), *Discovering history through language: Papers in honour of Malcolm Ross* (Pacific Linguistics, Vol. 605, pp. 61–71). Canberra: Research School of Pacific and Asian Studies, Australian National University.
Blust, R. (2013) *The Austronesian languages* (Rev. ed.). Asia-Pacific Linguistics Open Access Monographs 008. Canberra: Research School of Pacific and Asian Studies, Australian National University.
Braudel, F. (1949). *La Méditerranée et la Monde Méditerranéan à l'Époche de Philippe II*. Three volumes. Paris: Armand Collin.
Carson, M. T. (2006). Samoan cultivation practices in archaeological perspective. *People and Culture in Oceania, 22*, 1–29.
Carson, M. T. (2008). Refining earliest settlement in Remote Oceania: Renewed archaeological investigations at Unai Bapot, Saipan. *Journal of Island and Coastal Archaeology, 3*, 115–139.
Carson, M. T. (2012). Evolution of an Austronesian landscape: The Ritidian Site in Guam. *Journal of Austronesian Studies, 3*, 55–86.
Carson, M. T. (2013a). Micronesian archaeology. In I. Ness & P. Bellwood (Eds.), *Encyclopedia of Human Migration* (Prehistory, Vol. 1, pp. 314–319). New York: Wiley-Blackwell.
Carson, M. T. (2013b). Austronesian migrations and developments in Micronesia. *Journal of Austronesian Studies, 4*, 25–52.
Carson, M. T. (2014). Paleoterrain research: Finding the first settlement sites of Remote Oceania. *Geoarchaeology, 29*, 268–275.
Carson, M. T., & Kurashina, H. (2012). Re-envisioning long-distance Oceanic migration: Early dates in the Mariana Islands. *World Archaeology, 44*, 409–435.
Coomans, P. (1997). History of the Mission in the Mariana Islands: 1667–1673. In R. Lévesque (Ed. & Trans.), *Occasional Historical Papers Series Number 4*. Commonwealth of the Northern Mariana Islands Division of Historic Preservation, Saipan.
Cunningham, L. J. (1992). *Ancient Chamorro society*. Honolulu: Bess Press.
DeFant, D. G. (2008). Early human burials from the Nation Beach Site, Tumon Bay, Island of Guam, Mariana Islands. *Journal of Island and Coastal Archaeology, 3*, 149–153.
Denham, T. P., Christopher, B. R., & Jim, S. (2012). Dating the appearance of Lapita pottery in the Bismarck Archipelago and its dispersal to Remote Oceania. *Archaeology in Oceania, 47*, 39–46.

Dixon, B., Walker, S., Golabi, M. H., & Manner, H. (2012). Two probable latte period agricultural sites in northern Guam: Their plants, soils, and interpretation. *Micronesica, 42*, 209–257.

Driver, M. G. (1993) *Fray Juan Pobre in the Marianas, 1602*. Miscellaneous Series Number 8. Micronesian Area Research Center, University of Guam, Mangilao.

Farrer, D. S., & Sellmann, J. D. (2014). Chants of re-enchantment: Chamorro spiritual resistance to colonial dominion. *Social Analysis, 58*, 127–148.

Fritz, G. (1904). Die Chamorro: Eine geschichte und ethnographie der Marianen. *Ethnologisches Notizblatt, 3*, 25–100.

Fritz, G. (2001). *The Chamorro: A History and Ethnography of the Mariana Islands* (E. Craddock, Trans.). Occasional Historical Papers Series, Number 1. Commonwealth of the Northern Mariana Islands Division of Historic Preservation, Saipan.

Garcia, F. (2004) The Life and Martyrdom of the Venerable Father Diego Luis de San Vitores of the Society of Jesus (J. A. McDonough Ed. & M. Margaret, F. P. Higgins, & J. M. H. Ledesma, Trans.). Translated and edited version of original published in 1684. Monograph Series Number 3. Micronesian Area Research Center, University of Guam, Mangilao.

Hanson, D. B., & Butler, B. M. (1997). A biocultural perspective on Marianas prehistory: Recent trends in bioarchaeological research. *American Journal of Physical Anthropology, 104*, 271–290.

Hanson, D. B., & Pietrusewsky, M. (1997). Bioarchaeological research in the Mariana Islands of the western Pacific: An overview. *American Journal of Physical Anthropology, 104*, 267–269.

Hezel, F. X. (2000). *From conquest to colonization: Spain in the Mariana Islands 1690 to 1740* (2nd ed.). Occasional Historical Papers Series, Number 2. Commonwealth of the Northern Mariana Islands Division of Historic Preservation, Saipan.

Hornbostel, H. (1925). *Unpublished field notes, 1921–1924*. Record on file at Library of Bishop Museum, Honolulu.

Horrocks, M., Peterson, J. A., & Carson, M. T. (2015). Pollen, starch and biosilicate analysis of archaeological deposits on Guam and Saipan, Mariana Islands, northwest Pacific: Evidence for Chamorro subsistence crops and marine resources. *Journal of Island and Coastal Archaeology, 10*, 97–110.

Hung, H.-C., Carson, M. T., Bellwood, P., Campos, F., Piper, P. J., Dizon, E., et al. (2011). The first settlement of Remote Oceania: The Philippines to the Marianas. *Antiquity, 85*, 909–926.

Ikehara-Quebral, R., & Douglas, M. T. (1997). Cultural alteration of human teeth in the Mariana Islands. *American Journal of Physical Anthropology, 104*, 381–391.

Intoh, M. (1997). Human dispersals into Micronesia. *Anthropological Science, 105*, 15–28.

Kirch, P. V. (1997). *The Lapita peoples: Ancestors of the Oceanic world*. Cambridge: Blackwell.

Kirch, P. V., & Green, R. C. (2001). *Hawaiki, Ancestral Polynesia: An essay in historical anthropology*. Cambridge: Cambridge University Press.

le Gobien, C. (1700). *Histoire des Îles Mariannes Nouvellement Converties à la Réligion Chrétienne*. Paris: N. Pepie.

Lévesque, R. (Ed. & Trans.). (1992–1999). *History of Micronesia: A Collection of Source Documents*. Twenty volumes. Éditions Lévesque, Montréal.

Lum, J. K., & Cann, R. L. (1998). MtDNA and language support a common origin of Micronesians and Polynesians in Island Southeast Asia. *American Journal of Physical Anthropology, 105*, 109–119.

Lum, J. K., & Cann, R. L. (2000). MtDNA lineage analyses: Origins and migrations of Micronesians and Polynesians. *American Journal of Physical Anthropology, 113*, 151–168.

Madrid, C. (2006). *Beyond distances: Governance, politics and deportation in the Mariana Islands from 1870 to 1877*. Saipan: Northern Mariana Islands Council for the Humanities.

Matisoo-Smith, E. (1994). The human colonisation of Polynesia, a novel approach: Genetic analyses of the Polynesian rat (*Rattus exulans*). *Journal of the Polynesian Society, 103*, 75–87.

McKinnon, J., Mushynsky, J., & Cabrera, G. (2014). A fluid sea in the Mariana Islands: Community archaeology and mapping the seascape of Saipan. *Journal of Maritime Archaeology, 9*, 59–79.

Peterson, J. A. (2012). Latte villages in Guam and the Marianas: Monumentality or monumenterity? *Micronesica, 42*, 183–208.

Pietrusewsky, M., Douglas, M. T., & Ikehara-Quebral, R. M. (1997). An assessment of health and disease in the prehistoric inhabitants of the Mariana Islands. *American Journal of Physical Anthropology, 104*, 315–342.

Pregill, G. K., & Steadman, D. W. (2009). The prehistory and biogeography of terrestrial vertebrates in Guam, Mariana Islands. *Diversity and Distributions, 15*, 983–996.

Rainbird, P. (1994). Prehistory of the north-west tropical Pacific: The Caroline, Mariana, and Marshall Islands. *Journal of World Prehistory, 8*, 293–349.

Rainbird, P. (2004). *The archaeology of Micronesia*. Cambridge: Cambridge University Press.

Reid, L. A. (2002). Morphosyntactic evidence for the position of Chamorro in the Austronesian language family. In R. S. Bauer (Ed.), *Collected papers and southeast Asian and Pacific languages* (Pacific Linguistics, Vol. 530, pp. 63–94). Canberra: Department of Linguistics, Research School of Pacific and Asian Studies, Australian National University.

Ross, M. D. (1997). Social networks and kinds of speech-community event. In R. Blench & M. Spriggs (Eds.), *Archaeology and language I: Theoretical and methodological orientations* (pp. 209–261). London: Routledge. One World Archaeology.

Ross, M. D. (1998). Sequencing and dating linguistic events in Oceania: The linguistics/archaeology interface. In R. Blench & M. Spriggs (Eds.), *Archaeology and language II: Correlating archaeological and linguistic hypotheses* (pp. 141–173). London: Routledge. One World Archaeology.

Russell, S. (1998). *Tiempon i Manmofo'na: Ancient Chamorro Culture and History of the Northern Mariana Islands*. Micronesian Archaeological Survey Report Number 32. Commonwealth of the Northern Mariana Islands Division of Historic Preservation, Saipan.

Safford, W. E. (2009). *Useful Plants of the Island of Guam*. Facsimile reprint of 1902 original publication by G. P. Putnam and Sons. Hagatna, Guam: Guamology Publishing.

Sather, C. (1995). Sea nomads and rainforest hunter-gatherers: Foraging adaptations in the Indo-Malaysian Archipelago. In P. Bellwood, J. J. Fox, & D. Tryon (Eds.), *The Austronesians: Historical and comparative approaches* (pp. 229–268). Canberra: Department of Anthropology, Research School of Pacific and Asian Studies, Australian National University.

Simiand, F. (1903). Méthode historique et science social. *Revue de Synthèse Historique, 16*, 113–137.

Stanley, H. E. J. (Ed.). (1874). *The first voyage round the world by Magellan, translated from the accounts of Pigafetta and other contemporary writers*. London: The Hakluyt Society.

Steadman, D. W. (1995). Prehistoric extinctions of Pacific Island birds: Biodiversity meets zooarchaeology. *Science, 267*, 1123–1131.

Summerhayes, G. R. (2007). The rise and transformations of Lapita in the Bismarck Archipelago. In S. Chiu & C. Sand (Eds.), *From Southeast Asia to the Pacific: Archaeological perspectives on the Austronesian expansion and the Lapita cultural complex* (pp. 141–184). Taipei: Center for Archaeological Studies, Research Center for Humanities and Social Sciences, Academia Sinica.

Thompson, L. (1932). *Archaeology of the Mariana Islands*. Bishop Museum Bulletin 100. Honolulu: Bishop Museum Press.

Topping, D. M., Pedro, M. O., & Bernadita, C. D. (1975). *Chamorro-English dictionary*. Honolulu: University of Hawai'i Press.

Van Peenen, M. W. (2008). *Chamorro Legends on the Island of Guam*. English-Spanish edition of 1945 original publication. Publication Series Number 4. Micronesian Area Research Center, University of Guam, Mangilao.

Vilar, M. G., Chan, C. W., Santos, D. R., Lynch, D., Spathis, R., Garruto, R. M., & Koji Lum, J. (2013). The origins and genetic distinctiveness of the Chamorros of the Mariana Islands: An mtDNA perspective. *American Journal of Human Biology, 25*, 116–122.

Wickler, S. K. (2004). Modelling colonisation and migration in Micronesia from a zooarchaeological perspective. In M. Mondini, S. Munoz, & S. K. Wickler (Eds.), *Colonisation, migration and marginal areas: A zooarchaeological approach* (pp. 28–40). Oxford: Oxbow Books.

Zerega, N. J. C., Diane, R., & Timothy, J. M. (2006). Breadfruit origins, diversity, and human-facilitated distribution. In J. M. Timothy & N. J. C. Zerega (Eds.), *Darwin's harvest: Origins, evolution, and conservation of crop plants* (pp. 213–238). New York: Columbia University Press.

Zerega, N. J. C., Ragone, D., & Motely, T. J. (2004). Complex origins of breadfruit (*Artocarpus altilis*, Moraceae): Implications for human migration in Oceania. *American Journal of Botany, 91*, 760–766.

Zobel, E. (2002). The position of Chamorro and Palauan in the Austronesian family tree: Evidence from verb morphosyntax. In F. Wouk & M. D. Ross (Eds.), *History and typology of western Austronesian voice systems* (Pacific Linguistics, pp. 405–434). Canberra: Department of Linguistics, Research School of Pacific and Asian Studies, Australian National University.

Zumbroich, T. J. (2008). The origin and diffusion of betel chewing: A synthesis of evidence from South Asia, Southeast Asia and beyond. *Electronic Journal of Indian Medicine, 1*, 63–116.

Zumbroich, T. J. (2011). To strengthen the teeth and harden the gums: Teeth blackening as medical practice in Asia, Micronesia and Melanesia. *Ethnobotany Research and Applications, 9*, 97–113.

Chapter 6
Range of Archaeological Material Culture

In order to build a chronological narrative, the basic archaeological materials need to be considered, following two key questions. Firstly, what do the artefacts, structural features, and other physical vestiges reveal about the past? Secondly, how can these material findings be organised into a sensible chronological order? This chapter covers the range of known archaeological objects in the Mariana Islands, with attention to the kinds of information that can be learned and pursued in later chapters.

The archaeological record is composed of the durable material remains of past cultural activities, often described as "material culture". The surviving material offers direct yet incomplete clues about the past as manifest in long-lasting creations of stone tools, pottery, and other preserved items. Anyone who peaks into a modern rubbish landfill can find valuable yet imperfect information about the society that created the piles of trash. Likewise, the material record of an archaeological site requires controlled interpretation for making cautious statements about an ancient cultural group.

When collected from the context of a known place and time period, sets of artefacts and other material culture can represent several aspects of the cultural group who produced this material. Archaeologists tend to discuss their findings as representing ancient people who made certain forms of pottery, lived in different forms of houses, used specific ecological zones, and so on. These conclusions necessarily are based on technological and economic aspects of material objects that are easily identified from simple observation, whereas other aspects of social life, political organisation, religion, and cultural beliefs would require complicated and intensely debated theoretical arguments not always upheld by the available material evidence. As illustrated in this book, findings of archaeological material culture can be interpreted in variably convincing ways.

M.T. Carson, *Archaeological Landscape Evolution*,
DOI 10.1007/978-3-319-31400-6_6

In order to address this book's central theme of evolution or co-evolution of societies with their landscapes, the available material evidence must be assessed for the ability to provide relevant and sufficient data. What kinds of artefacts and other materials can provide useful information about how people lived in the past and how they interacted with their landscapes? Are the sources of information flimsy and speculative, or do they support logically compelling arguments though robust hard data?

Artefacts

Certain types of artefacts can endure for thousands of years in archaeological sites, where today they bear diagnostic signs of deliberate manufacture by people, and often they exhibit material proof of their past use. Marianas archaeological sites are most easily defined by large quantities of broken earthenware pottery, variable amounts of shell and stone stools, in some cases shell ornaments, and rare instances of bone tools. Other perishable items, such as those made of wood and plant fibres, tend to decay too quickly for possible preservation in ancient site deposits.

The surviving artefact record represents an important yet understandably limited window into the past, in some ways like tracing the clues of a "cold case" in police work. Rather than unveiling the truth about how a specific murder, theft, or other known event was committed, however, archaeologists struggle to learn the basics of what happened at a site and when those events occurred. Knowing in general terms how people create and use material objects in their lives, archaeologists can comprehend the original contexts of the artefacts that have survived. The physical properties of pottery, stone tools, shell ornaments, and other objects provide a basis for ascertaining how they were made, what happened during their use-life cycle, and how they related with other aspects of a past society and environment.

Without any doubt, broken pottery comprises the most abundant type of artefact throughout the Mariana Islands. Dozens or sometimes hundreds of pieces could be shattered from a single pot, and most sites contain the dispersed remains from several original pots (Fig. 6.1). Many sites are defined solely by concentrations of potsherds. The intensity of past cultural activity often can be estimated by the density of potsherds in a measured space.

Prior to European and Asian ceramic imports in the Marianas, all pottery was made of earthenware. Earthenware was fired at lower temperature than the harder stoneware and porcelain products that later gained prevalence after Spanish contact. It was baked in open fires without kilns, using heaps of burning wood and probably palm fronds as fuel.

Marianas earthenware was made of local clays, tempered with varieties of sand-sized particle inclusions, most often classified as volcanic, quartzose, and calcareous beach sands or combinations of these (Dickinson 2006; Dickinson et al. 2001). Clays could be found in almost all of the islands in variable quantity and quality, but certain sources probably were favoured over others for clay particle size and purity. A study of the chemical composition of pottery from Guam concluded that the clay recipes were made from a distinctive set of natural resources that could be discerned clearly

Fig. 6.1 Potsherds in a surface-visible concentration near the eastern end of Ritidian, Guam. Scale bar is in 20-cm increments

from the clay recipes in other islands of Micronesia (Descantes et al. 2001). Natural inclusions of grit potentially could be desirable, but generally the temper inclusions of regulated size and material were added artificially for controlling the workability of clays, resistance to fracture during firing, and practical strength of the final products.

Given the balance of artistic and technical choices involved in making pottery, the resulting traditions are sensitive to variation in a large-scale view of time and space (Fig. 6.2). Over time, Marianas earthenware became larger, thicker, more coarsely made, and generally less elaborately decorated. Along with overall thickening of the vessel walls, the rims and lips became especially thickened after A.D. 1000. The earlier thinner vessels were made with very fine beach sand temper, whereas the later thicker vessels were made with medium and large-grain volcanic sand temper inclusions. Quartzose tempers were used in varying frequencies but generally declining over time. Red slip dominated in earlier vessels, later replaced by non-slipped surfaces. Decoration was rare but exquisite in earlier pottery, whereas it was represented commonly in coarse combing or brushing in later pottery.

As should be clear here and elaborated in later chapters, no single trait in pottery alone can diagnose its original time period and use context, but rather a combination of factors can give the best insight. The noted trends in Marianas earthenware occurred as parts of gradual processes that overlapped with one another, but they did not all coincide precisely or suddenly. Additionally, some geographic variation can be detected, such as the more pronounced rim thickening in Guam as compared to other islands during the later periods.

Pottery is joined by several other artefacts in the Marianas repertoire, including types of pounding and processing tools, chopping implements, cutting and slicing tools, personal adornments, fishing gear, and weapons (Fig. 6.3). Raw materials for these objects most commonly are stone, shell, and bone. Baked clay was used almost exclusively for pottery, but in rare cases it was used for net sinkers and perhaps a few other items. The specific forms and styles will be discussed in reference to their known time periods in the following chapters.

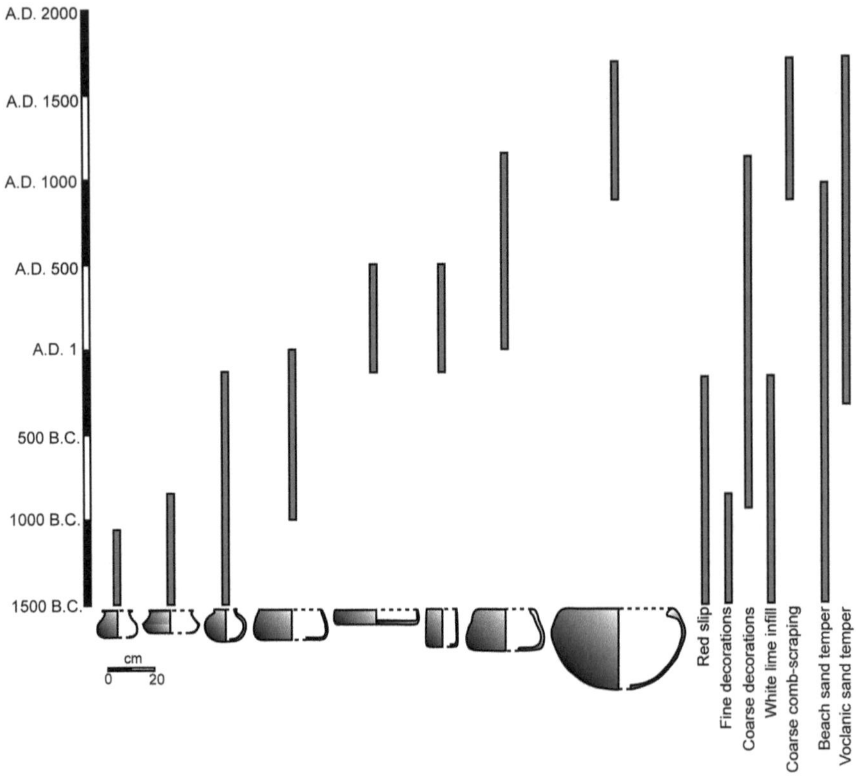

Fig. 6.2 Pottery chronology, noting major vessel shapes and trends

These broadly defined function-oriented categories of artefacts must be recognised as heuristic devices for the purpose of facilitating discussion and further study. This scheme is based primarily on technological or economic function, but others follow material type or objective physical shape. In any case, the complexities and nuances of an artefact's function cannot be reduced to any solitary overarching categorising scheme, no matter how unifying the scheme may appear to be.

Pounding and processing tools in the Marianas include items for grinding, mashing, and pulverising food and other materials. Most common are the grinding basins, known as *lusong*, and the associated pounder or pestle tools, known as *lummok*, found usually in surface-visible contexts at sites of the *latte* period. *Lusong* typically are made in portable or semi-portable boulders at residential sites, but in several cases they are manufactured into the natural bedrock shelves near cave entrances. The narrow-type of *lummok* pestle is found primarily in later-aged sites, but some larger-sized pounders have been found in earlier sites. Many pounders, basins, and boards likely were made of wood that never created any clear archaeological trace. Formalisation into longer-lasting stone, instead of perishable wood, may have applied to pounders and grinding basins just as it did to house posts since the beginning of the *latte* period about A.D. 1000.

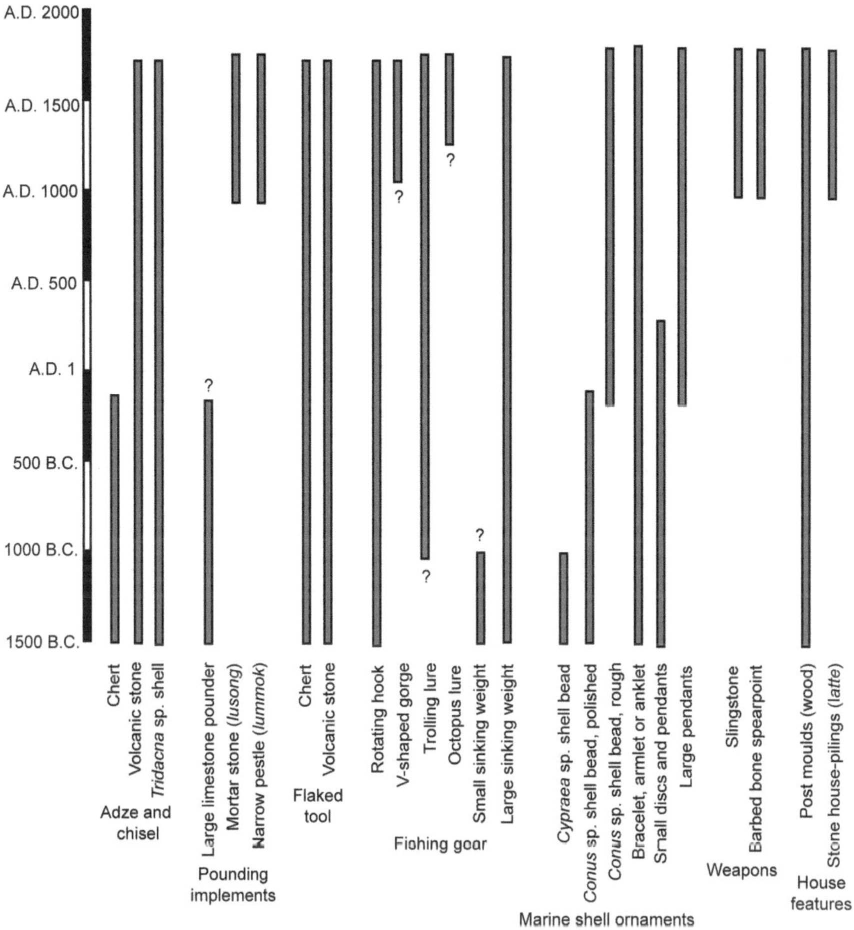

Fig. 6.3 Stone, shell, and bone artefact chronology

Chopping implements generally included adzes, chisels, and rare axes. Adzes and axes (but not always chisels) originally were hafted onto a handle, for different purposes of controlled chopping at an angle back towards the user (with an adze), direct percussive force moving forward into a narrow point (with a chisel), or heavy-duty chopping by side-swinging motion (with an axe). Of these, the adze consistently was the most popular throughout the Marianas chronological sequence, most often made of the shell of the giant clam (*Tridacna* sp.), also of volcanic stone where it was available, rarely of chert in the earlier sites, and so far in just one case of the horned helmet shell (*Cassis cornuta*.). Chisels suggest finer wood-working and carving, and these so far have been found in small numbers in sites of all ages. Axes are very rare, so far documented most clearly in the oldest occupation layer at House of Taga in Tinian, where they were made of chert and varieties of limestone.

Cutting and slicing tools are poorly understood in the Marianas archaeological record. They are known primarily from use-modified flakes, flaking debris, and flaking cores of sharp stone and shell material. Some of these items may have been held–held, while others may have been inset into wood handles. From a core, flakes were removed, and the flakes subsequently were utilised directly or else refined further into finished tools. During the manufacturing process, small bits of debris or debitage typically are discarded, but precious raw materials tend to be conserved as much as possible. The apparently preferred material was chert, but varied quantities of volcanic stone, limestone, or hard shell (usually of *Tridacna* sp.) were used. Chert tends to have been most popular in earlier sites and then losing popularity over time, possibly due to dwindling of the most preferred raw material sources. The use of a variety of materials, ever since the first settlement period, stresses the familiarity of the first settlers with exploring and exploiting their natural environment as part of the successful settlement of the Mariana Islands.

Examples of personal adornments in Marianas archaeological sites predominantly are made of marine shells, and their shapes can be described as beads, pendants, bangles, and possibly other categories. The most popular material was *Conus* spp. shell, especially for beads, which tends to be highly polished in earlier sites as compared to coarse and often non-polished in later sites. Additionally, a rare type of *Cypraea* sp. shell bead has been found in the earliest-period sites. Beads of orange coloured *Spondylus* sp. shells have gained great popularity today, and some have been found at *latte* period sites. A few forms of pendants are found repeatedly, such as discs made from the tops of *Conus* spp. shells at the earliest sites, but others tend to be highly individualised and made of various shells with attractive colour or lustrous character. Bangles often were made of larger *Conus* spp. shells, possibly used as bracelets but potentially as anklets or armbands.

Fishing gear is represented most often by a simple rotating hook found in sites of all time periods, but a V-shaped gorge was popular at some coastal sites of the *latte* period (Sinoto 2007). For both rotating hooks and V-shaped gorges, nearly all items were fashioned from *Isognomon* sp. shells and sometimes from *Turbo* spp. shells, along with plentiful debitage by-products bearing diagnostic traits of these shells. Rare items of bone were made probably from human bone. Pieces of trolling lures were made of *Isognomon* sp. shells, but so far they have been very rare in any sites pre-dating the *latte* period. Shell components of octopus lures were made from *Cypraea* sp. shells, so far evident only in the *latte* period. Net sinkers have been found so far in two types, including baked clay for larger items and limestone cave flowstone crystals for smaller items, suggestive of two different kinds of netting practice.

Weapons so far have been identified only in sites of the *latte* period, and these include slingstones and spearpoints. Wooden spears and bamboo knives may have been popular weapons or general-purpose tools, but they did not survive in archaeological sites. Slingstones were made of both limestone and volcanic stone. Spearpoints were made of human bone and carved with barbs, considered to be imbued with magic for inflicting both physically and spiritually deadly results (McNeill 2005). A few pieces of cut human bone points have been identified in older site layers, but the intended usage is not yet clarified.

For each of these types of artefacts mentioned here, considerable more research can be anticipated for learning how they were made and used. Experiments with replicas and raw materials surely can provide useful context for understanding how people worked with stone, shell, bone, and clay. Similar experiments hold great potential for examining patterns of use-wear in replicas of artefacts, as compared to the patterns seen on authentic archaeological specimens, best aided by microscope analysis for detecting the faint traces of use-wear. Further, residues of plant starch and other material have been observed on many artefacts, but only rarely have these been studied due to a need for developing suitable reference collections for identifying the residues in question.

Midden

A curious category of "midden" may be seen in the pages of archaeological literature, including this book, although the word has different meanings and connotations. Midden in this book refers to the preserved remains of plants and animals in a site deposit, including any such material that has not been modified into a definable human-made artefact. This definition can encompass remains of animal bones and shellfish, residues adhering to pottery and stone tools, and particles of bone, shell, burned wood, pollen, or other remains of plants and animals that have become preserved in sedimentary layers.

This broadly defined midden category is compatible with the notion of "ecofacts" as important elements of the past ecology or environment, although these materials were not transformed by human action into formal artefacts such as pottery, stone stools, or shell ornaments. Some of the key aspects of midden were described in Chap. 5, specifically in reference to faunal records and botanical records. These materials can be found within traditional archaeological sites, but they also can be found in non-site or off-site areas where they have accumulated naturally and without any human involvement.

Structural Features

Structural features include the ruins and remnants of houses and other constructions, representing large or immovable entities. *Latte* house pillar stones, post moulds, hearths, trampled living floors, pebble-cobble floor pavings, cobble-boulder alignments, rubbish pits, mounded heaps, and other features are considered more permanent or long-lasting than small pieces of broken artefacts and midden. These remnants are most easily recognised in surface contexts where the components are fully visible, but they can be difficult to identify when exposed only partially or after some degree of disturbance in buried contexts unearthed during excavations.

These kinds of features play special roles in the interpretations of archaeological sites. Worldwide, many archaeological chronologies are built primarily if not entirely on the contents of structural features, because sealed or intact features of discrete episodes offer the most confident contexts for precise and reliable dating, for example referring to the burial of a single person, cooking of one particular meal, installation of a house post, or discard of an individual broken pot. Furthermore, structural features enable insights into specific activity areas and cultural use of space within their datable contexts, directly relevant for addressing several research questions.

Collections of features can be interpreted as representative of a geographic region, cultural area, or time period. In the Mariana Islands, the *latte* period of approximately A.D. 1000 through 1700 can be defined most clearly by features of megalithic house post ruins, concentrations of thickened-rim earthenware pottery, and stone grinding basins plus often by burial pits, hearths, and other features. Varied forms and combinations of features potentially can characterise earlier periods, but they are known solely through subsurface discoveries that require different forms of documentation and interpretation than in surface-visible structural ruins, as will be discussed in later chapters.

Rock Art

Rock art constitutes a unique form of immovable archaeological feature, made by artistic expression on the surface of permanent rock. Rock art most often is enclosed inside a cave, but occasionally it can be found on other natural surfaces in open-air settings. In the Mariana Islands, most rock art images are pictographs, made by applying pigments of black (using charcoal), red (haematite or other iron-rich material), or white (calcium carbonate from slaked lime) mixed with some sort of binding agent that can make the material cohesive and adhere to the rock surface. Very rarely in the Marianas, petroglyphs are reported, made by etching or pecking of a shape into the rock.

Rock art images represent human figures, body parts, animals, and a number of enigmatic shapes (Fig. 6.4). The most frequently identified images are of human figures, drawn in various forms and positions. Others are real hand prints of individual people, as well as representational diagrams of hands, arms, and possibly other anatomical elements. Animal depictions are uncommon, but they include turtles and others that are difficult to specify. The various enigmatic shapes include collections of dots and arrangements of lines that can be interpreted in endless ways.

Marianas rock art sites evoke notions of past activities that probably occurred in or around these locations, but they are notoriously difficult to interpret. For example, where headless human figures are depicted on a cave's walls, the cave likely was related to a tradition of removing the head from a deceased individual or perhaps to the notions of spirits as humans without heads (Cabrera and Tudela 2006). In any case, questions linger about the cave as the actual place of witnessing or thinking about spirits, head removal, burial of certain individuals, specialised training or learning, or other activities. Similarly, hand prints indisputably represent actions of

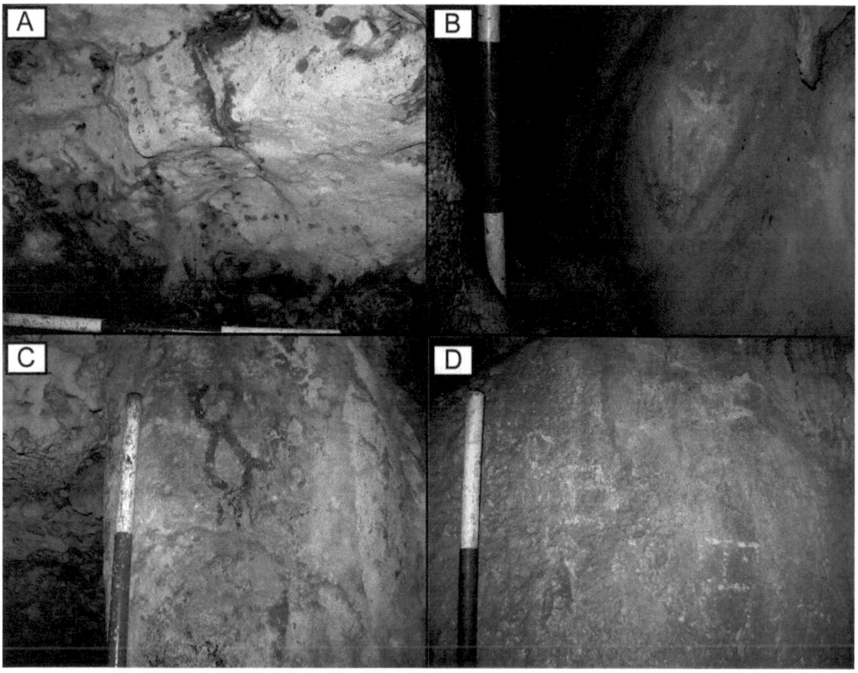

Fig. 6.4 Examples of rock art images at Ritidian. (**a**) *reddish brown dots* in vertical and horizontal lines. (**b**) *red pigment* hand print above *red pigment line*-drawn representation of a hand or other figure. (**c**) *black pigment* of unspecified image. (**d**) *white pigment* of human figures, including one headless figure at the *bottom right side* of the image

individual people who pressed their hands against a cave wall, but the meaning is open to speculation. Clusters of dots and various geometric shapes may have been intuitively understood by their makers, but this knowledge has become unclear today.

In addition to the partially or ambiguously understood meanings of rock art designs and symbols, dating has been vague and often not even considered. In most cases, nearby cultural deposits include thickened-rim pottery of the *latte* period and even more recent materials of Spanish-era and World War II contexts, so cultural awareness of the locations can be inferred possibly as early as A.D. 1000 but continuing into the recent past. The dates of cultural deposits may or may not have been confirmed by radiocarbon, but in any case this kind of dating does not necessarily apply to the rock art itself.

Direct dating of rock art is possible through radiocarbon and other techniques (Rowe 2009), but it requires good control of several factors that are not always controllable. Tiny bits of charcoal or other organic residue potentially can be isolated inside a sample of a pigment, but the sample potentially could incorporate inbuilt old age of an old clay deposit, burned fossil shell, or other material within the pigment recipe. Other problems involve possible contamination of the pigment from the geologically much older rock surface, dripping of water with constituent minerals through porous limestone, or algal growth over the rock surface.

Rock art bears significant potential for new research in the Mariana Islands, with only very few published studies at this time. Several sites are known throughout the islands, but they mostly are described as parts of much larger regional synthesis reports and studies without focusing on a research methodology for the highly specialised qualities of rock art (Hornbostel 1925; Spoehr 1957; Thompson 1932). Kalabera Cave in Saipan is one of the best known rock art sites, currently being developed as a tourist attraction, where so far the most observations have been recorded and have supported interpretations of the rock art as representing various events in Chamorro cultural history (Cabrera and Tudela 2006). At Tarague, within Andersen Air Force Base in northern Guam, several pictographs have been recorded in terms of their basic physical attributes (April 2006), although further studies have not yet been pursued. At Ritidian also in northern Guam, rock art has been recorded in a few caves (Carson 2012a), and a pilot study successfully identified the physical composition of pigments with possible organic binding agents needing closer analysis (Peterson 2014).

Caves

Caves can be defined generically as cavities in the ground, where people can enter. In some definitions, a true cave includes a dark zone where no natural light can enter, whereas smaller cavities are called rockshelters or other terms if they consist entirely of light zones and twilight zones without interior fully dark zones. In other views, cavities can be classified according to their objectively measurable attributes of internal volume, number of chambers, or combined factors of length, width, and height. Still other schemes refer to the number of openings, direction of access horizontally or vertically, the ability of torchlight to illuminate different surfaces, acoustical properties, and other factors that influence human behaviours inside these unique spaces (Pastoors and Weniger 2011).

As can be claimed in any region of the world, Marianas cave archaeology presents a mix of benefits and frustrations. Caves often are perceived as having potentially excellent preservation of rare items that cannot be found in most other settings, but meanwhile they are vulnerable to numerous types of disturbance, inter-mixing, and other compromising factors (Strauss 1990). Of special concern in the Mariana Islands, many caves were used as defensive positions during World War II by Japanese soldiers, often modified by tunnelling (Dixon et al. 2012). Nearly all of these caves and tunnels suffered intensive damage during and after wartime, involving disturbance, destruction, or removal of whatever archaeological contents may have once existed there. Further problematic for cave archaeology, caves tend to be places of rock art and different kinds of activities that do not occur regularly in open-air or above-ground sites, so the material records in caves likely reflect a narrow and somewhat skewed sampling of the past.

All caves embody special cultural places, principally because human beings ordinarily live above ground and require compelling motivation to engage in activities inside subterranean cavities. Contrary to popular notions of a "cave man" tradition in the human past, people did not in fact dwell in the deep and dark zones of caves, at least not for prolonged periods (Moyes 2012). People routinely used the open light and twilight portions of caves and rockshelters, while the remote dark zones were traversed only rarely and under specific circumstances (Clottes 2012). For these reasons, caves can be described as areas of specialised activities, although the "specialised" quality may be unclear and variable from one case to another.

Caves certainly are unique places, as can be appreciated when comparing archaeological materials inside and outside caves. A contrast of cave versus non-cave space is seen repeatedly along the limestone cliff-line of Ritidian of northern Guam (Carson 2012a). In these caves that escaped severe wartime damage, ancient cultural activity can be confirmed through rock art and accumulation of sediments with charcoal, ash, and limited amounts of discarded marine shells. Dense concentrations of artefacts and midden tend to be moderate in the light and twilight zones while notably sparse or totally absent in the dark zones, in contrast to their great abundance directly outside the same caves. At many of these caves, grinding basins (*lusong*) can be found carved into the bedrock outside the entrance, further enhancing a distinction between the interior and exterior space.

Some caves contain human grave features, but most of the known ancient graves were at individual houses and residential areas, at least during the well-documented *latte* period since A.D. 1000 (Carson 2012b). Limited studies at Ritidian in northern Guam indicate cave burials within the same general *latte* period, later disturbed and displaced by rainwater drainage, trampling, and other actions in the sensitive cave sediments (Carson 2012a). Although the findings so far are limited, the cave burials could represent individuals who for whatever reasons could not be interred at a *latte* house or village complex.

Each cave's past cultural use potentially can be understood in reference to the surrounding landscape and settlement system, as Dixon and Schaefer (2014) have proposed for caves in the Mariana Islands. Furthermore, the cultural use of cave and non-cave space may have varied through time. During any given time period, perhaps people followed an idealised mental template of how to use caves versus non-cave spaces, but inevitably each case reflects some degree of localised patterns in natural landforms and individual cultural expression.

Dating of cave use can be complicated by the paucity of reliable dating material in secure stratigraphic positions, but presumably people were aware of caves and made some use of them throughout the sequence of human presence in the Mariana Islands. According to a recent review of cave dating in Guam and Tinian, cultural activities rarely have been preserved older than A.D. 1000, but a few radiocarbon dates could be as old as 500 B.C. (Dixon and Schaefer 2014). Findings at Kalabera Cave in Saipan, however, indicate cave use there at least as early as 1000 B.C. (Swift et al. 2009). New discoveries at Ritidian in northern Guam indicate early red-slipped pottery at the entrance dripline of one large cave, prior to 1418–1144 B.C. and perhaps as old as 2097–1722 B.C. (Carson 2014), among the oldest of any site dating in the Marianas.

Landscapes

In a practical sense, all archaeological sites are found in landforms, and each landform can be associated with an ecological setting or landscape context (Stafford 1995). In this view, the archaeological landscape encompasses the geographic distribution of sites across the available ecological zones or landscape (Stafford and Hajic 1992). The inter-connections among sites can be interpreted as part of a complex landscape ecology, but the situation is more complicated when accounting for chronological change in the landscape itself and dynamic human use of it.

A landscape is comprised of the terrain and other attributes of an environmental and social setting within which sites can be situated, so that any single site can be appreciated in its surrounding landscape context. The "land" portion of the word "landscape" should not limit the ability to accommodate elements of land, sea, forests, animal life, human behaviours, and other components of the context in which sites originally were inhabited. A residential site may be situated on a ridgetop, possibly near a few other residences scattered along the same ridge, all overlooking the adjacent hill slopes and the stream valleys at the base of the sloped terrain. A group of travellers could make repeated use of a campsite on slightly mounded land within view of a riverbank. Inter-related elements thus form an essential character of any landscape.

Landscapes embody more than just physical terrain and ecological zones, and they imply a sense of connection between people and place. These connections are possible only when human beings conceptualise of their landscapes as inhabited environments, meaning that people live actively within their landscapes rather than just gaze at their surrounding scenery (Thomas 2012). Cultural landscapes can be expressed not only through distributions of archaeological sites but also through place names, stories associated with these place names, and traditional activities in places known for certain kinds of shellfish, therapeutic plants, or other resources.

Perhaps most importantly, landscapes are inhabited and experienced places. People are engaged in and with landscapes, rather than objectively on or alongside them as if they were detached entities (Ingold 2000). Most islanders understand that they live *in* the realms of their island landscapes, including the land mass, forests, surrounding ocean, and sphere of cultural influence. This notion contradicts the views of people living *on* islands.

In the remotely situated Mariana Islands, the landscape necessarily involves land and sea, but further it involves an ever-changing network of natural and cultural elements of a social-ecological system that has evolved over at least 3500 years. This liberal scope of a landscape is explored throughout this book. It is built through substantive datasets of archaeological research and other lines of evidence as outlined briefly above and discussed in more detail in the following chapters.

References

April, V. N. (2006). Talagi Pictograph cave, Guam. *Micronesian Journal of the Humanities and Social Sciences, 5*, 53–69.

Cabrera, G., & Tudela, H. (2006). Conversations with i man-aniti: Interpretation of discoveries of the rock art in the Northern Mariana Islands. *Micronesian Journal of the Humanities and Social Sciences, 5*, 42–52.

Carson, M. T. (2012a). Evolution of an Austronesian landscape: The Ritidian site in Guam. *Journal of Austronesian Studies, 3*, 55–86.

Carson, M. T. (2012b). An overview of latte period archaeology. *Micronesica, 42*, 1–79.

Carson, M. T. (2014). Contexts of natural-cultural history: A 3500-year record at Ritidian in Guam. In M. T. Carson (Ed.), *Guam's hidden gem: Archaeological and historical studies at Ritidian* (British archaeological reports international series, Vol. 2663, pp. 1–43). Oxford: Archaeopress.

Clottes, J. (2012). Ritual cave use in European Paleolithic caves. In H. Moyes (Ed.), *Sacred darkness: A global perspective on the ritual use of caves* (pp. 15–26). Boulder: University Press of Colorado.

Descantes, C., Neff, H., Glascock, M., & Dickinson, W. R. (2001). Chemical characterization of Micronesian ceramics through instrumental neutron activation analysis: A preliminary provenance study. *Journal of Archaeological Science, 28*, 1185–1190.

Dickinson, W. R. (2006). *Temper sands in prehistoric Oceanian pottery: Geotectonics, sedimentology, petrography, provenance*. Special Paper 406. Boulder, CO: Geological Society of America.

Dickinson, W. R., Butler, B. M., Moore, D. R., & Swift, M. (2001). Geological sources and geographic distribution of sand tempers in prehistoric potsherds from the Mariana Islands. *Geoarchaeology, 16*, 827–854.

Dixon, B., Gilda, L., & Bulgrin, L. (2012). The archaeology of World War II Japanese stragglers on the island of Guam and the Bushido Code. *Asian Perspectives, 51*, 110–127.

Dixon, B., & Schaefer, R. (2014). Archaeological investigation of caves and rock shelters and Guam and Tinian: A synthesis of their use through time. *Journal of Pacific Archaeology, 5*, 52–74.

Hornbostel, H. (1925). *Unpublished field notes, 1921–1924*. Record on file at Library of Bishop Museum, Honolulu.

Ingold, T. (2000). *The perception of the environment: Essays on livelihood, dwelling and skill*. London: Routledge.

McNeill, J. R. (2005). Putting the dead to work: An examination of the use of human bone in prehistoric Guam. In G. F. M. Rakita, J. E. Buikstra, L. A. Beck, & S. R. Williams (Eds.), *Interacting with the Dead: Perspectives on Mortuary Archaeology for the New Millennium* (pp. 305–315). Gainesville: University Press of Florida.

Moyes, H. (Ed.). (2012). *Sacred darkness: A global perspective on the ritual use of caves*. Boulder: University Press of Colorado.

Pastoors, A., & Weniger, G.-C. (2011). Cave art in context: Methods for the analysis of the spatial organization of cave sites. *Journal of Archaeological Research, 19*, 377–400.

Peterson, J. A. (2014). Community archaeology at Ritidian: University of Guam archaeological field schools and contributions to community history and cultural resource management. In M. T. Carson (Ed.), *Guam's Hidden gem: Archaeological and historical studies at Ritidian* (British archaeological reports international series, Vol. 2663, pp. 84–95). Oxford: Archaeopress.

Rowe, M. (2009). Radiocarbon dating of ancient rock paintings. *Analytical Chemistry, 81*, 1728–1735.

Sinoto, Y. (2007). A study of gorges from the Gognga Cove Beach Site, Tumon, Guam. In A. Anderson, K. Green, & F. Leach (Eds.), *Vastly ingenious: The archaeology of Pacific material culture in honour of Janet M. Davidson* (pp. 209–215). Dunedin: Otago University Press.

Spoehr, A. (1957). *Marianas prehistory: Archaeological survey and excavations on Saipan, Tinian and Rota* (Fieldiana: Anthropology, Vol. 48). Chicago: Chicago Natural History Museum.

Stafford, C. R. (1995). Geoarchaeological perspectives on paleolandscapes and regional subsurface archaeology. *Journal of Archaeological Method and Theory, 2*, 69–104.
Stafford, C. R., & Hajic, E. R. (1992). Landscape scale: Geoenvironmental approaches to prehistoric settlement strategies. In J. Rossignol & L. A. Wandsnider (Eds.), *Space, time, and archaeological landscapes* (pp. 137–161). New York: Plenum Press.
Strauss, L. G. (1990). Underground archaeology: Perspectives on caves and rockshelters. In M. B. Schiffer (Ed.), *Archaeological method and theory* (Vol. 2, pp. 255–304). Tucson: University of Arizona Press.
Swift, M. K., Randy A. H., & Michael A. F. (2009). *Archaeological assessment: Kalabera Cave development project, Marpi, Saipan.* Report prepared for Herman B. Cabrera and Associates. Swift and Harper Archaeological Research Consulting, Saipan.
Thomas, J. (2012). Archaeologies of place and landscape. In I. Hodder (Ed.), *Archaeological theory today* (2nd ed., pp. 167–187). Cambridge: Polity Press.
Thompson, L. (1932). *Archaeology of the Mariana Islands.* Bishop Museum Bulletin 100. Honolulu: Bishop Museum Press.

Part II
Chronological Sequence

Chapter 7
Building an Archaeological Chronology

Chronology is essential for examining landscape evolution, but the methods of building such a chronology are not always clearly understood. In the present study, the specific findings of archaeological sites and palaeoenvironmental records are examined in reference to their definable time intervals. In terms that may be familiar to archaeologists but not many others, the time intervals in Marianas archaeology are ascertained through stratigraphic position, associated materials, and radiocarbon dating. Even among jargon-conversant archaeologists, the dates assigned to a time period routinely are questioned, and the existence of a time period in itself may be challenged in fierce debates.

Archaeological chronologies continually are refined, as new information and new techniques inevitably become available through ongoing research. Generally, the chronology of any region is divided into smaller and smaller units as archaeologists accrue more and more data over decades of investigations. Initial chronologies may begin with general periods that span some centuries or even millennia, and later efforts provide finer units or sub-units often on the order of a few centuries each. Additionally, the dates of transitioning between one period and another may be vague at first, followed by more confident estimation from investigations specifically of these transitional periods.

An archaeological time period implies a beginning and ending date of a set of traits that belong to this period. For instance in the Mariana Islands, a distinctive form of thickened-rim pottery is found in contexts dated by radiocarbon as early as A.D. 1000 and extending into the early Spanish occupation period, often in association with megalithic *latte* residential sites dated as late as A.D. 1700. These and other traits may be viewed as belonging to a period of roughly A.D. 1000 through 1700, although this broadly defined period potentially can be sub-divided in two or more units depending on the available materials and the limits of radiocarbon dating for ascertaining a fine scale of chronological discrimination.

M.T. Carson, *Archaeological Landscape Evolution*,
DOI 10.1007/978-3-319-31400-6_7

Prior to the 1950s, the entirety of Marianas archaeology was perceived as extending no earlier than the *latte* period, referring to a singular native Chamorro culture that preceded Spanish imperial transformations. The region's first archaeologists, like Antoine-Alfred Marche (1889, 1982) and Hans Hornbostel (1925), excavated at surface-visible megalithic *latte* sites, where they necessarily focused on information from the cultural deposits associated directly with these sites. Notably, radiocarbon dating was not available prior to the 1950s, so the ages of *latte* sites were unknown except as most likely involving several centuries in order to account for the sedimentary deposits and abundant archaeological materials.

In the post-World War II years of the 1940s and continuing into the 1950s, Alexander Spoehr (1957) aimed to build a basic archaeological chronology in the Mariana Islands, using stratigraphic excavations in combination with the newly invented technique of radiocarbon dating. Spoehr (1957) documented the *latte* period approximately as early as A.D. 1000, but he found deeper and older cultural layers as well. These deeper and older cultural layers were dated at least as early as 1000 B.C., and they contained red-slipped pottery very much different from the thickened-rim and coarse pottery of the *latte* period. Additional excavations, inland from the House of Taga Site in Tinian, recovered a larger sample of the earliest pottery, including several examples with red slip and finely executed decorations (Pellett and Spoehr 1961), but no radiocarbon dating was attempted for this particular site at that time.

A comprehensive history of Marianas archaeology has been compiled elsewhere (Carson 2012), but the decades since the 1950s can be described as having concentrated strongly on the *latte* period, while only very rarely were any discoveries made of more than 2000 years of older archaeological material. Over several decades, the *latte* period became well documented and increasingly important as a symbol of native Chamorro heritage. Meanwhile, everything pre-dating the *latte* period was poorly understood and by default described as "pre-*latte*".

Currently, the Marianas archaeological record is known to extend at least as old as 1500 B.C., and not surprisingly it encompasses a series of changing conditions in natural and cultural history. Today, the *latte* period typically is divided into subunits of pre-Spanish and post-Spanish influence, but arguably it can be sub-divided according to slight change in climate conditions or according to distributions of radiocarbon dates associated with different cultural layers. Perhaps, the most productive advance however has been in refining the chronological sequence prior to the *latte* period, including a number of distinguishable types of pottery, shell beads, and other materials as mentioned in Chap. 6 and explored fully in the next chapters. Moreover, these different assemblages of artefacts can be associated with periods of changing sea level, coastal ecology, plant communities, and other attributes of an evolving landscape.

Within the Marianas archaeological sequence, the clearest evidence of chronological change can be seen in the forms and styles of earthenware pottery. This material is greatly abundant, and it happens to involve multiple opportunities for technical and artistic choices that potentially can result in considerable variation in the site collections both geographically and chronologically. As mentioned above, a thickened-rim form gained overwhelming popularity during the *latte* period, but other

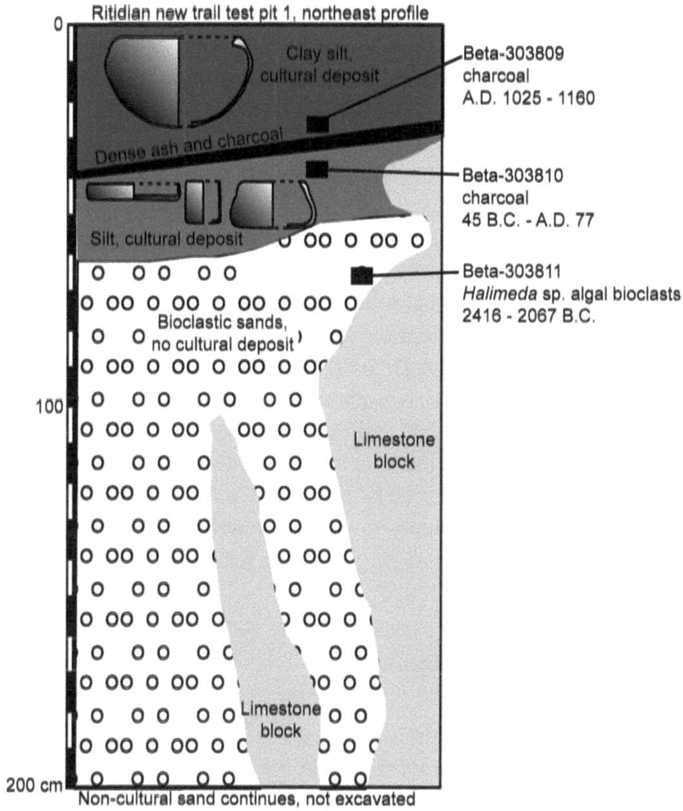

Fig. 7.1 Relative sequence of stratigraphic layers, with associated radiocarbon dating results. Example at New Trail Test Pit 1, outside "Upper Cave" at Ritidian, Guam. Modified from Carson (2012)

traits were popular during earlier periods. Variation over time is noted in vessel forms, decorative style, wall thickness, material composition, and other aspects of earthenware pottery, as well as to a lesser extent in stone, shell, and bone artefacts.

When a specific form or style of pottery or other artefact is found in a definable cultural layer, then it potentially can serve as an indicator of the associated time period (Fig. 7.1). Red-slipped pottery, for example generally pre-dates 200 B.C. in the Marianas, and those with thinner vessel walls tend to be much older. The exact date range can be ascertained through radiocarbon dating of the layer, ideally bracketed by dating in older layers and younger layers for refining the beginning and ending of the period when the cultural layer in question was formed. The dating may vary from one site to another, and some of the transitions of artefact types occurred over longer or shorter periods than others.

Looking at the full range of material culture, as introduced in Chap. 6, several different forms and styles of pottery and other objects are noted over time in the Marianas archaeological sequence (see Figs. 6.2 and 6.3). Some items endured over several centuries, but others gained and lost popularity within comparatively brief flashes of just a few centuries. A small and simple bowl, about the size of a coconut shell, fulfilled general utility in innumerable contexts throughout most of the chronology, although change may be noted in colour, friability, wall thickness, and other traits. By comparison, decorative styles in colour and design were more variable in pottery over time. Likewise, shell beads consistently were produced in all time periods, but their specific forms varied.

In terms of building a chronological sequence, the transitions from one period to another are of key importance, but they are not always rigidly defined. The ubiquitous thickened-rim pottery in the Marianas generally can be dated as early as A.D. 1000, but in some localities its first appearance may have been a century or two earlier or later. Instead of an exact date of A.D. 1000, a more cautious approach is to assign a potential range of A.D. 800–1000, but others may prefer A.D. 900–1000, A.D. 800–1200, or other refinements. As for an ending date of the thickened-rim trait, it appears to have continued until the end of the Spanish-imposed *reducción* period, overall complete by A.D. 1700, although it varied by as much as a few decades from one place to another. The same ambiguities apply to other artefact types and especially in older cultural layers with less intensive research than the *latte* period.

Use of Radiocarbon Dating

Within the time depth of a few thousand years in Marianas archaeology, radiocarbon dating is well within its limits that now have been extended to 50,000 years (Reimer et al. 2013). The dating is based on measuring ratios of different carbon isotopes preserved in a sample of organic material. When any carbon-based plant or animal organism dies, some of its carbon isotopes remain stable while others slowly decay at a predictable rate over time. By measuring the ratios of these isotopes, the time of death of the organism can be calculated within a margin of error, sometimes up to a few centuries for any single dating result.

The choice of organic material for dating can affect the result, for example if the material happens to incorporate carbon that is older or younger than the targeted point of archaeological interest. Soil layers continue to integrate organic material (and therefore carbon) from ongoing soil-formation processes, but they also could retain older carbon from variable sources. Burned wood (charcoal) offers one reliable material for radiocarbon dating, most confidently if it is from a short-lived specimen free of any in-built older age, for example from a nutshell or twig that had lived for just a few years while avoiding the deep interior old growth of a large tree trunk. Marine shell is another reliable material for radiocarbon dating, but the results need to be adjusted for the older carbon retained in the world's oceans, now very well documented in a marine calibration curve (Reimer et al. 2013). Even more important for the Marianas region, a locally specific marine reservoir correction has been calculated and proven accurate for *Anadara* sp. shells (Carson 2010, 2014), and it appears accurate when applied to archaeological samples such as *Halimeda* sp. algal bioclasts that absorbed their carbon in similar nearshore shallow-water settings (Carson and Peterson 2012).

Concerns sometimes arise about the archaeological context of a dating sample. Typically, the dated material comes from a cultural layer. Ideally, a reliable sample of short-lived charcoal or age-correctable shell is retrieved from a securely situated feature such as a hearth that is very closely associated with the artefacts and other materials of interest.

Radiocarbon dates must be acknowledged as ranges of probability, within which any single specific point could be correct. The results typically span some centuries when calibrated into conventional calendar years at the 2-Sigma (95.4% confidence) margins of accuracy, as reported throughout this book. For instance, a range of 1600–1400 B.C. conceivably could point to an event that occurred during any one of those 200 years.

When multiple radiocarbon dating samples independently provide age ranges that overlap with one another, then their cumulative probabilities indicate an increased chance of being accurate. Potentially, the results can overlap in such a way that helps to refine the total age range, for example with probabilities of 1700–1500 B.C., 1600–1400 B.C., and 1550–1350 B.C. for three separate parts of a single archaeological layer, each cross-confirming each other and thus increasing the overall probability that the archaeological layer was used by people during the range of 1550–1500 B.C. On the other hand, the three dating samples could be perceived as referring to three separate events of slightly different ages within a few continuous centuries when the archaeological layer developed over a period of time, in principle extending potentially as much as from 1700 B.C. to 1350 B.C. Especially in the case of a long potential date range, the end points can be constrained at least partially by dates from overlaying later-aged contexts or from underlying older-aged contexts.

Radiocarbon dating as a technique is reliable, but archaeologists criticise each other's work when specific samples are obtained from vague or insecure contexts. In the Mariana Islands, one such problem related to the early site dating at Unai Bapot in Saipan (Fig. 7.2), as reviewed in detail elsewhere (Carson 2014:38–40, 109–113; Carson and Kurashina 2012:428–430). One investigation obtained dates of two different *Anadara* sp. shells from a small ash pile slightly older than 1500 B.C., reinforced by a date on a piece of charcoal from a super-imposed layer that was in fact quite dense with charcoal about 1100–1000 B.C. (Carson 2008). Another investigation questioned the early dating and instead preferred to obtain dates of small particles of carbonised nutshells and flecks of short-lived wood taxa (Clark et al. 2010), but these tiny particles produced essentially the same age of 1100–1000 B.C., regardless of their stratigraphic position throughout more than 1 m of changing pottery types and sedimentary contexts. The dating results were precisely on their targets for each sample, but those targets had been displaced from their original contexts of the charcoal-rich layer of a stable backbeach dated 1100–1000 B.C., filtered downward in the sedimentary column into the underlying unit of unstable beach sands that had accumulated during an older period of higher sea level. Furthermore, the dating on *Anadara* sp. shells could be refined according to a Marianas-specific correction for these shells (Carson 2010), whereas other taxa of dated marine shells at Unai Bapot proved unreliable due to their incorporation of environmental carbon of variable ages and unpredictable dating results.

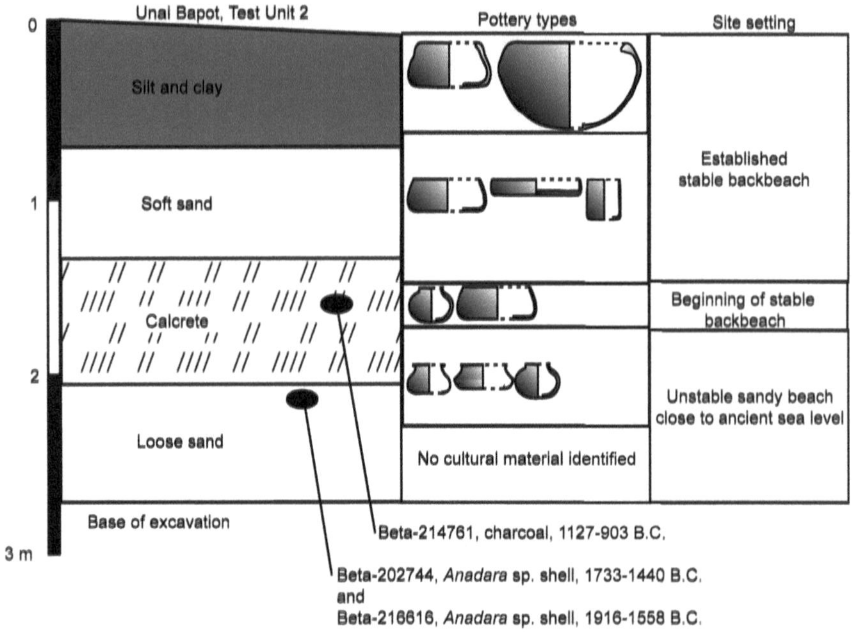

Fig. 7.2 Radiocarbon dating in relation to stratigraphic layers, pottery types, and environmental setting at Unai Bapot, Saipan. Based on data from Carson (2008, 2014: 109–113)

The radiocarbon dating at Unai Bapot gained special significance because of the unexpectedly old dating, slightly older than 1500 B.C. By choosing unreliable materials or samples from insecure contexts, an incorrectly younger age would be concluded. Within the Marianas and generally within the Pacific Islands, the difference of just a few centuries can be extremely important for dating a "pottery trail" linked with migrations of people from Southeast Asia through the Pacific (Carson et al. 2013). As reviewed in Chap. 5, the earliest red-slipped and decorated pottery in the Mariana Islands at 1500 B.C. slightly pre-dated similar pottery in the Bismarck Archipelago east of New Guinea and very certainly pre-dated other related pottery in the remote islands of Southern Melanesia and West Polynesia after 1200 B.C.

The technical and interpretive issues of site dating apply to any part of a chronological sequence, not just the earliest dating as illustrated at Unai Bapot. Transitions from one type of artefact to another cannot always be dated directly, but more realistically these transitions are dated by associated layers that often have overlapping ranges of radiocarbon dates. Exact pin-pointing of a transition rarely is possible. Although rapid change may have occurred in some cases as tightly constrained events, other transitions unfolded over a few human generations or even a few centuries. A time range of few centuries can be clarified if the radiocarbon samples in question refer to reliable materials in secure contexts, and often this range of a few centuries is more than sufficient within the full scope of an archaeological chronology covering some millennia.

Marianas Chronological Outline

In the next chapters, time periods of the Marianas landscape are identified according to radiocarbon dates of sedimentary layers or other horizons, and these temporal units then are described in terms of their material contents. These temporally constrained materials refer to certain types of artefacts, food refuse, and ecological conditions. Many of these existed primarily within their associated time periods but not appreciably earlier or later. Others were sustained more or less consistently across two or more chronologically definable time periods. In a few cases, the material findings may have existed during very short time spans that cannot yet be detected through the available dating.

The current 3500-year-long Marianas chronology includes seven broadly defined periods, each varying in length up to as much as 700 years (Fig. 7.3). As noted, these periods are known primarily by available dating, whereas their associated material findings are variable and not always congruent with the given parameters of the

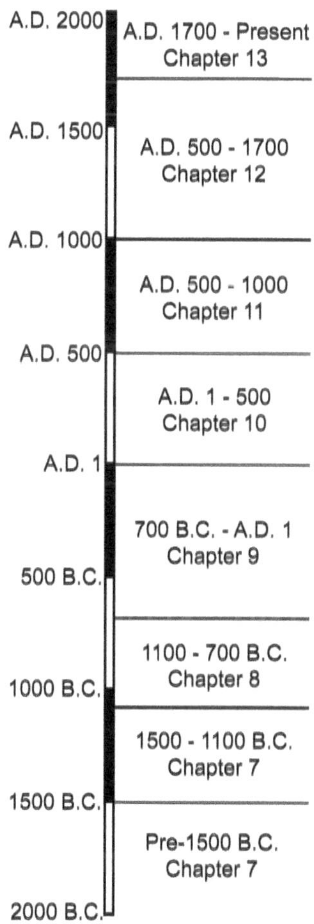

Fig. 7.3 Currently definable major periods of Marianas landscape chronology

discerned time units. The precise beginning and ending dates of every period surely will be refined after future research, and perhaps other internal sub-divisions will be identified.

As presented here, the Marianas chronology transcends changing conditions of the natural environment and cultural setting. The changing conditions are known from archaeological artefacts (see Figs. 6.2 and 6.3), as well as from records of local and regional sea level (see Fig. 3.8), archaeo-faunal remains (see Fig. 5.9) preserved botanical materials (see Fig. 5.11), and other evidence as discussed in the next chapters. Collectively, this information illustrates the natural–cultural landscape during each time period, and change over time can be tracked through the chronological sequence as a whole.

In this book, time periods are referenced by their date ranges, although other systems have assigned nomenclature for periods according to site names, types of pottery, or other factors (Hunter-Anderson and Butler 1995; Reinman 1977; Russell 1998). In the present narrative, names of periods are avoided as potentially misleading or over-emphasising certain aspects of the past at the expense of others. For instance, Hunter-Anderson and Moore (2001) proposed an "*unai* period" referring to the "sand" or "sandy beach" settings of earliest habitation sites, but research now reveals that the broad sandy beaches (*unai*) did not exist at the time of first settlement about 1500 B.C. (Carson 2011). Another example is the conventionally accepted *latte* period following A.D. 1000, although the *latte* house ruins are only one of many aspects of a setting that experienced changing conditions of climate, population size and density, and material culture expressions. The *latte* period nonetheless retains its significance for reasons of cultural heritage, and it does appear to be a strongly defined archaeological period whose name continues to be used.

The periods are not named as in prior schemes, because the present study considers multiple lines of evidence about landscapes that inherently contradict over-simplification. If a specific name or phrasing is offered, then it must be understood as merely a convenient label allowing for multiple interpretations and open to other labels. Naming a period after a particular site would cause unnecessary problems in the Marianas, because many sites contain evidence from multiple time periods. Naming after a pottery type, condition of coastal morphology, population size, or any other single factor necessarily would detract from the many different kinds of information about the associated time period.

All of the time periods as presented in this book witnessed variable degrees of change in every component of the natural and cultural landscape, thus enabling a discussion of how the landscape has evolved through time. In this view, time periods are artificial constructions for the sake of analysis and discussion, in a sense constituting the analytical framework of a model system, much like observations can be made of a biological organism during a succession of hours, days, or weeks as convenient time markers for keeping track of ongoing change. In this case, time intervals of hundreds of years each are presented as overall summaries of the landscape during measurable spans of time, together forming a continuous sequence of more than three millennia.

In the following chapters, each time period is described towards the goal of comparing the results throughout the chronological sequence. In each case, the dating parameters and most informative known sites are discussed, and then key questions are addressed about the ancient landscape. What were the available landforms and resource zones of the inhabited landscape during each identifiable time interval? What did the archaeological sites look like, and what kinds of artefacts and other materials characterised these sites? How do these results compare with the findings in other areas of the Asia-Pacific region?

References

Carson, M. T. (2008). Refining earliest settlement in Remote Oceania: Renewed archaeological investigations at Unai Bapot, Saipan. *Journal of Island and Coastal Archaeology, 3*, 115–139.

Carson, M. T. (2010). Radiocarbon chronology with marine reservoir correction for the Ritidian Archaeological Site, northern Guam. *Radiocarbon, 52*, 1627–1638.

Carson, M. T. (2011). Palaeohabitat of first settlement sites 1500–1000 B.C. in Guam, Mariana Islands, western Pacific. *Journal of Archaeological Science, 38*, 2207–2221.

Carson, M. T. (2012). History of archaeological study in the Mariana Islands. *Micronesica, 42*, 312–371.

Carson, M. T. (2014). *First settlement of Remote Oceania: Earliest sites in the Mariana Islands*. New York: Springer.

Carson, M. T., Hung, H.-c., Summerhayes, G., & Bellwood, P. (2013). The pottery trail from Southeast Asia to Remote Oceania. *Journal of Island and Coastal Archaeology, 8*, 17–36.

Carson, M. T., & Kurashina, H. (2012). Re-envisioning long-distance Oceanic migration: Early dates in the Mariana Islands. *World Archaeology, 44*, 409–435.

Carson, M. T., & Peterson, J. A. (2012). Radiocarbon dating of algal bioclasts in beach sites of Guam. *Journal of Island and Coastal Archaeology, 7*, 64–75.

Clark, G., Petchey, F., Winter, O., Carson, M., & O'Day, P. (2010). New radiocarbon dates from the Bapot-1 Site in Saipan and neolithic dispersal by stratified diffusion. *Journal of Pacific Archaeology, 1*, 21–35.

Hornbostel, H. (1925). *Unpublished field notes, 1921–1924*. Record on file at Library of Bishop Museum, Honolulu.

Hunter-Anderson, R. L., & Butler, B. M. (1995). *An overview of Northern Marianas prehistory*. Micronesian Archaeological Survey Report Number 31. Commonwealth of the Northern Mariana Islands Division of Historic Preservation, Saipan.

Hunter-Anderson, R. L., & Moore, D. R. (2001). The Marianas pottery sequence revisited. *Presentation prepared for Ceramic Tradition Workshop at International Symposium on Austronesian Cultures: Issues Relating to Taiwan*. Taipei: Academia Sinica.

Marche, A.-A. (1889). Rapport general sur une mission aux Îles Mariannes. *Nouvelles Archives des Missions Scientifiques et Littéraires, Nouvelle Série, 1*, 241–280.

Marche, A.-A. (1982). *The Mariana Islands* (R. D. Craig Ed. & S. E. Cheng Trans.). Publication Series Number 8. Micronesian Area Research Center, University of Guam, Mangilao.

Pellett, M., & Spoehr, A. (1961). Marianas archaeology: Report on an excavation on Tinian. *Journal of the Polynesian Society, 70*, 321–325.

Reimer, P. J., Bard, E., Bayliss, A., Beck, J. W., Blackwell, P. G., Bronk, R. C., et al. (2013). INTCAL13 and MARINE13 radiocarbon age calibration curves, 0–50,000 years cal BP. *Radiocarbon, 55*, 1869–1887.

Reinman, F. R. (1977). *An archaeological survey and preliminary test excavations on the Island of Guam, Mariana Islands, 1965–1966*. Miscellaneous Publication Number 1. Micronesian Area Research Center, University of Guam, Mangilao.

Russell, S. (1998). *Tiempon i Manmofo'na: Ancient Chamorro culture and history of the Northern Mariana Islands*. Micronesian Archaeological Survey Report Number 32. Commonwealth of the Northern Mariana Islands Division of Historic Preservation, Saipan.

Spoehr, A. (1957). *Marianas prehistory: Archaeological survey and excavations on Saipan, Tinian and Rota* (Fieldiana: Anthropology, Vol. 48). Chicago: Chicago Natural History Museum.

Chapter 8
1500–1100 B.C., Initial Settlement

First human settlement definitely occurred by 1500 B.C. in the Mariana Islands. Some evidence could suggest a slightly older date of human arrival, but 1500 B.C. can be accepted as the time when an established population lived in the islands and interacted significantly with this Remote Oceanic landscape. The oldest human presence is recorded in red-slipped pottery and other materials of at least eight sites of three separate islands, as well as in a horizon of human-caused impact on the native vegetation communities beginning at this same time. The first settlement period has been discussed at length in a separate book (Carson 2014b), so only the most relevant points about ancient life and landscape will be considered here, along with informative updates and clarifications.

The oldest known sites in the Marianas shared several characteristics of their ecological settings, artefact forms, and midden compositions that were sustained over a few centuries, approximately 1500–1100 B.C. Future research may push the oldest dating farther back by as much as a few centuries, but any significantly older dating possibly would refer to a context of very few people living in the region and with a significantly different landscape at that time. In its current definition, this earliest period ended about 1100 B.C. with the beginning of a drawdown in sea level, change in shellfish taxa in site middens, number and distribution of habitation sites, and transitions in the forms and styles of pottery, shell beads, and other artefacts.

Site Inventory and Dating

The early-period Marianas archaeological sites and palaeoenvironmental records have been confirmed in the larger southern-arc islands of Guam, Tinian, and Saipan (Fig. 8.1). The ancient environmental context was most thoroughly studied at Ritidian in Guam (Carson 2012, 2014c), as well as in several natural archives of lake-bottom and swamp-bottom deposits (Athens and Ward 1998, 1999, 2004,

M.T. Carson, *Archaeological Landscape Evolution*,
DOI 10.1007/978-3-319-31400-6_8

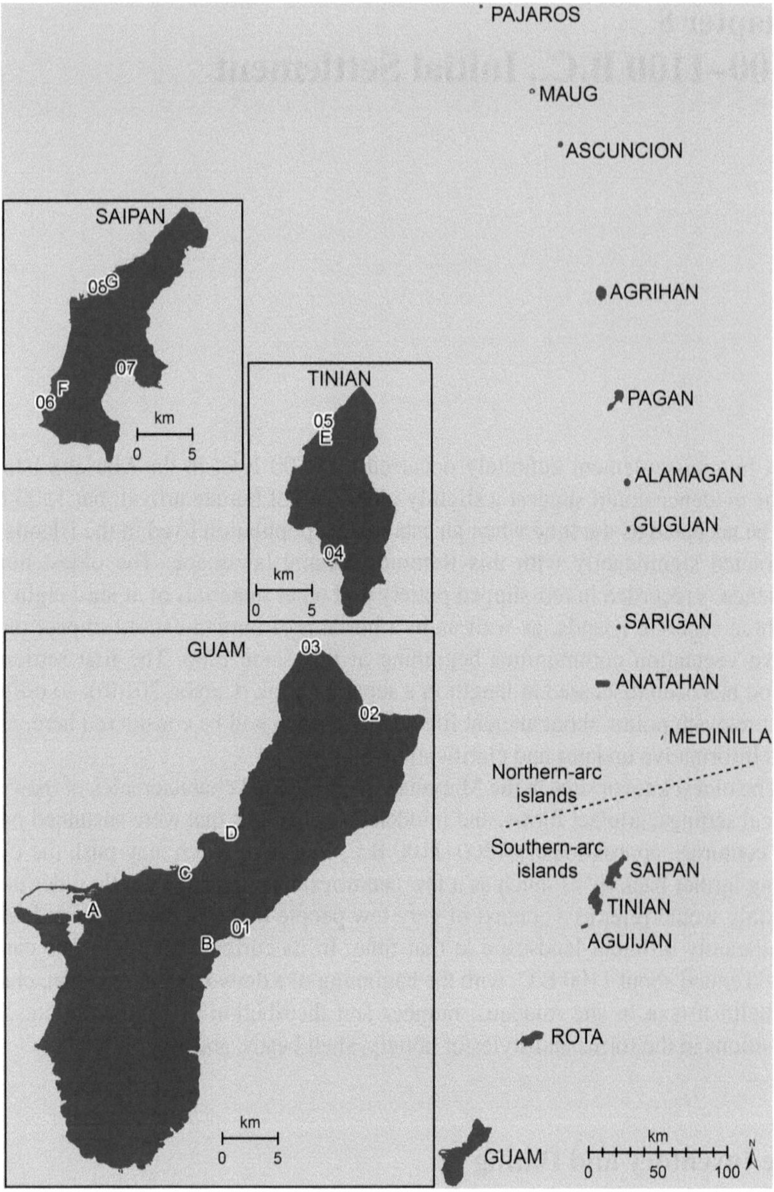

Fig. 8.1 Major site locations, 1500–1100 B.C. 01–08: Archaeological sites. 01: Mangilao. 02: Tarague. 03: Ritidian. 04: House of Taga. 05: Unai Chulu. 06: Chalan Piao. 07: Unai Bapot. 08: Achugao. A–G: Other location of reference. A: Laguas. B: Pago. C: Hagatna. D: Tumon. E: Hagoi. F: Susupe. G: San Roque

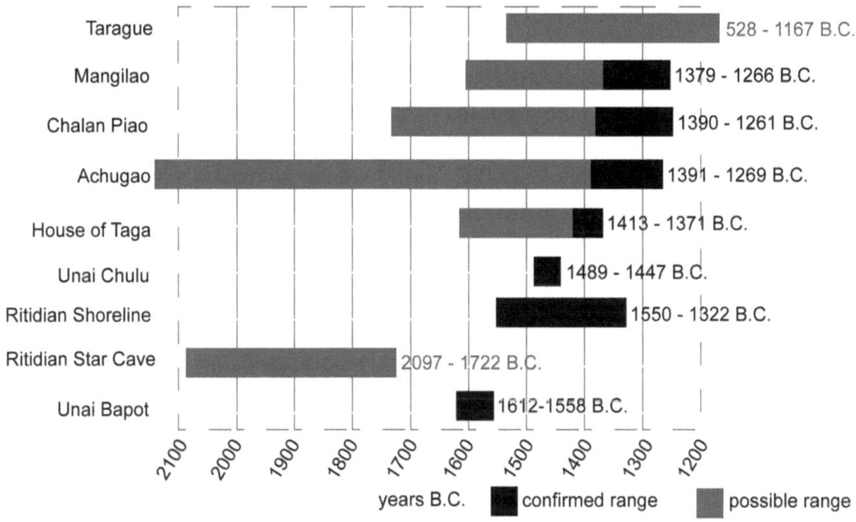

Fig. 8.2 Earliest site dating summary. Modified and updated from Carson (2014b, c) and Carson and Kurashina (2012)

2006; Athens et al. 2004). The material culture was best documented at House of Taga in Tinian (Carson 2014b:119–134). The oldest validated dating so far has been at Unai Bapot in Saipan (Carson 2008), but other sites may yet be proven as even older (Carson and Kurashina 2012).

The initial cultural occupations occurred in shoreline-oriented settings, now obscured beneath more than three millennia of sedimentary layers and stranded far inland from today's shorelines. Given the depth typically of 2+ m and sometimes sealed beneath solidified calcrete (Carson and Peterson 2011), the ancient cultural layers are known mostly from narrow sample windows of small test pits (Carson 2014a). Only very recently have excavations exposed larger contiguous areas of the ancient living surfaces, such as the largest so far about 90 m^2 at House of Taga in Tinian (Carson and Hung 2014).

A date of 1500 B.C. refers to the time when people indisputably were living in the Mariana Islands, but the precise timing differed slightly from one site to another (Fig. 8.2). According to a critical review of the early-period radiocarbon dating (Carson and Kurashina 2012), using redundant overlap of multiple radiocarbon dating samples, a range of approximately 1500–1300 B.C. applies for the oldest cultural activity at most sites, but more realistically people lived in these sites over a period of time with gradual accumulation of material over a few centuries, extending both earlier and later than the redundant overlap range of the radiocarbon samples. The oldest singular site dating was confirmed by overlap of three dating samples at 1612–1558 B.C. at Unai Bapot in Saipan, but the non-confirmed date range at the same site extends as early as 1916 B.C. (Carson 2014b:38–40). Other non-confirmed date ranges are potentially as old as 1616 B.C. at House of Taga in Tinian (Carson 2014b:123–124), 1741 B.C. at Chalan Piao in Saipan (Moore et al. 1992), 2097 B.C.

at Ritidian Star Cave in Guam (Carson 2014c), and 2133 B.C. at Achugao in Saipan (Butler 1994). Moreover, palaeoenvironmental records have been interpreted to indicate human-caused forest-clearing and introductions of overseas taxa as early as 2200–2000 B.C. (Athens and Ward 2004; Athens et al. 2004) although a more cautious interpretation points closer to 1500 B.C. (see Fig. 5.11).

Dating slightly older than 1500 B.C. would not be too terribly objectionable, but archaeologists working in Pacific Oceania have grown cautious of accepting old dates. Programs of "chronometric hygiene" have revealed that many sites do not necessarily date as early as may have been claimed, and overall younger date ranges are evident when using stricter controls of secure sample contexts, choice of short-lived datable specimens, and multiple cross-confirming dates (Spriggs and Anderson 1993). Accordingly, many island chronologies have been re-adjusted to accommodate the youngest verifiable dating in a "short chronology", with the possibility of eventually accepting older dating in a "long chronology". A short chronology is inherently easy to prove and treads into self-fulfilling logic simply by discounting anything older than a pre-decided acceptable age, while a long chronology undergoes close scrutiny. Older dates have potentially large impacts and implications, so they bear heavy burdens of needing to be proven against intense criticisms, whereas younger dates typically do not attract this kind of attention for debate.

As depicted in Fig. 8.2, a restrained short chronology begins by 1500 B.C. in the Marianas, but a liberal long chronology extends a few centuries earlier. Each singular radiocarbon date range encompasses the full probability within which any particular point could be correct, often spanning a few centuries. If multiple dates are available for the same cultural layer in a site, then their overlapping date ranges can cross-confirm each other as the most likely portion of the fuller ranges. Additional dating samples can increase the probability of their overlapping range as the most likely of the total possibilities. The results further can be refined by obtaining dating samples from clearly older and younger layers in the stratigraphic profile of a site, for example, with dates of underlying non-cultural coral reef or with dates of overlaying cultural deposits containing diagnostically later pottery and other materials. Ultimately, the most convincing dates are based on samples from secure contexts in cultural layers, preferably using short-lived specimens for producing two or more date ranges that can constrain each other for a more refined result.

The dating overlap can be useful for specifying the time when a site very certainly was occupied, but it potentially misrepresents the age of a cultural layer that accumulated over a longer period of time. In the Mariana Islands, the oldest buried cultural deposits reflect gradual sedimentary build-up, implying a length of time to account for their formation. Within a single layer, radiocarbon dates can be arranged in order from oldest to youngest, covering a number of centuries from beginning to end. At Mangilao in Guam, eight radiocarbon dates from the lowest cultural layer span from as early as 1601–1266 B.C. through as late as 1111–811 B.C. (Dilli et al. 1998), thus representing a few centuries of continuous cultural activity, within which the majority of dating results reveal a redundant overlap at 1379–1266 B.C. as the time when people absolutely must have been living at the site (Carson 2014b:29). The same approach can be applied to 16 radiocarbon dates securely from the oldest cultural layer at Unai Chulu in Tinian, in total reflecting continuous habitation from as early as 1581–1186 B.C. through as late as 1291–916 B.C. (Craib 1993; Haun et al. 1999), although the oldest cross-confirmed overlap indicates 1413–1371 B.C. for the age of definite habitation in a conservative short chronology (Carson 2014b:34–35).

Earliest Marianas site dating cannot be understood without reference to the findings in adjacent regions, mentioned only briefly here and revisited later in this chapter and elsewhere. Quite simply, when considering a date of 1500 B.C. or earlier, very few places in the Asia-Pacific region could have been the homeland of the people who sailed across the ocean to the Mariana Islands. The red-slipped and finely decorated pottery of the Marianas appears convincingly similar to the traditions known in the Philippines and parts of Indonesia by approximately 1500 B.C. (Carson et al. 2013), but the only known examples clearly pre-dating 1500 B.C. were in the northern and central Philippines as early as 2000–1800 B.C. (Hung 2008; Hung et al. 2011). No pottery of any kind appeared in any part of Island Southeast Asia before 2200 B.C. (Bellwood and Dizon 2013).

Most instructive for clarifying the Marianas homeland and its potential oldest dating has been the practice of applying white lime-infill inside the decorated designs on red-slipped pottery. This trait did not occur by accident, and it most reasonably can be attributed to a context where people already were using slaked lime (a white powder, made from heating coral, limestone, or marine shell) as one ingredient in preparing betel nut quids (Fitzpatrick et al. 2003). Betel nut (from the *Areca catechu* palm tree) acts as a narcotic stimulant when chewed as a quid with slaked lime powder and the leaf of a *Piper betle* shrub, wherein all three ingredients interact for this effect (Zumbroich 2008). The biological origin of *Areca catechu* has been traced to the Philippines, prior to a much broader dispersal probably aided by human travellers to various destinations throughout Asia and the West Pacific (Zumbroich 2008). Similarly, the oldest known shell container with residue of slaked lime has been found in the Philippines, specifically in Duyong Cave of Palawan, dated by association with a human burial feature as old as 2700 B.C. and a charcoal-rich hearth as old as 3700 B.C. (Fox 1970:62–65), by far pre-dating the emergence of pottery-making in the region. The oldest verifiable usage of white lime-infill in pottery decoration was much later, about 2000–1800 B.C. in the Cagayan Valley of the northern Philippines (Hung 2008). Further intriguing, pollen of the betel nut tree (*Areca catechu*) suddenly appeared in the Marianas within an unmistakeable horizon of initial forest-clearing and burning, as seen in the most recent analysis of a swamp-bottom coring record from Susupe in Saipan (Athens and Ward 2006), argued in some views to date as early as 2200 B.C. but probably in fact dating closer to 1500 B.C.

Considering the dates of red-slipped and white lime-infilled pottery in the Philippines, an oldest age of 2000–1800 B.C. appears acceptable for the origin of this unique combination of traits that eventually occurred in the Mariana Islands. A slightly older date of 2200 B.C. potentially can be entertained when considering the oldest evidence of pottery-making in the region, prior to the confirmable dating of lime-infilled pottery decoration. These dates can be compared with the ages of the corals directly beneath the initial habitation layers in the Mariana Islands, where the corals last lived at some time prior to the emplacement of the overlaying cultural deposits, dated 2455–2068 B.C. and 1929–1644 B.C. at Ritidian Shoreline in Guam (Carson 2012), as well as 3031–2731 B.C. at House of Taga in Tinian (Carson 2014b:34–35). Curiously, dating as old as 2580–2043 B.C. was reported for an enigmatic subsurface feature at Tumon in Guam (Bath 1986), described as a hearth but lacking any artefacts or midden or even an associated cultural layer, later found to have been submerged beneath a sub-tidal zone of Tumon Bay until the formation of a sandy beach by A.D. 100–200 (Carson 2011, 2014b:30–31).

Possible sites significantly prior to 1500 B.C. may relate to a different cultural context than was evident in the material outlined here as characterising the period 1500–1100 B.C. The available evidence allows a long chronology in the Marianas perhaps as early as 2000 B.C., but no archaeological site in the Marianas so far has yielded artefacts, midden, or other hard evidence confidently dated to such as early age. In the absence of relevant data, imaginations may conjure scenes of roaming sea nomads, hapless castaways lost far from home, and other visitors who left behind only ephemeral traces on the Marianas shores, but these unsubstantiated speculations are impossible to confirm or deny at this time.

Landforms

Landforms in the Marianas have been affected by changing sea level, both directly and indirectly, most notably in coastal zones. This information has been essential for making sense of the earliest settlement in relation to a very different coastal ecology of that time as compared to modern conditions (Fig. 8.3; see also Fig. 3.8). As has been introduced in Chap. 3, during the time when people first settled in the islands, the ocean level itself stood about 1.8 m higher than today throughout this sector of the Pacific Ocean (Dickinson 2000, 2001, 2003). A period of sea-level drawdown began about 1100–1000 B.C., eventually reaching a stable point about A.D. 100–200 before additional drawdown. The most recent trend has been a rising sea level although today's level still is 1.8 m lower than it was during the first settlement period in the Marianas. These rates and magnitudes of sea-level change apply in the Mariana Islands, but they are known to vary across different parts of the world's oceans and further according to localised tectonic activities.

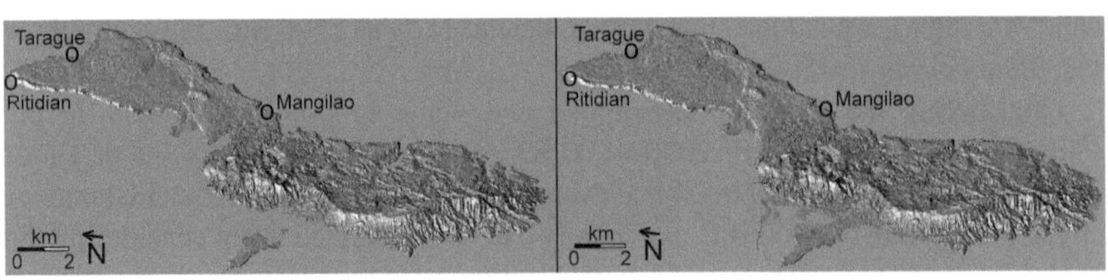

Fig. 8.3 Palaeo-terrain model of Guam, 1500–1100 B.C., compared to modern setting. Based on data from Carson (2011, 2014a)

The sea-level history in the Marianas has been outlined more confidently than so far has been possible in other regions. In the southern-arc islands, limestone formations retain notches where higher stands of sea level cut into the limestone, exacerbated by additional weathering from marine organisms and other factors. The elevations of these tidal notches can be measured, and they can be matched with portions of coral reefs that have become exposed above sea level. Importantly, those portions of emerged reef can be dated directly by radiocarbon. Moreover, new efforts at deep excavations have discovered old coral reefs in subsurface contexts, for example, at Ritidian and Hagatna in Guam (see Figs. 3.12 and 3.16), where absolute elevations were measured and coordinated with radiocarbon dates of the last time when those corals were living. Later stages of sea-level drawdown and temporary stability gain support from findings of ancient beach surfaces, again with verified elevations and dates, for example, in the successive layers of beach formation at Tumon in Guam (see Fig. 3.11).

Even a small fluctuation in sea level can make a significant difference in island and coastal settings, not only in the shapes of coastlines but also in the behaviours of people who live closely with the sea. A change of 1.8 m of sea level potentially can submerge or expose very broad sections of lowland terrain, in total comprising several square kilometres when considering the Mariana Islands as a whole. As illustrated in Fig. 8.3, the configuration of coastal landforms in Guam was substantially different for the first settlers at 1500 B.C. than can be seen in the present-day conditions.

During the earliest settlement period, a higher sea level touched on or near the cliff faces and hill slopes of the islands, with little or no opportunity for the low-lying coastal terrain that can be seen today. In particular, broad sandy beaches did not exist in the Marianas, but instead a few narrow beach fringes, ridges, and berms could be found in scattered locations. These ancient sandy formations later were covered by thick and wide deposits of more recent sediments. The middle to outer portions of coral reefs later were exposed above the lowering sea level, while the inner reef zones and lagoons were buried beneath accumulations of storm-surge sands. Slope-eroded sediments eventually filled over valley floors and coastal terraces, but these zones mostly were watery without dry land when the first people lived in the Marianas.

According to the oldest known site locations and their original contextual settings, the earliest settlers occupied shoreline zones within easy reach of a range of nearshore resources. These areas were the only options at the time for people who wished to live close to the seashore. Otherwise, habitations could be installed in areas somewhat removed from the coasts, for instance in elevated limestone plateau or in volcanic hilly terrain, where the known sites mostly post-date A.D. 1000 and very few date as early as 500 B.C.–A.D. 100.

A shoreline-oriented character of early settlement matches most expectations of the early settlement period, but to some extent the records sealed beneath beach sands could be over-represented in comparison to sites in other landforms with poor preservation conditions. People surely knew about the volcanic hills and limestone plateau terrain, but cultural layers are unlikely to have survived in the thin, rocky, and acidic silts and clays. If people ventured into these areas and left behind any material traces, then the material evidence has become lost due to natural chemical and mechanical weathering, disturbance by vegetation roots and rainwater, and cultural reworking of the thin and rocky sediments over several centuries. As noted, the known site remnants in volcanic hills and limestone plateau landforms so far refer only to the later components of the Marianas archaeological sequence.

Resource Zones

For people seeking new habitable locales in the remote Mariana Islands, the conditions at 1500 B.C. would guide them to a number of scattered places in the larger southern-arc islands, where fresh water and other resources could be acquired most easily. These larger islands offered greater diversity of habitats and overall more biomass, but they also offered reliable access to essential life-supporting fresh water in aquifers exposed inside deep caves, in seeps draining near the coastlines, and in streams especially in southern Guam. Otherwise, capture of rainwater was possible in any of the islands, but it was most predictable in the southern islands with slightly more rainfall year-round. The accessible groundwater sources could tip the balance in favour of survival during rainless periods. A number of sparse sandy beach patches were habitable in scattered points around the edges of the larger islands, close to coral reefs and other shallow-water swamp-like zones with abundant and diverse shellfish and other important resources. These aspects of the natural ecosystem were not so attractive in the smaller volcanic-cone formations of the northern-arc islands.

The higher sea level supported lenses of fresh water at a slightly more elevated level than today (see Fig. 3.21), and some of these lenses were more easily accessible during the period 1500–1100 B.C. than can be seen now. In particular, pools of fresh water stood in the bottoms of caves at the level of the natural aquifers of the larger southern-arc islands, plus the largest caves in limestone formations provided greater surface area for collecting ceiling-drip water. These essential resources very likely were major criteria when people were selecting the first habitation sites of the region. Accordingly, large caves are found near some of the earliest sites, such as at Ritidian in Guam, where the cave floors were very close to the elevation of the ancient freshwater lens. Importantly, a freshwater lens could be accessed at these precise elevations above the zone of mixing with underlying brackish and salty water. Today, many of these caves no longer contain freshwater pools, due to lowering of the freshwater lens over a falling base of sea level after 1100 B.C.

The earliest Marianas sites were situated at the ancient shorelines of their time, implying that people targeted specific niches with access to resources on land and at sea. The cultural deposit at Ritidian Shoreline in Guam had formed inside an intertidal or shallow sub-tidal zone, with meadows of *Halimeda* sp. algae overlaying a floor of *Heliopora* sp. coral, measured at 1.8 m above today's sea level and thus matching the local sea-level history (Carson 2012). Another cultural deposit at Ritidian Star Cave was emplaced inside a different portion of the widespread bed of *Halimeda* sp. algal bioclasts, directly outside the entrance to a large cave complex (Carson 2014c). The oldest cultural layers at Achugao and Chalan Piao, both on the west-facing leeward coast of Saipan, have been described as situated on small berms or ridges of sands, surrounded by shallow-water zones and detached from the main land mass of the island (Butler 1994; Moore et al. 1992). Other early site deposits must have been very close to their contemporary shorelines, according to their absolute elevations in comparison to the local sea-level history, additionally reinforced by the presence of reef-derived detritus, lagoon clasts, or other indicators of the ancient localised beachfront conditions (Carson 2014a).

The exposed beachfront settings prior to 1100 B.C. are clarified by findings of definitively stable backbeach conditions in the immediately post-dating superimposed layers. These over-capping layers of sands and silts after 1100 B.C. contain abundant evidence of in situ burning of hearths and various pits, whereas burning episodes were preserved less commonly in the older settings subjected to tidal washing. In sites such as Ritidian in Guam, the diagnostic lagoon facies sands of *Halimeda* sp. bioclasts are overlain by entirely different sandy compositions. Furthermore, the later stable conditions coincided with a sharp decline or total disappearance of the earliest forms of very thin red-slipped pottery and other artefacts, instead associated with thicker pottery and other materials, indicative of a transition in the cultural setting while the coastal zones began to transform substantially.

The stable backbeach zones began forming about 1100 B.C. in the sites where dating is available, such as at Ritidian Shoreline and Mangilao in Guam (Carson 2012; Dilli et al. 1998), at House of Taga and Unai Chulu in Tinian (Carson 2014b:34–35; Haun et al. 1999), and at Unai Bapot in Saipan (Carson 2008). Especially interesting are dense concentrations of branch coral debris in whole layered deposits, most likely broken from their reef habitats and thrown ashore during major events of periodic typhoons or other storm-surge actions. In order to create the evident thick layers instead of just isolated pieces of branch coral debris, these events must have occurred during the last time when the site surfaces were exposed and prior to a drawdown of sea level approximately around 1100–1000 B.C. As the sea level lowered, the ancient habitation sites were stranded somewhat inland and at higher elevation from the new active shoreline. Prior to very much drawdown in the sea level, while those ancient sites still were within reach of storm-surge, broken bits of branch corals covered the oldest cultural deposits and in some cases disturbed them, as at Mangilao and Unai Chulu. In one instance, a sample of *Acropora* sp. branch coral was dated 1364–1050 B.C. at Ritidian Shoreline, directly overlain by a cultural deposit dated 1056–842 B.C. (Carson 2012, 2014c). A slightly earlier age of this transition in coastal morphology is recorded at the nearby Ritidian Star Cave, where a layer of *Halimeda* sp. algal bioclasts containing extremely thin red-slipped pottery was dated 2097–1722 B.C., overlaid by a formation of silty sand containing a later type of slightly thicker red-slipped pottery dated 1418–1144 B.C. (Carson 2014c).

The active shoreline contexts in some cases have contributed to disturbance of the ancient cultural deposits, but these instances should not be misconstrued as indicating hopelessly damaged sites. Some of the deposits contain pottery fragments with rounded edges, characteristic of rolling in the water, particularly at Tarague in Guam (Kurashina and Clayshulte 1983a, b), the most seaward area of the ancient deposit near House of Taga in Tinian (Pellett and Spoehr 1961), and various areas of Achugao in Saipan (Butler 1994). Other sites contain pottery with non-eroded edges, and sometimes the broken pieces can be refitted, as at the inter-tidal zone of Ritidian Shoreline in Guam (Carson 2012) and the landward portion of the most ancient habitation layer near House of Taga (Carson 2014b:119–134).

The oldest cultural layers in the Mariana Islands very certainly were formed close to the ancient shorelines, thus stirring practical questions about how the sites were occupied in contact with the tidal waters. Where structural features have been documented, post moulds suggest that houses were raised on stilts made of tree trunks generally 20–30 cm in diameter but as large as 40 cm in diameter, for instance at House of Taga (Carson and Hung 2014), Unai Chulu (Haun et al. 1999), and Achuago (Butler 1994). In the largest so far exposed ancient living surface of about 90 m^2 near House of Taga, the post moulds were surrounded by bracings of stone cobbles, and very tellingly all of the hearth features were situated at the more inland side of the habitation zone and farthest from the active shoreline (Carson 2014b:119–134).

Similar to the case in the Marianas about 1500–1100 B.C., stilt-raised houses stood over inter-tidal or shallow sub-tidal zones in sites dated as early as 1500–1350 B.C. in the Bismarck Archipelago (Kirch 1997, 2001) and 1100–900 B.C. in Fiji (Nunn 2007). This form of habitation may have been a preferred way of life for many coastal people of the greater region at this time. The tradition of stilt-raised or pile-raised housing later was formalised into stone pillars during the *latte* period after A.D. 1000 in the Mariana Islands, but these later sites were inhabited in stable backbeach zones and inland terrain. The immediate proximity with active shorelines was strongly associated with the oldest sites, but this practice diminished over the course of a few centuries, after the sea level began its drawdown and coastal ecologies transformed following approximately 1100 B.C. in the Marianas.

As may be expected, the shoreline habitations contain dense shell middens. Shellfish naturally were easily accessible, and accordingly the selected taxa can reveal what sorts of habitat zones were available or preferred for shellfish-collection by the local residents. The archaeo-shellfish records are well preserved in the earliest sites, where the shells were discarded beneath and around the house structures and quickly accumulated in thick heaps.

Most abundant in the oldest habitation sites are shells of *Anadara* sp., an ark clam (of the Family Arcidae) that prefers to live in swampy or other shallow-water habitats such as with sea grass beds or meadows. A swampy or marshy habitat presumably could be found in the immediate vicinity of sites such as Chalan Piao and Achugao in Saipan, as well as probably Unai Chulu in Tinian, where wetland sediments currently exist in close proximity and suggest even larger zones in the past during a period of higher sea level than today's conditions. A broad expanse of shallow water can be verified at Ritidian in Guam, characterised by meadows of algal bioclasts (Carson 2012, 2014c). Similar supporting habitats likely existed at sites such as Unai Bapot in Saipan, where large-sized *Anadara* sp. shells were overwhelmingly abundant (Carson 2008). This same kind of shallow-water zone, with very little turbidity, likely supported growth of seaweeds, for which the nutritional and culinary values cannot be overlooked, but material traces of seaweeds are unlikely to have survived in the habitation deposits after human consumption.

After the first few centuries of habitation in the Marianas, the archaeo-shellfish records indicate a nearshore resource depression, resulting from natural sea-level drawdown combined with human-caused harvesting (see Fig. 5.9). Notably, *Anadara* sp. shells declined markedly in frequency after 1100–900 B.C. (Carson 2008), and today they are extremely rare in the Marianas except in a few places of mangroves and other swamps (Amesbury 1999, 2007). Meanwhile, inter-tidal rock-clinging and reef-clinging taxa such as sea urchins, limpets, and chitons were depopulated locally at the oldest habitation sites although they continued to exist in healthy numbers in other places until eventually people made new residences and affected those resource zones as well (Carson 2012, 2014b:93–96).

As the supplies of *Anadara* sp. diminished, along with other taxa affected by the transforming nearshore zones, the archaeo-shellfish records reflect a steady reliance on shellfish from the middle and outer reef zones. These taxa may have been more resistant to the changing conditions as the sea level underwent a period of draw-down after 1100 B.C. Taxa such as *Trochus* spp. and *Turbo* spp. are particularly well represented and steady throughout the chronological sequence. Only much later, after A.D. 1000, did a major shift occur in harvesting large amounts of *Strombus* sp. gastropods from the newly formed reef and lagoon ecosystems.

Evidence of a nearshore resource depression is clear by 1100 B.C., but the contributing factors of human harvesting must have begun earlier and during the first settlement period. This scenario would suggest a heavy reliance on the shallow-water and inter-tidal zones very close to the oldest habitation sites, perhaps not surprising at all but certainly important for understanding the composition of those habitats. In particular, the targeted habitats appear to be in places with direct access to ample provisions of *Anadara* sp. shells, thus limiting the choice of residential sites to these kinds of zones.

Regarding the kinds of mangrove swamps and other shallow-water habitats that once supported *Anadara* sp. shellfish, potential candidates can be found scattered around the coastlines of the larger southern-arc islands, but only a few so far have been verified to contain evidence of early-period habitation (Carson 2011, 2014a). In Guam, for example, only three of several possible such locales have been found to contain earliest habitation sites (see Fig. 8.3). Practically speaking, additional sites may yet be discovered after more thorough subsurface explorations, but at least some of the potentially attractive ecological zones do appear to have been uninhabited.

Large uninhabited zones may have been due to the small numbers of people residing in the Mariana Islands during the first few centuries of human presence. Long stretches of coastlines, nearly all inland forests, vast expanses of the sea, and entire islands were not inhabited during the earliest settlement period, and many of these were not inhabited until several centuries later. Whether intentionally or not, these uninhabited zones may have acted as social buffers between groups, and they certainly allowed people at each separate site to recruit their resources from very broad areas and multiple habitat zones.

While people lived mostly near the seashore, at least some people ventured into adjacent and farther resource zones. People needed to search inland areas and other places removed from the primary habitation sites in order to find suitable clays or making pottery and specific types of stone (notably chert) for making tools. Meanwhile, lakes and swamps received flecks of charcoal inflowing from burning of the native forests, and pollen records show a decline in native forests combined with increase of economically useful taxa such as coconut and betel nut palms (Athens and Ward 2006). Although not necessarily as an intensive endeavour, people were accessing inland forests, finding clay and stone sources, and manipulating the environment since the beginning of human settlement in the Mariana Islands.

The effects on native forests deserve more discussion, due to the basic human nutritional need for plant foods that were naturally abundant in Island Southeast Asia but not in the Remote Oceanic islands such as the Marianas. Human-caused clearing of island forests therefore gains special significance because it implies deliberate removal of native vegetation while encouraging the growth of culturally useful plants. A definite horizon of impact is attested in the palaeo-botanical records (see Fig. 5.11), showing an influx of charcoal from burning that never happened for thousands of years until suddenly at one point. This same initial burning coincided with a decline in native palms, increase in grasses and ferns that typically intrude into newly opened lands, significant rise in coconut palms as one of the most useful of all plants in the region, and first appearance of both betel nut palms and ironwood trees (Athens and Ward 2004, 2006; Athens et al. 2004).

The most useful plant foods could not grow directly at the beachfront sites composed of calcareous substrates, but rather they were accessed in nearby terrain with adequate soil development. In these locales, native forests could provide coconuts, a type of local seeded breadfruit (*Artocarpus mariannensis*), and possibly nuts and starch from assorted palms and cycads. These resources may have been sufficient to allow initial founding groups of people to survive, at least long enough to encourage growth of more of these same plants while nurturing the plantings of new items from overseas.

The definite overseas imports of plants included the betel nut palm (*Areca catechu*), ironwood tree (*Casuarina equisetifolia*), non-seeded breadfruit (*Artocarpus altilis*), bananas, yams, and varieties of taro (except perhaps a native swamp taro). Of these, only the betel nut palm and ironwood tree have been dated in pollen records within the initial cultural impact horizon in the Marianas (Athens and Ward 2006). Neither of these plants provided direct dietary nutrition, but rather they were useful for ingredients in narcotic stimulants (betel nut) and for supplying extremely hard and dense wood (ironwood). All other suspected early plant imports are undated so far, due to poor preservation of their remains in the lake-bottom and swamp-bottom coring records. However, preserved starches and other residues have been identified on artefacts in the lowest cultural layer at House of Taga in Tinian, and very likely at least some of these can be verified as belonging to imported starchy foods, pending taxonomic identifications with reference collections still in progress.

Additional clues about the cultural use of resource zones are found in animal bones of the oldest sites, so far difficult to interpret due to the small numbers of bones found in these sites. They are scarce, but these bones consist primarily of small pieces from fish, turtle, and bird without any indication of imported or domesticated species. The depositional contexts in active beachfronts were less than ideal for preserving small bone fragments, especially the narrow and hollow bird bones that easily can disintegrate or simply float into disappearance with the tides. Of the surviving fish bones, nearly all are taxonomically ambiguous, other than a few mouth parts belonging mostly to reef-grazing fish, but vertebrae sizes reflect a broad range of small, medium, and large species. Turtle bones are common in the earliest cultural layers, although they decline rapidly thereafter, likely as a result of people interfering with habitual feeding grounds and nesting areas. Bird bones have been very sparse in the oldest sites, partly due to the contexts of deposition and preservation as noted, but other factors may be considered about the processing of small bird bones before, during, and after meals. In any case, more studies with larger samples of animal bones will be needed for clarifying the kinds of animals that were captured and from what range of habitats.

Tentatively, the records of animal bones suggest very little adverse impact on the native vertebrate fauna during the earliest settlement period, other than apparently chasing away the turtles from the immediate vicinities of habitation sites. As noted, future studies may yet discover de-populations of birds, as has been found in numerous other Pacific Islands (Steadman 1995). So far, an avifaunal extinction horizon has not been identified in the Marianas until much later in the cultural sequence, coincidentally after the first arrival of rats about A.D. 1000 (Pregill and Steadman 2009). Additionally, imported domestic animals were absent in the Mariana Islands until after Spanish contact following Magellan's arrival in A.D. 1521 although domesticated pigs, dogs, and chickens played important roles in most other islands of Remote Oceania (Wickler 2004).

Overall, the preserved food remains depict an emphasis on protein from the immediately accessible nearshore zones at the first habitation sites, with a range of plant foods from nearby interior island terrain. The protein appears to have been dominated by shellfish, supplemented by variable amounts of fish, turtle, and bird. The plant foods are not yet fully documented, but coconuts certainly were important among a suite of others as mentioned. Starches were retained on surfaces of stone pounding and slicing tools likely used during food preparation, and pottery vessels definitely were used for cooking assorted meals.

The recruitment of diverse resources intuitively can be perceived as managing a set of immediate, adjacent, and distant resource zones. The oldest habitation sites were situated with immediate access to nearshore resources of shellfish, probably seaweeds, certain kinds of generally smaller reef fish, unsuspecting turtles, perhaps some seabirds, and freshwater sources in the terrain just above sea level. The same sites were adjacent to other resource zones within easy reach of walking, paddling, swimming, or wading for just a few minutes, including edges of forests, lower portions of elevated terraces and other upper terrain, caves near some sites, and the outer reef zones near all sites. More distant resource zones required at least a little pre-planning and implied a degree of personal skill for accessing deeper interior parts of the forests, upland terrain, places for obtaining raw materials of clays and tool-making stones, the ocean outside the reef barriers, and farther islands.

The configuration of landforms and natural resource zones clearly affected the possible choices of where people first settled in the Mariana Islands, but human populations very quickly began to affect these resources and the ecological balance that had existed prior to human arrival. The Marianas landscape overall could absorb these impacts or adjust with some time, but soon a lowering sea level added to the ecological imbalance and began one of a series of long-term transformations of the island ecologies, especially in the nearshore niches that happened to be most essential for human survival. The apparent targeting of certain shoreline sites may have been productive at first, but it was not sustainable through irreversibly changing conditions.

Material Culture

First human settlement indelibly changed the Marianas landscape, bringing new dynamics of human–environment interactions that did not exist here previously. People in a sense became an invasive species in an isolated environment that had been evolving for thousands of years without any human presence. Both intentionally and unintentionally, the first settlers created a new landscape system in the Mariana Islands, in some ways imposing cultural notions on the environment but in other ways fitting into the available niches.

The material record of first settlement provides a substantive chronicle of the process of creating a new cultural landscape, perhaps more accurately understood as a natural–cultural landscape system. Broken pottery and other artefacts reveal the technological, economic, and other aspects of a society interacting with the Remote Oceanic environment for the first time in human history. Missing, though, is any written account or other direct register of the thought process when people assigned names, meanings, and stories to the islands, mountain peaks, reefs and lagoons, strange new plants, locally unique stones, and other aspects of the landscape. Toward understanding these processes, a number of inferences can be developed from observations of the surviving archaeological materials, but a comprehensive view first requires an introduction to the assemblages of the earliest Marianas settlement period.

The oldest material culture assemblages are most abundantly characterised by thin-walled and red-slipped pottery, but of course several other items are essential for a fuller picture of the scene. Among the red-slipped pottery, rare decorated pieces are extremely important for understanding the cultural context of the earliest period. Additional rarities in the pottery collections are black-burnished fragments and other paddle-impressed pieces. Along with the pottery are varied forms of shell beads and ornaments, as well as shell and stone tools, in some cases exclusive to the earliest cultural period but in other cases continuing longer in the sequence. Occasional structural features have been scarcely documented, except in the larger excavation exposures, where the findings have included hearths, post moulds, rubbish pits, and stonework ruins such as cobble pavings.

The most commonly found early-period artefact material consists of broken earthenware pottery, generically categorised as a component of the "Marianas Red" series, but the assemblages include a number of variants that are not always red in colour (Fig. 8.4). As first described by Alexander Spoehr (1957), Marianas Red

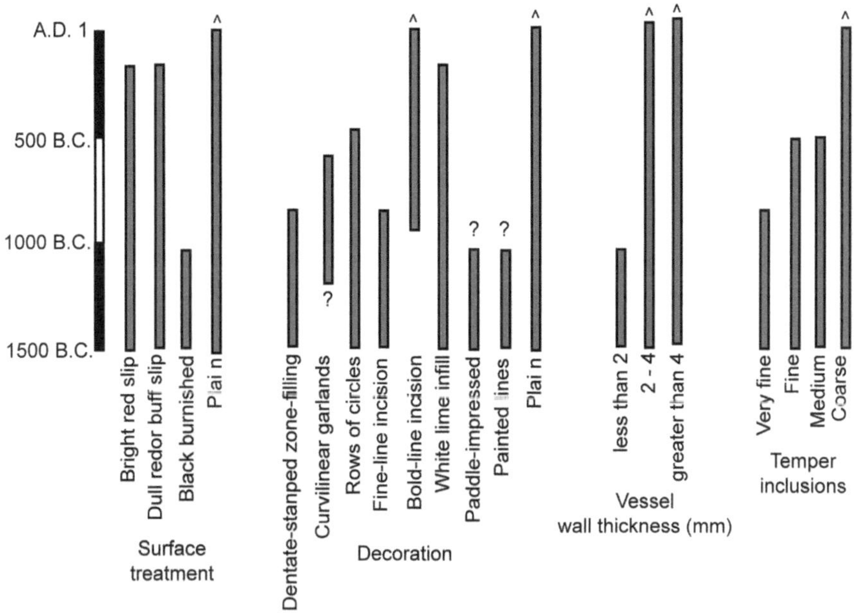

Fig. 8.4 Variable traits of "Marianas Red" pottery

refers to a broad range of pottery pre-dating the preponderance of thick, coarse, and generally non-slipped earthenware of the *latte* period. The first descriptions were vague, due to the small numbers of potsherds found in very few early-period sites. Additional research clarified the thin-walled, red-slipped, finely decorated, and black-burnished varieties of the earliest pottery landward of House of Taga in Tinian (Pellett and Spoehr 1961). The narrowed rim profiles (A-type rims) contrasted obviously against the thickened rims (B-type rims) of the *latte* period, previously recognised by Laura Thompson (1932) although without the chronological control of later excavations. Fred Reinman (1977) further noted that the thickened-rim (B-type) pottery tended to contain coarse volcanic-sand temper in the clay paste, whereas presumably older pottery with narrow A-type rims tended to contain calcareous beach-sand temper, thus leading to categories of volcanic-sand temper (VST), mixed sand temper (MST), and early calcareous ware (ECW).

The oldest Marianas pottery was well made by experts, from the beginning of island habitation, despite necessary adjustments according to local raw materials and possible bottleneck of cultural knowledge and skills. Although in theory a few pots likely were transported across the ocean from a distant homeland, the continued pottery tradition in the Mariana Islands required use of local clays, temper inclusions, and of course skilled potters. The predominant beach sand temper cannot be linked definitively with any specific geological province, as calcareous material mostly appears identical cross-regionally, but trace amounts of andesitic and dacitic aggregates appear consistent with local geological sources in the Marianas (Dickinson 2006:143). In the habitation deposits, small pieces of haematite (a reddish mineral

Fig. 8.5 Unfired and partially worked clay with fingertip impressions, landward of House of Taga

form of iron) likely were used as the colouring agent for preparing a red slip, but they also may have been used for pictographs and other applications of red-colouring besides just pottery-making. Nodules of unfired clay more clearly support a scenario of local pottery manufacture within at least some of the habitation sites. Particularly informative is a lump of partially worked yet never fired clay retaining fingertip impressions, found near House of Taga in Tinian (Fig. 8.5).

The first generations of Marianas potters maintained traditions of red-slipped vessels, in some cases with distinctive incised and stamped decorations highlighted by white lime-infill. As noted elsewhere, these traits of the end products match with contemporaneous and older findings in the Philippines of Island Southeast Asia, but the techniques of manufacture varied in terms of how to achieve those final products. Such variation may be expected, according to differences in raw materials, personal preferences, interplay of supply and demand, and specialised skillsets. In the archaeological assemblages, these differences are detectable through analysis of vessel wall thickness, rim and lip profiles, material composition, traces of clay-forming techniques, and comparison of decorative motifs.

In one comprehensive approach, the unique characteristics of the oldest Marianas pottery were examined via the technical and artistic choices involved in a multi-step process from collection of raw materials, through stages of forming and shaping, to the final products (Carson 2014b:53–68). By keeping the end products in mind, the choices made along the way can be more reasonably ascertained and appreciated.

This approach leverages the observations of thousands of individual pottery fragments toward understanding the cultural processes involved in making the final products of whole vessels.

So far the most locally distinctive trait of earliest Marianas pottery, not yet documented in similar-aged sites in Island Southeast Asia, was the production of impressively thin walls for the earthenware vessels. In many cases, the body portions were sometimes only 0.5 mm, but generally they were 1–2 mm in thickness, with tendency for slightly thicker portions at critical points of curvature needing more strength. In order to achieve these thin-walled products, other choices were made of small and fine temper inclusions, tight compaction of the clay paste during initial and secondary hand-forming, additional processing by beating and trimming, thorough firing, and overall small vessel size. In combination, these choices resulted in a set of localised material traits in the Marianas early-period pottery, and furthermore this combination of traits began to change noticeably after approximately 1100 B.C.

The early-period Marianas pottery consisted mostly of red-slipped but rarely black-burnished thin and small vessels of about 20 cm or less in diameter, often with a carination or angled shoulder, and very seldom with dentate-stamped, circle-marked, and fine-line incised decorations (Fig. 8.6). The vessel forms were rather simple, without any elaborate anatomy. Only two examples of handles have been found in the House of Taga assemblage. Decoration was restricted to the upper and most visible portions, but it was very rare in less than 1 % of the total pottery fragments.

The decorated varieties were described in two major categories of "Achugao Type" and "San Roque Type", named after the two sites where they were documented in large collections by Brian Butler (1994, 1995). Both types of designs were made with very fine precision, impressed or incised into leather-hard clay before firing and highlighted by white lime-infill. The Achugao Type referred to dentate-stamping in rectilinear zone-filling patterns, and it was produced at least as early as 1500 B.C. as seen in the oldest cultural layers most clearly at House of Taga, Unai Chulu, Unai Bapot, and Achugao. The San Roque Type referred to curvilinear garlands in horizontal bands, and it gained popularity slightly later but certainly prior to 1100 B.C. Both included rows of circle-marking, made either by prepared circle-shaped stamps or by hand-drawn circles. Rows of circles, without distinctive dentate-stamping or other decorative elements, have been observed in potsherds from the oldest layers at Ritidian and Tarague in Guam.

Although quite rare, the decorations on early pottery serve as excellent markers of the time period. They further contribute invaluably in cross-regional comparisons, as will be discussed later. Finely made incision and stamping, highlighted by white lime-infill, produced patterns of lines, circles, rows of repeated motifs, and juxtaposed zones of filled and non-filled elements (see Fig. 8.6). After 1100 B.C., the decorative output shifted in favour of coarser and bolder incisions and larger stamped circles, while the finer dentate-stamping, line incisions, and smaller circles disappeared from the sequence. The vessels meanwhile became thicker and slightly larger.

In the voluminous collections now available from House of Taga and also now found at the Ritidian Star Cave and Ritidian Beach Cave, three very rare forms been added to the repertoire of early-period pottery (see Fig. 8.6). First, a thin coating of red

Fig. 8.6 Pottery types, 1500–1100 B.C.

paint or pigment was applied on the surfaces of certain potsherds after firing, for an unknown purpose but possibly for aiding in measuring the quantity of material inside the vessel. Second, a carved or wrapped paddle created markings on the surface of very few pots, likely as a by-product of paddle-and-anvil final shaping of the clay that otherwise would have been erased during the finishing stages, but it was retained in rare cases for artistic or other values. Third, handle appendages have been found in just three instances of this earliest period. These extremely rare findings provide further insight into the inventory of skills and techniques of the early-period Marianas potters, as well as their possible links with pottery-making traditions in other regions.

In addition to the copious pottery fragments, other artefacts characteristically were made only during the earliest cultural period and then never again thereafter, while others diminished in popularity over some centuries or longer (see Fig. 6.3). The early-period assemblages contain large pounding implements, medium to heavy-duty chopping tools, small flaked tools for finely controlled cutting and slicing, shell beads and other ornaments, and fishing-gear (Fig. 8.7). These basic functional categories and their variants will be considered here, with attention to the characteristics that define the early-period Marianas material culture.

Of the very few known pounding tools, consistently these were made of limestone, including varieties with greenish hue and others made from crystallised flowstone, most likely collected from caves. These objects are considerably heavier, longer, and wider than the later occurrences of pestle-pounders in the *latte* period, so a different cultural usage is implied for more heavy-duty processing. Slight battering damage is noted at one end or sometimes at two ends of the long axis. Conceivably, volcanic stone would have been suitable for these tools, but so far no pounders have been discovered of volcanic stone in the earliest cultural layers in contrast to their greater prevalence made into the thinner pestles of the *latte* period.

From the earliest cultural period, the chopping and slicing tools preferentially were made of chert, sometimes of hard shell, occasionally of volcanic stones, and rarely of limestone. The most easily recognisable tools are adzes, intended for wood-chopping or other wood-working, with fully polished blades and often with polish extending around other surfaces. Other items are flaked pieces, often with visible use-wear on the sharp edges, possibly gripped directly by hand or attached to a stick of wood. Manufacture or maintenance occurred within the habitation sites, reflected in small bits of debitage by-products and flaking cores with clear scars where the flakes were removed.

The choice of stone or shell for different items reveals a familiarity with the local environment, as well as an ability to work effectively with a range of materials. The early preference for chert obviates an awareness of the local geology from the very beginning of human settlement in the Mariana Islands, but later the use of chert would dwindle, perhaps related to limited sources of the raw materials. Volcanic stone was used rarely during the earliest period, perhaps related to the distribution of oldest sites in areas of limestone terrain, but people did at least utilise these sources and further made more extensive use of the sources of chert as noted. The usage of shell may be interpreted as efficient in an island setting, knowing that the raw materials could be found near almost any patch of coastline. Some items may have been fashioned from wood (such as the imported hard and dense ironwood) or bamboo (which can be made into a sharp-edged or pointed tool), but these have not survived in the archaeological assemblages.

Shell adzes in the Marianas have attracted some attention for possibly reflecting cross-regional connections in traditions of using shell rather than stone. Most common was the use of the durable shell of giant clams (Family Tridacnidae), cut and

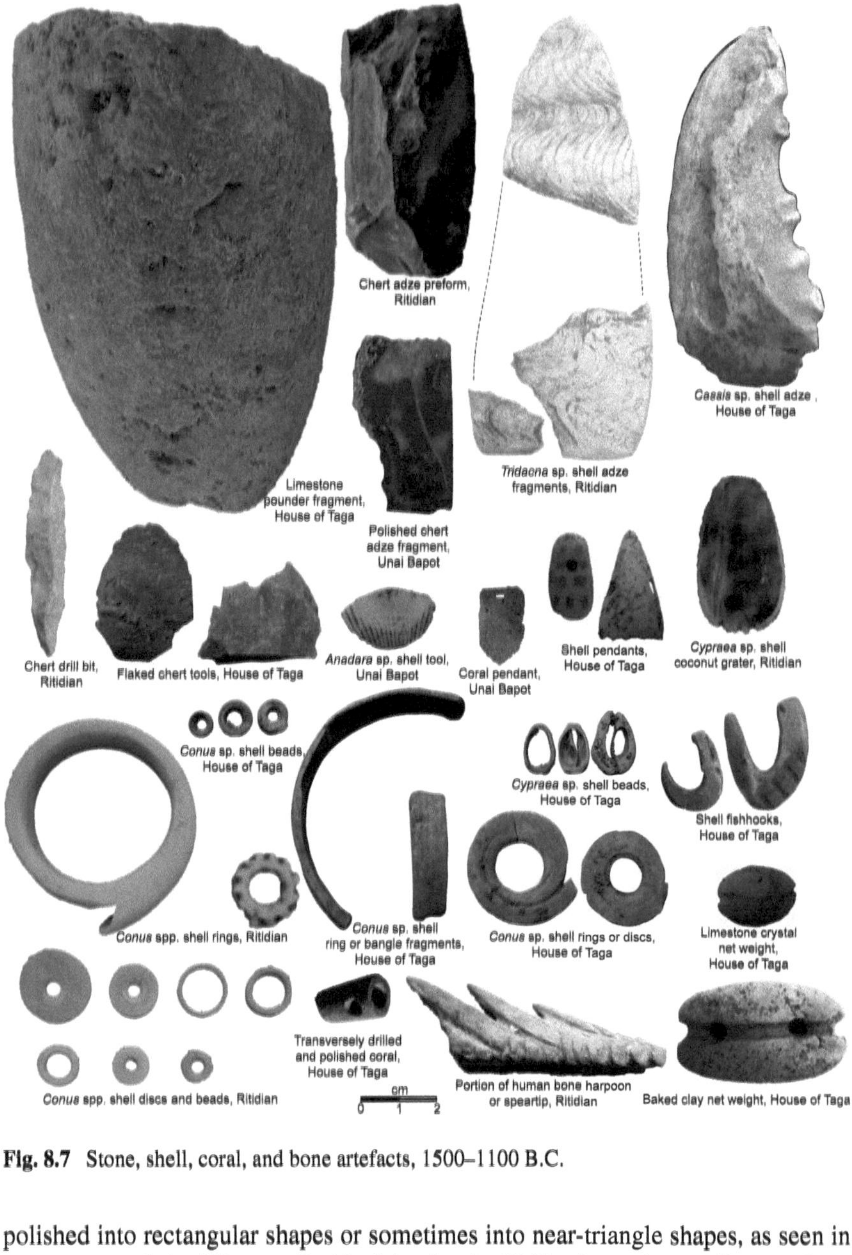

Fig. 8.7 Stone, shell, coral, and bone artefacts, 1500–1100 B.C.

polished into rectangular shapes or sometimes into near-triangle shapes, as seen in many parts of the Western Pacific Islands, the Philippines, the smaller southern islands of Ryukyu, and the Ogasawara Islands of southern Japan (Kidder et al. 1994). The taxonomic species is not always identifiable in the finished products, but most appear to be *Tridacna* spp. while *Hippopus* sp. has not been clearly identified in the Marianas. In other areas, *Hippopus* sp. adzes are verified taxonomically, for example, in Fiji where their numbers vastly decreased with sea-level drawdown after about 900 B.C. (Seeto et al. 2012). Gustav Paulay (1996) suggested that *Hippopus* sp. giant clams may have suffered a similar fate in the Mariana Islands, so in theory some shells of this taxon may have been utilised by the first settlers during the centuries prior to sea-level drawdown.

Besides the common findings of *Tridacana* sp. shell adzes, one different type of shell adze has been discovered in the oldest cultural layer near House of Taga in Tinian, in this case made of the lip of a horned helmet shell, *Cassis cornuta*. This occurrence could be the oldest of its type, certainly prior to 1100 B.C. and possibly close to 1500 B.C., compared to others dating no earlier than A.D. 100–200 in other parts of Micronesia and Indonesia (Bellwood 1997). The same later age applies to rare specimens found in Tikopia of the Solomon Islands (Kirch and Yen 1982), as well as in the Marquesas Islands of East Polynesia (Suggs 1961).

Along with the large and durable shells used for adzes, other smaller shells were fashioned into beads, bangles, and pendants during the earliest Marianas settlement period. *Conus* spp. shells were most common for all of these body ornaments, ranging in size from small and polished beads through larger pendants and bangles. Rarely, *Cypraea* sp. shells were cut and polished into beads that could be sewn into cloth or hair, found exclusively in the early-period sites and never again in the Marianas. Similarly, the tiny polished beads, wafer-thin discs, and ringed shapes of *Conus* spp. shells quickly declined in popularity after this earliest cultural period although a few of these forms continued to be produced in small numbers for another few centuries (see Fig. 6.3). Numerous pendants, discs, and rings were fashioned from a wide variety of shells, often with unique shapes and patterns of colours indicative of individual personality or expression. At least one piece of colourful *Cypraea* sp. shell was made into a small coconut grater at Ritidian Beach Cave, hinting at the artistic qualities of practical daily-use items.

Rare instances of pink coral tube-like artefacts have been found landward of House of Taga in Tinian, likely intended to be strung together as links in a necklace or other body ornament. These objects technically are not true tubes, because the interiors are not hollow as cylinders, but rather the two ends were drilled transversely. The transverse end-drilling may have provided an alternative solution instead of drilling a complete cylinder that would have required a specialised technology. Long cylinder drilling evidently was not practiced in the Marianas. Even in much later periods after A.D. 1000, shorter-depth transverse end-drilling was applied for pieces of *Tridacna* sp. shells, worn in links of a necklace known today as *sinahi* (Chap. 13).

Fishing-gear must have been essential for people living at the earliest Marianas habitations in shoreline niches, but only small amounts of artefacts clearly can be identified as fishing related. Small rotating hooks were made of shiny nacreous shell, mostly of *Isognomon* sp. and rarely of *Turbo* spp. shells whose diagnostic parts can be seen in discarded debitage. Sharp-angled jabbing hooks, V-shaped gorges, and composite-piece lures so far have not been found in the earliest deposits. Fishing hooks presumably were used for carnivorous species that could be taken by a baited hook, whereas most reef-dwelling fish more likely were caught by spearing or netting. Prior to the appearance of carved human bone speartips in the much later *latte* period, a few pieces of cut human bone points could represent generic point tools or other implements in progress, and a singular item at Ritidian Beach Cave resembles a portion of an ornately carved harpoon made from human bone.

Net weights or sinkers were identified in the oldest cultural layer near House of Taga, including larger types of baked clay and smaller types of crystallised limestone. The smaller and lighter sinkers could relate to nets tossed by hand toward a targeted area of the water. The heavier baked clay weights were more suitable for sinking from the side of a boat or otherwise dropping into the water without tossing.

The early-period artefacts supported sets of activities at residential habitations near the ancient shorelines, but only recently have excavations unearthed informative views of what those residential sites may have looked like. Beyond the usual constraints of just a few small test pits and narrow trial trenching, numerous scattered blocks and trenches were opened at Achugao (Butler 1994), a contiguous area of 16 m^2 was uncovered at Unai Chulu (Haun et al. 1999), and slightly more than 90 m^2 of ancient living surface was exposed near House of Taga (Carson 2014b:119–134). Each of these larger excavations revealed arrangements of post moulds and other features, related to the locations of shoreline-oriented houses. Prior to knowing about these findings, the first Marianas settlement period presented an enigma of a full material culture repertoire made by people who explored the regional landscape and invested in using its various resources, yet the actual housing structures of these people were not documented in the limited windows of small test excavations.

The larger excavation exposures now leave no doubt that the earliest Marianas sites were residential habitations, involving stilt-raised houses in shoreline zones. The largest contiguous excavation, near House of Taga in Tinian, showed the remnants of substantial house posts, often braced by cobbles, among living surfaces with hearths, pits, stonework pavings, and other features (Fig. 8.8). Most of the stone bracings around the posts were made by a few stones lining the perimeter, but a few consisted of more extensive stonework that created larger paved surfaces (Fig. 8.9). Artefacts tended to be more abundant within and around the stonework, whereas food-refuse midden tended to be more common outside the immediate vicinities of structural features and especially dense in the seaward portion of the site deposit.

Hearths were constructed beneath or between the houses, evident in the landward portions of the excavated areas at slightly higher elevation and farther from the active tidal zone. These features were made of limestone cobbles, arranged in rectangular or oval patterns. The cobbles had been heated, resulting in sooty coating and chalky texture. Tiny specks of carbonised wood and plant fibres were retained in most of the hearths, along with soot-coated remains of shellfish and sometimes the bones of fishes, turtles, and birds.

Alignments of stones appear to mark edges of activity areas, possibly related to the boundaries of individual houses. In one spot, three upright cobbles were arranged at equal distance along a row, on the seaward side of a set of house posts but landward of the pits closer to the active tidal wash zone (Fig. 8.10). Other instances were composed of cobbles and small boulders positioned flat on the ground, where the hearths consistently were on the landward side of these alignments.

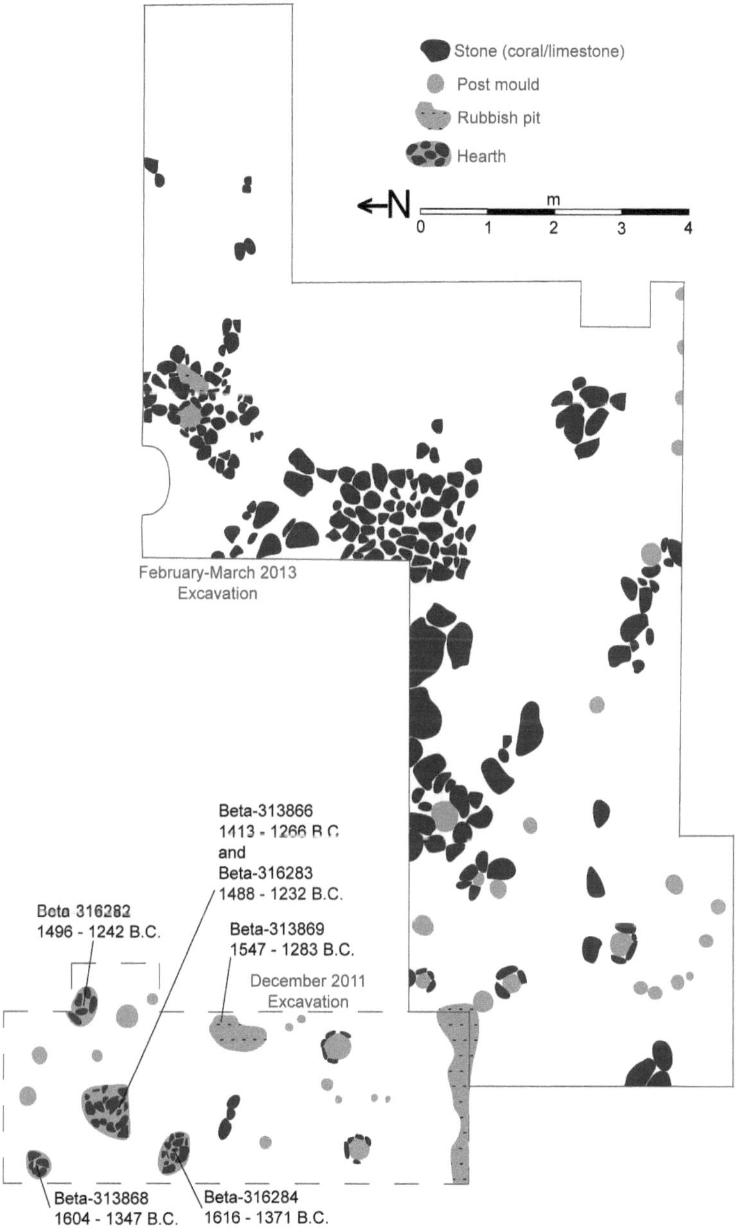

Fig. 8.8 Structural features and dating in the lowest cultural layer of approximately 1500–1100 B.C., landward of House of Taga

Fig. 8.9 Stonework feature, landward of House of Taga. Scale bars in the upper right background are in 20-cm increments

Fig. 8.10 Alignment of three upright stones, landward of House of Taga. Scale bars are in 20-cm increments

The total layout of cultural space is not yet known at these oldest sites, but the findings so far indicate that houses were distributed along an axis following the length of the associated shoreline at each site. If a person could walk from the seashore toward the inland back side of the settlement, then he or she would walk past one or a few houses. Walking along the length of the shore, however, a person would see several houses, one after another. At Unai Bapot in Saipan, the habitation zone covered probably no more than 20 m on the landward-seaward axis, compared to at least 30 m and possibly much longer in its axis parallel with the shoreline. Near House of Taga, the landward-seaward axis revealed its densest in situ habitation over less than 20 m, with another 10 m length of material mixed into the tidal wash zone on the seaward end as compared to extremely sparse material distributed farther inland, but the length in parallel with the shoreline is not yet fully explored beyond a single exposure over 17 m. At least a few sites have survived in much smaller areas, possibly due to their original contexts in patches of inter-tidal sands, for example, covering less than 20 by 20 m at Ritidian Shoreline.

These oldest habitation sites were situated for optimal use of specific shoreline niches, perhaps best illustrated in a modern-day example of "floating villages" in Southeast Asia (Fig. 8.11). The most popular settings are at the edges of mangroves and other swamps, often bordering rivers and lakes but not necessarily facing the ocean. These locales are ideal for access to diverse shellfish, swimming fish, birds, trees, and other resources. Every house has at least one boat or canoe that can be tethered beneath it, and a significant amount of activity every day takes place in the water, often in watercraft but also by wading or swimming.

Fig. 8.11 Modern example of a "floating village" at the edge of the Tonle Sap or Great Lake in Cambodia

In the modern floating villages of Southeast Asia, the primary residential stilt-raised houses are built directly over inter-tidal and shallow sub-tidal zones, where the maximum level of water tends to be about 3 m. This positioning results in elongate rows of houses, all installed within a narrow range of preferred habitat. The stilts or pile risers typically lift a house 3 m but often higher, in order to accommodate highest water levels of lakes and rivers during the tropical monsoon wet seasons. A similar height may have been useful for the earliest Marianas settlements on seashores with regular ocean tidal fluctuations and vulnerability to frequent storm-surge events.

In the modern examples, people live close with the water, but they routinely take advantage of land exposed during low tides and dry seasons for making temporary shelters, cooking sheds, open-air workshops, and other activity spaces. Many of the on-ground activities leave behind clear material traces in the more stable ground surfaces, but most often the material remnants are disturbed or removed after a number of washings beneath periodically higher water levels. Furthermore, people need to obtain at least some of their starchy plant foods and other resources from places outside the immediate habitats of the floating villages. Most villages are within a few hours of travel by boat or by foot to lands with fertile soils for growing rice and other crops, tended by people from the same or sometimes different villages.

At the earliest Marianas sites, similar formats of floating villages likely were situated at the shores of certain ecological niches (Fig. 8.12). The houses were built near coral reefs and lagoons, particularly in the places that were close to mangroves or shallow-water sea grass habitats. The landward edges of the housing zones were near the bases of limestone cliffs in some cases such as Mangilao, Ritidian, Tarague, and Unai Bapot. They were backed by swamps or wetlands in other cases such as Unai Chulu, Chalan Piao, and Achugao. The terrain near House of Taga was an apparent exception of a broad and very gently rising surface of an ancient and weathered limestone terrace.

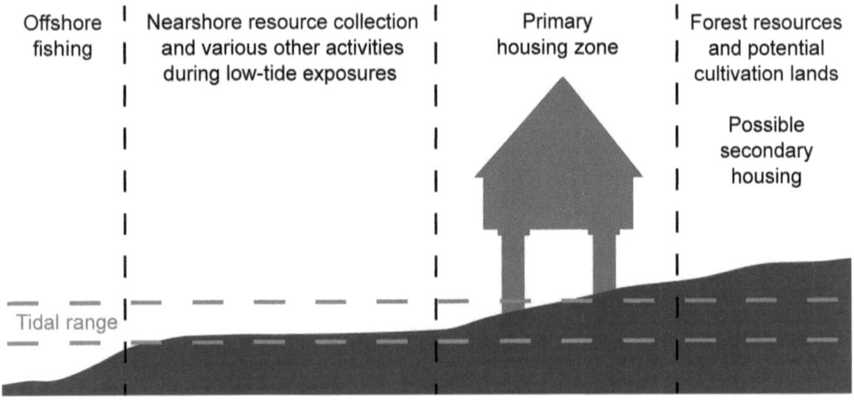

Fig. 8.12 Major habitat zones related to a floating village settlement. Schematic diagram is not to scale

Archaeological deposits resulted from decay of the stilt-raised houses into the underlying unstable beach sands, mixed in some areas with remnants of activities that occurred directly on the ground. Materials from both on-ground and above-ground contexts and from slightly varied ages potentially were inter-mixed within the gradual build-up of sediments, but overall the materials have been found in stratigraphic order when considering large scales of sedimentary units. Such a scenario accounts for the spotty irregularities in the stratigraphy and dating of sites such as Achugao in Saipan (Butler 1994). Near House of Taga in Tinian, the earliest cultural deposit consisted of only the lowest and oldest remnant of the original living surface, and the upper portions of the same deposit contained a mixture of materials from episodes of different ages, in total capped by an entirely later-aged sedimentary unit.

Given that early-period Marianas settlement is known almost exclusively from the remnants of stilt-raised communities, some aspects of ancient life inevitably have been under-represented or entirely missed in the archaeological record. The surviving site deposits relate only to the activities that occurred directly in the shoreline-oriented housing zones, but many other activities occurred elsewhere when people accessed forests, oceans, and probably caves. The latest discoveries in early-period layers at Ritidian Beach Cave and Ritidian Star Cave very strongly reflect non-residential activities with specialised types of pottery, shell ornaments, and unusual food remains not at all typical of the residential activities as seen at Ritidian Shoreline and other habitation sites of the same age. Meanwhile, some of the off-site activities are inferred by the presence of chert tools, clay used in pottery-making, and post moulds made from large forest trees. Other non-residential activities are unknown archaeologically because they occurred in places that have not yet been discovered or in contexts involving little or so preservation of cultural material. Notably absent are records of ancient mortuary practice, otherwise greatly informative about social and religious life in many parts of the world yet so far unknown for the oldest Marianas settlement period.

The ancient floating villages were impractical for preserving grave interments, and accordingly human burial features have not yet been found in any Marianas site of this age. If the deceased were buried at all, then the graves may have been distant from the known habitation sites, in places that have not yet been discovered or with poor preservation qualities. For instance, a number of caves contain burial pits, often badly disturbed, but none so far are dated older than 500 B.C. while most post-date A.D. 1000. Burial beneath a stilt-raised shoreline house may have been technically plausible, but it may not have been preferred and certainly not long-lasting in this type of setting. Other possibilities may have involved exposure of bodies in natural elements, including perhaps the ocean, as well as potentially curating skulls and bones among the living without opportunities for long-term preservation inside sedimentary deposits.

Regional Context

Most important for the regional context at 1500–1100 B.C., the initial generations of settlers in the Mariana Islands were the first people ever to live successfully in the Remote Oceanic environment. This isolated settlement in the Marianas differed significantly from the surge of populations moving into a completely different part of Remote Oceania after 1200 B.C., when people spread rapidly throughout several dozens of separated archipelagos across Southern Melanesia and West Polynesia within a brief span of a few centuries, effectively all colonised by 800 B.C. (see Fig. 1.1). As has been mentioned, the first human encounters with the Remote Oceanic environment in the Mariana Islands took place more than 2000 km distant from the nearest conceivable homeland of people who made a diagnostic red-slipped and decorated pottery.

During the first centuries of Marianas settlement, the realm of Remote Oceania was only just beginning to take solid shape as a tangible entity in human experience. Eventually, after several centuries of explorations, Pacific Oceania would become populated far and wide, fostering deep connectivity among people living throughout a "sea of islands" (Hau'ofa 1994). Prior to the development of such an extensively inhabited seascape, the communities in the Mariana Islands indeed were isolated. A few sailors and sea nomads may have earned reputations of staying at sea for weeks or even months, and they may have shared stories about distant islands, shoals, emerging atolls, and portions of the ocean with different wind and wave patterns. The installation of a viable human population in Remote Oceania, however, was an entirely different endeavour that required successful overseas voyaging plus effective survival in a remote environment for the first time in human history. Moreover, it was at its time the world's longest distance of ocean-crossing human migration at 1500 B.C.

As outlined here, the Marianas archaeological record began with a full repertoire of material culture traits already in use. Necessarily, many of these components were inherited from an overseas homeland region, while a number of specific expressions likely developed uniquely within the Marianas context. Traditions of pottery-making must have been brought by the first immigrant settlers, but the choice of technical and artistic output in theory could be modified according to the new natural and social conditions, perhaps related to the production of notably thin-walled earthenware in the Marianas. Likewise, proficiency in making stone tools must have existed among at least some of the initial founding settlers, but specific outcomes in the Marianas depended partially on the availability of raw materials and the unique needs of the people living in this new environment.

The knowledge base and skillset of the first Mariana Islanders entailed much more than making pottery and stone tools. The successful settlement implies expertise in remote-distance sailing and presumably navigation, talents of reading the environment to find suitable habitation locales, the ability to identify native plant foods and other resources, and the ability to modify or create supplies that were naturally rare or absent in the Remote Oceanic environment. Other kinds of cultural knowledge very likely were maintained by specialists responsible for cooking recipes, artworks, family histories and genealogies, folklore, and religious practice.

In order to account for all the activities involved in an efficacious settlement of the Mariana Islands, the first generations of people most likely came from varied backgrounds. At a minimum, the founding population included people with long-distance seafaring skills, as well as people who knew how to live in an island environment. The seafaring abilities likely were provided by groups who lived most of their lives at sea, possibly including sea nomads who historically have roamed the waters around the Philippines and Indonesia in Island Southeast Asia (Chen 2002). Sea nomads are known for living in close symbiosis with land-based communities, each supplying the other with important foods, prestige goods, and sometimes marriage partners (Sather 1995). Sea nomads alone do not account for the material record of Marianas settlement, noting the formal house structures in multiple sites of separate islands, local pottery-making, access to inland resources of forest goods and tool-making stones, modification of native forests to accommodate new crop growth, and of course the sustained residential habitation and survival of a local population clearly outside the definitive scope of itinerant sea nomads.

Among the artefacts of early-period Marianas sites, red-slipped pottery provides the most crucial clue about a potential homeland region, further specified by the use of white lime-infill in finely decorated patterns of dentate-stamped, incised, and circle-marked designs. The unique combination of traits constitutes the diagnostic footprint of a "pottery trail" whose steps can be tracked from the Mariana Islands back to a slightly older source in the Philippines (Carson et al. 2013). The exact route across the ocean is unclear, and a precise point of origin is uncertain. A few sites in the northern and central Philippines have yielded this type of pottery, but it has been best documented and dated about 2000–1800 B.C. in the Cagayan Valley of northern Luzon (Hung 2008; Hung et al. 2011). Moreover, the Philippines region in a broad sense appears to have been the origin zone of betel nut-chewing and familiarity with slaked lime (Zumbroich 2008), necessary for producing the white lime-infill in the pottery and further for supplying the betel nut palms that were transported across the ocean to the Mariana Islands during the earliest settlement period (Athens and Ward 2006).

The distinctive forms of earliest Marianas pottery decorations so far have been documented in at least three sites in the northern and central Philippines. Two of these sites are in the Cagayan Valley of northern Luzon, including Magapit probably at least as old as 1400 B.C. (Aoyagi et al. 1993) and Nagsabaran about 2000–1800 B.C. (Hung 2008). The third site is in the central Philippines, where the decorated pottery is not well dated in the Batungan Caves of Masbate Island, apparently from a context pre-dating about 1000 B.C. (Solheim 1968). Of these three sites, currently the largest collections of pottery and the most secure early dating results have been at Nagsabaran, where the distinctive decorated pottery appeared certainly by 1800 B.C. but perhaps earlier.

Several sites across the Philippines and Indonesia have yielded red-slipped pottery dating at least as early as 1000 B.C. and likely some centuries older, but so far only very few of these sites appear to be good candidates as possible homelands of first Marianas settlement. In most cases, the pottery lacks the distinguishing traits of the earliest Marianas decorative system, but perhaps the lack of evidence in this case is due to sampling size when knowing that the decorations can occur on less than 1 % of a total collection. Radiocarbon dating reveals a gradual north-to-south gradient, beginning about 2200 B.C. in the Batanes Islands (Bellwood and Dizon 2013), continuing through northern Luzon in the range of 2000 through 1500 B.C. (Bellwood 1997; Hung 2008), and then expanding into numerous sites throughout the Philippines and Indonesia in the range of 1500 through 1000 B.C. (Simanjuntak

2008; Spriggs 2007; Tanudirjo 2001). Within this larger picture, the few sites with finely decorated pottery appear quite rare as documented at Magapit, Nagsabaran, and the Batungan Caves in the northern through central Philippines. The examples in Indonesia appear to be more coarsely decorated overall, and a few rare pieces with finer decoration are dated closer to 1000 B.C.

While the early-period Marianas pottery shows a strong link with slightly older traditions in Island Southeast Asia, not all aspects of material culture were inherited from this homeland source. Most notably in the Marianas, barkcloth beaters (for producing tree-bark cloth) and spindle whorls (for spinning textiles) are missing from the local archaeological record, even though they were popular throughout East Asia and Southeast Asia. Moreover, these clothing-related artefacts continued to be absent throughout the entire pre-European archaeological sequence. Early Spanish records refer to the Chamorro people as lacking clothing, but perhaps some of those accounts were exaggerated or sensationalised.

Although clothing production is not attested in the Marianas material record, some of the oldest shell ornaments match with the decorative pieces known in clothing of Austronesian aboriginal tribes of eastern Taiwan (Fig. 8.13). Tiny polished *Conus* sp. shell beads were used for decorative leg coverings, skirts, vests, and necklaces. Some of the shell discs resemble those adorning shoulder straps, while other discs and rings are reminiscent of the complex elements in head dressings. The larger *Conus* sp, shell rings likely were used as bangles to adorn wrists or ankles. The unique forms of sliced *Cypraea* sp. shells can be matched with the rows of the same types of shells sewn in rows into vests, attaching the threads over the central spires of the sliced shells for the appearance of floating over the vest fabric. The only one of these shell ornament types that continued after 1100 B.C. in the Marianas was the type of small cut and polished *Conus* sp. shell bead that persisted in small numbers most likely for usage in necklaces through 500 B.C. although the earlier usage at 1500–1100 B.C. had involved more numerous shells as suitable for decorating larger objects of leg coverings, skirts, and vests.

These earliest forms of shell artefacts ceased in production in the Mariana Islands shortly after first settlement, thus suggesting the loss of a cultural tradition after the first few generations of human settlement in these remote islands. The same traditions can be traced at least as early as 1500 B.C. in archaeological sites of eastern coastal Taiwan and nearby offshore islands (Hung 2008), continuing in use through the modern era as seen in the ethnographic collections. They may have existed in coastal sites of Island Southeast Asia with access to marine shells, but such coastal sites have not yet been well documented in comparison to the stronger reporting at inland river valley sites, where marine shell artefacts are notably sparse.

Looking beyond the material artefact collections, the earliest Marianas site settings were unique as locations where no other human beings had resided previously. The relevant contemporaneous and slightly older sites in Island Southeast Asia occurred in land masses where people already had been living for several thousands of years previously as hunters, fishers, and foragers. The critical new developments after 2000 B.C. involved formal village settlements, along with production of pottery, more numerous polished stone and shell tools, and other aspects of residential

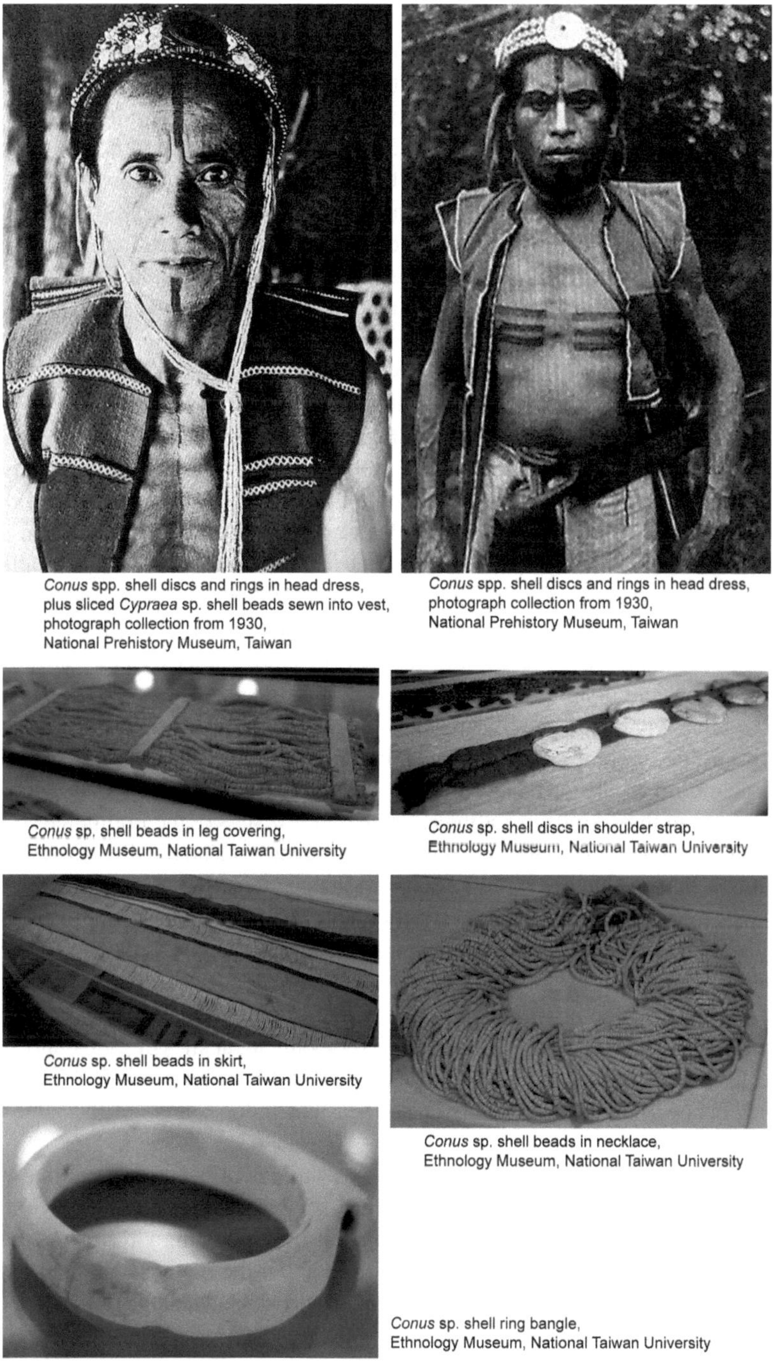

Fig. 8.13 Examples of shell ornaments similar to those found in early-period Marianas sites, documented ethnographically in eastern Taiwan

life and landscape. By 1500 B.C., this new lifestyle had been spreading and gaining popularity for several generations, and overall the populations were being resettled into the scattered shoreline niches of seashores and riverbanks.

At least some of the first Marianas settlers must have been familiar with coastal living in Island Southeast Asia, very likely in forms of stilt-raised floating villages along swampy shores, and this general lifestyle was repeated in the first Marianas sites. The evidence in the Mariana Islands points very specifically to ecological niches in unstable sandy patches, bordered by shallow-water lagoons or other swamp-like zones, with nearby coral reefs. At these sites, people consumed large amounts of *Anadara* sp. and lesser amounts of other shellfish, along with various fish, turtles, birds, and a number of starchy plant foods.

As in the earliest Marianas sites, so far no pottery-bearing habitation sites in Island Southeast Asia have revealed human burial features prior to about 700–500 B.C. Given the similar contexts of stilt-raised houses directly at shores of beaches and rivers, the ancient mortuary practice probably did not involve burial of individuals within these exact settings. Extended burials in formal pits have been documented in contexts with diagnostically later-aged materials and radiocarbon dates no earlier than 700–500 B.C., for example, as at Nagsabaran in Cagayan Valley of the Philippines (Hung 2008). Burials in these later periods occurred when the landforms such as at Nagsabaran already had become more stable above the level of the floodplain of the Cagayan River, thus referring to a different landscape context post-dating the ancient floating village settings.

The archetypal floating village niches had existed around 1500 B.C. and earlier along various seashores and riverbanks in the northern and central Philippines, as well as more broadly in Island Southeast Asia. Around 2000–1800 B.C., people were living in these kinds of settings in at least a few sites of the Philippines, where they produced not only dense heaps of shell middens but also the distinctive pottery traditions that eventually would be re-enacted in the Mariana Islands by 1500 B.C. By the time of first Marianas settlement, however, many of these preferred habitat zones in the Philippines were being filled by growing populations, and communities were being established in increasing numbers of shoreline niches farther afield in Island Southeast Asia. In particular, people were settling in parts of Indonesia and making red-slipped pottery, for example, at sites clustered along riverbanks in the Karama Valley of Sulawesi by 1500 B.C. (Simanjuntak 2008) and on the shores of a number of islands in northeastern Indonesia about the same time (Tanudirjo 2001).

In this larger regional view, first Marianas settlement occurred at a time when people in Island Southeast Asia were seeking new places to live, preferably on seashores and riverbanks suitable for installing stilt-raised floating villages. As the targeted ecological niches were being filled throughout the inter-visible land-masses of the Philippines and Indonesia, at least one group took advantage of new opportunities in unoccupied territory far overseas in the remote Mariana Islands. This event marked the first successful human settlement of Remote Oceania at least as early as 1500 B.C., when an overall cohesive lifestyle was sustained in at least eight sites in three of the larger southern-arc islands of the Marianas.

The first Mariana Islanders may not have been seeking remote islands per se, but rather they settled in niches where they could maintain lifestyles and landscape systems that were familiar to them. In addition to the allure of unoccupied land in the Marianas, social and cognitive factors may have instigated people to seek this unprecedented isolation. In theory, a long-distance migration may seem attractive for the youngest siblings in large families who did not inherit rights of land and wealth in their native homelands, but this motivation may not have been sufficient in itself to substantiate the longest ocean-crossing migration of its time in human history. Conceivably, some people sought isolation and freedom to pursue an independent lifestyle that was otherwise unpopular or punished in their homeland, for instance related to religious beliefs, choice of music or dance expression, or selection of marriage partners. Another possibility could involve a refuge colony for people defeated in warfare or other conflict. Realistically, the first immigrants to the Marianas acted on a number of different motivations, and no singular explanation ever may be satisfactory to account for the material record of successful settlement.

Regardless of the underlying motivations, the initial Marianas settlement successfully brought elements of Island Southeast Asian cultural landscapes into a new setting in the Mariana Islands, resulting in a specific archaeological signature starting from 1500 B.C. if not earlier. This new setting likely involved people from different backgrounds and traditions within a broadly similar homeland region, made distinctive by these unique combinations and further by localised developments within the isolated Marianas context. The inhabited Marianas landscape at first was based on residences in a few scattered shoreline niches, but soon a number of factors would begin to transform in the natural setting, in the cultural behaviours and expressions of the people living there, and in the interface between these people and their landscape. The beginnings of these transformations are attested in the material record around 1100 B.C.

References

Amesbury, J. R. (1999). Changes in species composition of archaeological marine shell assemblages in Guam. *Micronesica, 31*, 347–366.

Amesbury, J. R. (2007). Mollusk collecting and environmental change during the prehistoric period in the Mariana Islands. *Coral Reefs, 26*, 947–958.

Aoyagi, Y., Aguilera, M. L., Jr., Ogawa, H., & Tanaka, K. (1993). Excavation of hill top site, Magapit Shell Midden in Lal-lo Shell Middens, northern Luzon, Philippines. *Man and Culture in Oceania, 9*, 127–155.

Athens, J. S., Dega, M. F., & Ward, J. V. (2004). Austronesian colonisation of the Mariana Islands: The palaeoenvironmental evidence. *Bulletin of the Indo-Pacific Prehistory Association, 24*, 21–30.

Athens, J. S., & Ward, J. V. (1998). *Paleoenvironment and prehistoric landscape change: A sediment core record from Lake Hagoi, Tinian, CNMI* (Report prepared for United States Naval Facilities Engineering Command). Honolulu: International Archaeological Research Institute.

Athens, J. S., & Ward, J. V. (1999). *Paleoclimate, vegetation, and landscape change on Guam: The Laguas core* (Report prepared for United States Naval Facilities Engineering Command). Honolulu: International Archaeological Research Institute.

Athens, J. S., & Ward, J. V. (2004). Holocene vegetation, savannah origins and human settlement of Guam. In V. Attenbrow & R Fullagar (Eds.), *A Pacific Odyssey: Archaeology and anthropology in the Western Pacific: Papers in Honour of Jim Specht*, Records of the Australian Museum, Supplement 29 (pp. 15–30). Sydney: Australian Museum.

Athens, J. S., & Ward, J. V. (2006). *Holocene paleoenvironment of Saipan: Analysis of a core from Lake Susupe* (Micronesian Archaeological Survey Rep. No. 35). Saipan: Commonwealth of the Northern Mariana Islands Division of Historic Preservation.

Bath, J. E. (1986). *The San Vitores Road Project, Part 1* (Report prepared for Maeda Pacific Corporation by Joyce E. Bath. Document on file at Micronesian Area Research Center). Mangilao: University of Guam.

Bellwood, P. (1997). *Prehistory of the Indo-Malaysian Archipelago* (Rev. ed.). Honolulu: University of Hawaii Press.

Bellwood, P., & Dizon, E. (Eds.). (2013). *4000 Years of migration and cultural exchange: The archaeology of the Batanes islands, Northern Philippines*, Terra Australis 40. Canberra: Australian National University E Press.

Butler, B. M. (1994). Early prehistoric settlement in the Mariana Islands: New evidence from Saipan. *Man and Culture in Oceania, 10*, 15–38.

Butler, B. M. (Ed.). (1995). *Archaeological investigations in the Achugao and Matansa areas of Saipan, Mariana islands* (Micronesian Archaeological Survey Rep. No. 30). Saipan: Commonwealth of the Northern Mariana Islands Division of Historic Preservation.

Carson, M. T. (2008). Refining earliest settlement in remote Oceania: Renewed archaeological investigations at Unai Bapot, Saipan. *Journal of Island and Coastal Archaeology, 3*, 115–139.

Carson, M. T. (2011). Palaeohabitat of first settlement sites 1500–1000 B.C. in Guam, Mariana Islands, Western Pacific. *Journal of Archaeological Science, 38*, 2207–2221.

Carson, M. T. (2012). Evolution of an Austronesian landscape: The Ritidian site in Guam. *Journal of Austronesian Studies, 3*, 55–86.

Carson, M. T. (2014a). Paleoterrain research: Finding the first settlement sites of Remote Oceania. *Geoarchaeology, 29*, 268–275.

Carson, M. T. (2014b). *First settlement of remote Oceania: Earliest sites in the Mariana islands*. New York: Springer.

Carson, M. T. (2014c). Contexts of natural-cultural history: A 3500-year record at Ritidian in Guam. In M. T. Carson (Ed.), *Guam's Hidden Gem: Archaeological and historical studies at Ritidian, British Archaeological Reports International Series 2663* (pp. 1–43). Oxford: Archaeopress.

Carson, M. T., & Hung, H.-C. (2014). Pacific islands: Finding the earliest sites. In C. Smith (Ed.), *Encyclopedia of global archaeology* (pp. 5693–5696). New York: Springer.

Carson, M. T., Hung, H.-C., Summerhayes, G., & Bellwood, P. (2013). The pottery trail from Southeast Asia to remote Oceania. *Journal of Island and Coastal Archaeology, 8*, 17–36.

Carson, M. T., & Kurashina, H. (2012). Re-envisioning long-distance Oceanic migration: Early dates in the Mariana Islands. *World Archaeology, 44*, 409–435.

Carson, M. T., & Peterson, J. A. (2011). Calcrete formation and implications for buried archaeological deposits in the Mariana Islands, western Pacific. *Geoarchaeology, 26*, 501–513.

Chen, C.-Y. (2002). Sea nomads in prehistory on the southeast coast of China. *Bulletin of the Indo-Pacific Prehistory Association, 22*, 51–62.

Craib, J. L. (1993). Early occupation at Unai Chulu, Tinian, Commonwealth of the Northern Mariana Islands. *Bulletin of the Indo-Pacific Prehistory Association, 13*, 116–134.

Dickinson, W. R. (2000). Hydro-isostatic and tectonic influences on emergent Holocene paleoshorelines in the Mariana Islands, western Pacific Ocean. *Journal of Coastal Research, 16*, 735–746.

Dickinson, W. R. (2001). Paleoshoreline record of relative Holocene sea levels on Pacific islands. *Earth-Science Reviews, 55*, 191–234.

Dickinson, W. R. (2003). Impact of mid-holocene hydro-isostatic highstand in regional sea level on habitability of islands in Pacific Oceania. *Journal of Coastal Research, 19*, 489–502.

Dickinson, W. R. (2006). *Temper sands in prehistoric oceanian pottery: Geotectonics, sedimentology, petrography, provenance*, Special Paper 406. Boulder, CO: Geological Society of America.

Dilli, B. J., Haun, A. E., Goodfellow, S. T., & Deroo, B. (1998). *Archaeological Mitigation Program, Mangilao Golf Course Project Area, Mangilao Municipality, Territory of Guam, Vol. 1: Introduction, research design, and data recovery results* (Report prepared for Mr. Jetan Sahni). Hilo, HI: Paul H. Rosendahl, Ph.D.

Fitzpatrick, S. M., Nelson, G. C., & Reeves, R. (2003). The prehistoric chewing of betel nut (*Areca catechu*) in western Micronesia. *People and Culture in Oceania, 19*, 55–65.

Fox, R. B. (1970). *The Tabon Caves: Archaeological explorations and excavations on Palawan Island, Philippines* (Monograph No. 1). Manila: Philippines National Museum.

Hau'ofa, E. (1994). Our sea of islands. *Contemporary Pacific, 6*, 147–161.

Haun, A. E., Jimenez, J. A., Kirkendall, M. A., & Goodfellow, S. T. (1999). *Archaeological investigations at Unai Chulu, Island of Tinian, Commonwealth of the Northern Mariana Islands* (Report prepared for Naval Facilities Engineering Command, Pacific Division). Hilo, HI: Paul H. Rosendahl, Ph.D.

Hung, H.-C. (2008). *Migration and cultural interaction in southern coastal China, Taiwan and the northern Philippines, 3000 BC to AD 100: The early history of Austronesian-speaking populations*. Ph.D. Dissertation, Australian National University, Canberra.

Hung, H.-C., Carson, M. T., Bellwood, P., Campos, F., Piper, P. J., Dizon, E., et al. (2011). The first settlement of remote Oceania: The Philippines to the Marianas. *Antiquity, 85*, 909–926.

Kidder, J. E., Jr., Wilson, R. L., Koyama, S., Oda, S., Keally, C. T., Nakatsu, Y., et al. (1994). *Material culture along the black current: Prospectus*. Tokyo: Archaeology Research Center, International Christian University.

Kirch, P. V. (1997). *The Lapita peoples: Ancestors of the Oceanic world*. Cambridge, England: Blackwell.

Kirch, P. V. (Ed.). (2001). *Lapita and its transformations in near Oceania: Archaeological investigations in the Mussau islands, Papua New Guinea, 1985–88, Vol. 1: Introduction, stratigraphy, chronology* (Archaeological Research Facility Contributions Number 59). Berkeley, CA: University of California.

Kirch, P. V., & Yen, D. E. (1982). *Tikopia: The prehistory and ecology of a Polynesian Outlier* (Bishop Museum Bulletin 238). Honolulu: Bishop Museum Press.

Kurashina, H., & Clayshulte, R. N. (1983a). Site formation processes and cultural sequence at Tarague, Guam. *Bulletin of the Indo-Pacific Prehistory Association, 4*, 114–122.

Kurashina, H., & Clayshulte, R. N. (1983b), *Site formation processes and cultural sequence at Tarague, Guam* (Miscellaneous Publications Number 6). Mangilao: Micronesian Area Research Center, University of Guam.

Moore, D. R., Hunter-Anderson, R L., Amesbury, J. R., & Wells, E. F. (1992). *Archaeology at Chalan Piao, Saipan* (Report prepared for Jose Cabrera). Mangilao: Micronesian Archaeological Research Services.

Nunn, P. D. (2007). Echoes from a distance: Research into the Lapita occupation of the Rove Peninsula, southwest Viti Levu, Fiji. In S. Bedford, C. Sand & S. P. Connaughton (Eds.), *Oceanic explorations: Lapita and Western Pacific settlement*, Terra Australis 26 (pp. 163–176). Canberra: Australian National University E Press.

Paulay, G. (1996). Dynamic clams: Changes in the bivalve fauna of Pacific Islands as a result of sea-level fluctuations. *American Malacological Bulletin, 12*, 45–57.

Pellett, M., & Spoehr, A. (1961). Marianas archaeology: Report on an excavation on Tinian. *Journal of the Polynesian Society, 70*, 321–325.

Pregill, G. K., & Steadman, D. W. (2009). The prehistory and biogeography of terrestrial vertebrates in Guam, Mariana Islands. *Diversity and Distributions, 15*, 983–996.

Reinman, F. R. (1977). *An archaeological survey and preliminary test excavations on the Island of Guam, Mariana Islands, 1965–1966* (Miscellaneous Publication Number 1). Mangilao: Micronesian Area Research Center, University of Guam.

Sather, C. (1995). Sea nomads and rainforest hunter-gatherers: Foraging adaptations in the Indo-Malaysian Archipelago. In P. Bellwood, J. J. Fox, & D. Tryon (Eds.), *The Austronesians: Historical and comparative approaches* (pp. 229–268). Canberra: Department of Anthropology, Research School of Pacific and Asian Studies, Australian National University.

Seeto, J., Nunn, P. D., & Sanjana, S. (2012). Human-mediated prehistoric extinction in the tropical Pacific? Understanding the presence of *Hippopus hippopus* (Linn. 1758) in ancient shell middens of the Rove Peninsula, southwest Viti Levu Island, Fiji. *Geoarchaeology, 27*, 2–17.

Simanjuntak, T. (Ed.). (2008). *Austronesian in Sulawesi*. Yogyakarta, Indonesia: Center for Prehistoric and Austronesian Studies.

Solheim, W. G., II. (1968). The Batungan Cave Site, Masbate, Philippines. In W. G. Solheim, II (Ed.), *Asian and Pacific archaeology series, Vol. 2* (pp. 20–62). Honolulu: Social Science Research Institute, University of Hawaii.

Spoehr, A. (1957). *Marianas prehistory: archaeological survey and excavations on Saipan, Tinian and Rota, Fieldiana: Anthropology, Vol. 48*. Chicago, IL: Chicago Natural History Museum.

Spriggs, M. (2007). The Neolithic and Austronesian expansion within Island Southeast Asia and into the Pacific. In S. Chiu & C. Sand (Eds.), *From Southeast Asia to the Pacific: Archaeological perspectives on the Austronesian expansion and the Lapita cultural complex* (pp. 104–140). Taipei: Center for Archaeological Studies, Research Center for Humanities and Social Sciences, Academia Sinica.

Spriggs, M., & Anderson, A. (1993). Late colonization of East Polynesia. *Antiquity, 67*, 200–217.

Steadman, D. W. (1995). Prehistoric extinctions of Pacific Island birds: Biodiversity meets zooarchaeology. *Science, 267*, 1123–1131.

Suggs, R. C. (1961). *Archaeology of Nukuhiva, Marquesas Islands, French Polynesia* (Anthropological Papers, No. 1). New York: American Museum of Natural History.

Tanudirjo, D. (2001). *Islands in between: Prehistory of the northeastern Indonesian Archipelago*. Ph.D. Dissertation, Australian National University, Canberra.

Thompson, L. (1932). *Archaeology of the Mariana Islands* (Bishop Museum Bulletin 100). Honolulu: Bishop Museum Press.

Wickler, S. K. (2004). Modelling colonisation and migration in Micronesia from a zooarchaeological perspective. In M. Mondini, S. Munoz, & S. K. Wickler (Eds.), *Colonisation, migration and marginal areas: A zooarchaeological approach* (pp. 28–40). Oxford: Oxbow Books, England.

Zumbroich, T. J. (2008). The origin and diffusion of betel chewing: A synthesis of evidence from South Asia, Southeast Asia and beyond. *Electronic Journal of Indian Medicine, 1*, 63–116.

Chapter 9
1100–700 B.C., Changing Coastlines

The period 1100–700 B.C. witnessed sustained inhabitation of the Marianas landscape, while growing populations adapted with new conditions of falling sea level and changing coastal ecologies. Established communities took advantage of emerging beach ridges, berms, and other coastal landforms that provided more usable land area, but meanwhile the coastal zones were changing in their physical forms and their associated natural resource habitats. During these same centuries, several islands in other parts of Remote Oceania were being populated for the first time.

Site Inventory and Dating

A neatly bracketed date range of 1100–700 B.C. refers to a few centuries of the ancient landscape in the Marianas, continuing the shoreline orientation of prior centuries but with definite new adaptations to changing environmental and social conditions. Beginning around 1100 B.C., transformations are noticeable in sea-level drawdown, reconfiguration of costal landforms and resource zones, size and number of habitation sites, cultural harvesting of different compositions of shellfish and other natural resources, and transitions in the forms and styles of pottery, shell ornaments, and other material culture objects. Further substantive transformations occurred after 700 B.C. in these same categories plus others.

The archaeological sites of this period are found mostly at and around the same locations of the preceding centuries, as well as a few new places (Fig. 9.1). Overall the habitation zones became slightly larger while spreading over newly available coastal landforms with more stability than was possible in prior centuries. At Ritidian in Guam, a new beach ridge supported an elongate habitation zone over a length of about 150 m (Carson 2012, 2014a). Also in Guam, stable backbeach areas began to form in a few scattered locations, and some of these supported new habitation sites such as at Ypao in Guam (Leidemann 1980; Olmo and Goodman 1994). Lowered sea level meanwhile

M.T. Carson, *Archaeological Landscape Evolution*,
DOI 10.1007/978-3-319-31400-6_9

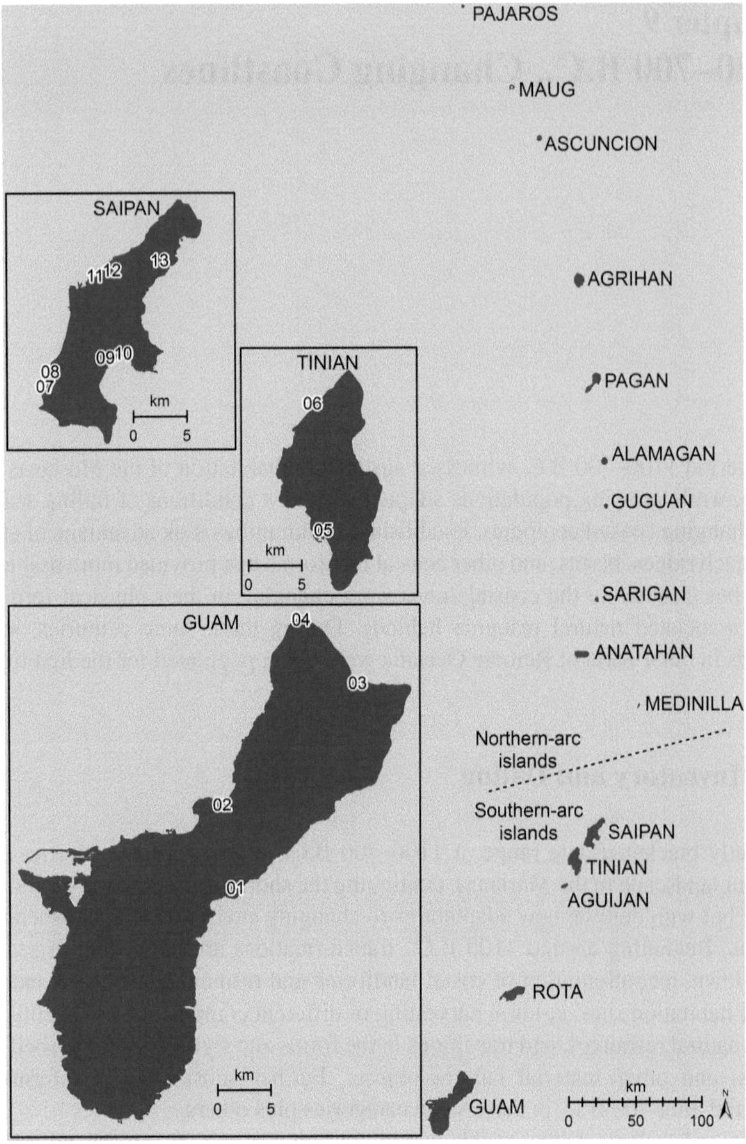

Fig. 9.1 Major site locations, 1100–700 B.C. 01: Mangilao. 02: Ypao. 03: Tarague. 04: Ritidian. 05: House of Taga. 06: Unai Chulu. 07: Chalan Piao. 08: Chalan Kanoa. 09: Laulau Rockshelter. 10: Unai Bapot. 11: Achugao. 12: San Roque. 13: Kalabera Cave

created several stranded mounds, berms, and other landforms in and around wetlands on the west coast of Saipan, thus creating more opportunities for patchworks of residences growing around the vicinities of the older sites of Chalan Piao (Athens and Ward 2006; Cleghorn and McIntosh 2002) and Achugao (Butler 1995).

An unexpected but important discovery has been a cultural layer at a large cave, Kalabera Cave in northern Saipan, at least as early as 1000 B.C. (Swift et al. 2009). This early age is unexpected because most other caves in the Marianas so far have not yielded direct evidence of cultural activity in the interior spaces prior to 500 B.C., instead showing cultural use only of the external areas such as at Ritidian in Guam. Such an early age does appear credible for the use of the external overhang areas at Ritidian Beach Cave and Ritidian Star Cave in Guam, as well as at Laulau Rockshelter in Saipan, where Alexander Spoehr (1957:52–58) reported a lower layer with red-slipped pottery, but these settings were clearly different from an enclosed dark zone of a cave such as at Kalabera Cave. Dense cultural layers did accumulate in the exterior lighted spaces directly outside some dark caves from an earlier date as noted at Ritidian, but curiously the early dated layer inside Kalabera Cave contains thickened-rim non-slipped pottery and other materials diagnostically belonging to much later-aged periods post-dating A.D. 1000. In this case, more explorations potentially can identify areas of greater and lesser disturbance in the cave floor sediments, and an intact early cultural deposit may yet offer more information about interior cave use at this early date.

The known sites of this period reflect a changing physical and cultural landscape, maintaining much the same basic framework of the pre-existing landscape system but forced to adapt to new conditions. People made and used different forms and styles of pottery and other artefacts, while they learned new ways of harvesting resources from their changing surroundings. The change may have been subtle at first, but it was noticeable in a number of different categories of the natural–cultural landscape. People still preferred to inhabit specific shoreline niches in the larger southern-arc islands at least for as long as the environment could support this lifestyle. After 700 B.C., the local way of life and landscape could no longer be sustained, resulting in more pronounced and systemic change.

Landforms

As the regional sea level began a period of drawdown, coastal zones of the Mariana Islands underwent a set of adjustments in terrain structure and associated habitat zones (Fig. 9.2). The pre-existing small pockets and fringes of unstable beach sands were stranded slightly above the falling sea level, while new shorelines formed slightly father away from the main land masses. In the new exposures of slightly elevated and broader coastal terrain, a number of landforms began to take shape.

Two major types of landform adjustments can be discerned during this time period, both involving land stranded above and behind the falling sea level. Near the seashore bases of steep limestone cliffs, such as Ritidian in Guam and Unai Bapot in Saipan, elongate beach ridges or emergent berms were formed. In wetland zones, such as around Susupe in Saipan and in the Hagatna Basin in Guam, the wetlands transitioned from marine to freshwater conditions, while scattered mounds and berms began to emerge above the lowering levels of the surrounding swamps.

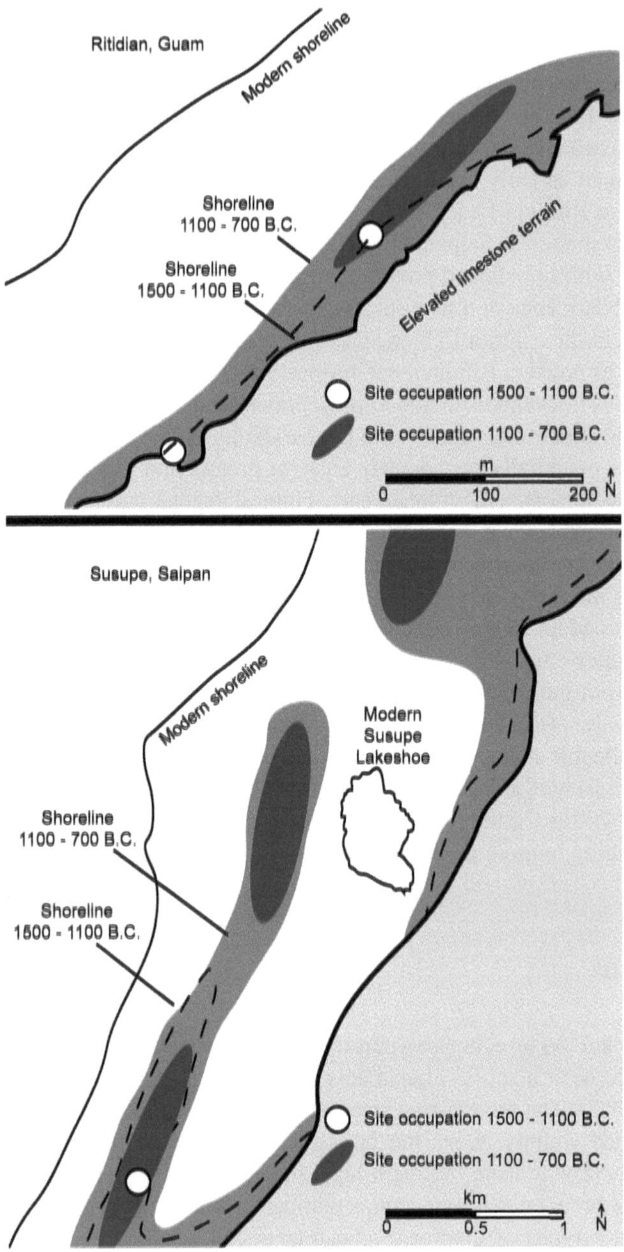

Fig. 9.2 Changing coastal habitats around 1100 B.C. Example at Ritidian is based on data from Carson (2011, 2012, 2014a). Example at Susupe is based on data from Athens and Ward (2006) and Carson (2014a, b:15–20)

Best documented at Ritidian, a shift in beach sand composition signalled an important change in local coastal ecology. The new sandy sediments at Ritidian were composed of pulverised storm-surge clasts that periodically were thrown over the emergent beaches, overlaying the deeper and older beds of subtidal lagoon facies that had been stranded there as the sea level began its drawdown. The locations of older habitations directly at the edges of lagoons and other shallow-water settings now were becoming covered by thicker and broader layers of pulverised storm-surge sands. The exact positions of shorelines clearly were shifting, and increasingly more dry ground surface was becoming available.

After 1100 B.C., the larger coastal landforms allowed more opportunities for on-ground activities than previously was possible. People evidently acted on these new opportunities, as seen in the thicker cultural deposits and better preservation of habitation debris over more spatially extensive areas. Hearths and pits were preserved over a significantly larger portion of stable backbeach at House of Taga, as compared to a previous restriction of hearths to the more landward portion of the site removed from the active tidal range. At both Ritidian and Unai Bapot, the newly stable land surfaces retained evidence of localised burning that had been absent in older layers of unstable sands, here diagnosed by slight reddening and hardening of sands around and beneath the burn spots. Moreover, these newer cultural layers contained widespread charcoal, trapped in place from numerous individual burning episodes while the sandy sediments accumulated without the prior conditions of repeated tidal washing.

Concurrent with the changing coastlines, inland and upland zones remained mostly stable as geological formations, yet modifications began to become more noticeable in forest clearing and patterns of slope erosion and re-deposition. Larger resident populations cleared more forests to accommodate growth of economically useful trees and root crops, and meanwhile grassland habitats expanded in areas of unattended land or secondary disturbance (Athens and Ward 2004). The change in vegetation cover generated greater amounts of slope erosion, especially in places where clays and silts were less protected by tree roots. Slope-eroded sediments began to accumulate in lower elevations, for example in the bottoms of lakes and swamps but also over the floors of stream-cut valleys.

Resource Zones

In the new configuration of coastal terrain after 1100 B.C., coral reefs in most cases adjusted their growth patterns at the lower sea level, but certain shellfish taxa suffered in the changing environment. Most striking was a decline in *Anadara* sp. shellfish that preferred mangrove swamps or other shallow-water settings, while exactly these supporting habitats were beginning to disappear. Meanwhile, *Gafrarium* sp. and other taxa proved more resilient to the changing nearshore conditions.

The nearshore natural resources appear to have been suffering at this time, most likely due to the compounded effects of falling sea level with increasing levels of harvesting by people. A region-wide decline in *Anadara* sp. can be attributed in large part to the naturally lowering sea level, because the same results occurred contemporaneously at each site. Wherever people established new residential sites after 1100 B.C., lesser and lesser amounts of *Anadara* sp. shellfish could be found in their irreversibly disappearing natural habitat, and eventually they became almost entirely unavailable in the Mariana Islands. Meanwhile, people still could find other nearshore taxa in impressive abundance when creating new sites along coastlines where sea urchins, limpets, and chitons had not yet been depleted. In the longer inhabited sites, however, these same shellfish taxa were scarce or non-existent in site middens after 1100 B.C., evidently because they were vulnerable to the impacts of harvesting by the resident human populations.

While the nearshore resources were depressed, more shellfish protein was being harvested from *Trochus* spp. and *Turbo* spp., commonly found in middle and outer reef zones. These findings could suggest that the coral reefs adjusted gradually with the sea-level drawdown, at least for a short time before the sea level dropped even more. Today, the ancient outer reef margins and algal ridges are seen exposed above sea level along some coastlines, but they have been buried beneath recent coastal sediments in other places.

The lowering sea level created a minor decrease in the elevations of freshwater lenses in coastal areas, but this minor physical change brought a potentially major cultural impact. Previously, a higher sea level supported lenses of fresh water, accessible in seeps at the tiny sandy beach fringes and in pools inside limestone caves of the larger southern-arc islands. As sea level began its drawdown after 1100 B.C., the freshwater sources lowered and became less and less voluminous in the places where people previously had found them. These sources still remained usable in most cases for up to a few centuries, but soon people would need to find other points of access, sometimes but not always within the immediate vicinity of a pre-existing residential site.

Material Culture

Beginning around 1100 B.C., transitions are evident in several aspects of material culture, overall continuing the traditions of preceding centuries yet with observable modifications and a few new inventions. Abundant earthenware pottery assemblages show change in vessel shape and thickness, as well as in decorative expressions. Certain types of shell beads and ornaments were no longer in use, but others continued to be produced in variable frequencies. Stilt-raised houses persisted as the preferred residential structures, newly constructed in slightly different locations as if chasing after the moving shorelines during the centuries following 1100 B.C.

In terms of the artefacts, probably most informative about the period 1100–700 B.C. are the changing attributes in the earthenware pottery (Fig. 9.3). The tradition of red slip continued, as did the use of beach sand as the predominant non-plastic temper inclusion when preparing the clay recipe. The very thin-walled pottery had ceased production, as did the rare instances of burnished blackware. The new products overall were thicker and coarser than in prior centuries, now mostly 2–5 mm and using less

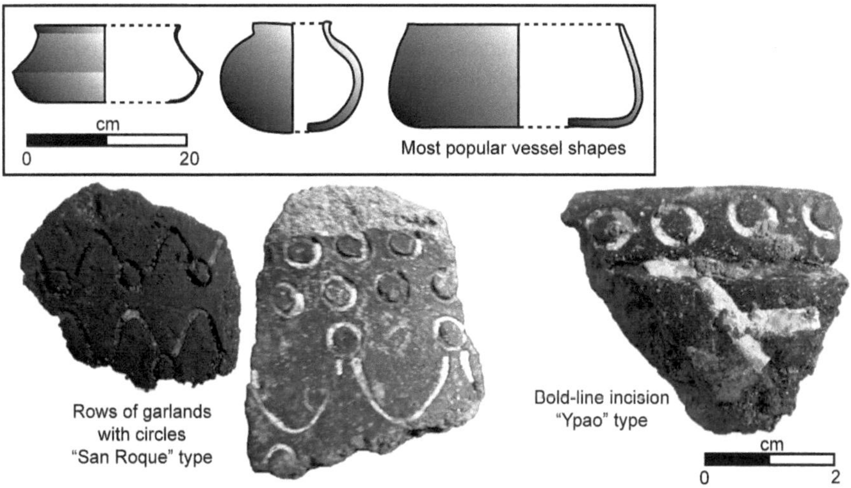

Fig. 9.3 Pottery types, 1100–700 B.C. All examples are from excavations landward of House of Taga, Tinian

refined clay paste with slightly coarser beach-sand tempers. These sturdier vessels were larger and more varied in shape than was possible with the older forms of very thin-walled earthenware. The finely dentate-stamped and zone-filled designs also had ceased by this time, but rows of curvilinear garlands and circles continued probably as late as 900–800 B.C. Meanwhile, a diagnostic new type of decoration entailed broad-line incisions in various angular patterns with large circle-stamped patterns, and these overall bolder designs were highlighted by white lime infill as in prior centuries.

The lingering curvilinear garland decorative style is best dated at sites where it appeared in the oldest cultural layers without any of the dentate-stamped zone-filling varieties. Such a case is most clear at San Roque in Saipan, where Brian Butler (1994, 1995) documented this decorative style in great detail and named it as the "San Roque Type". The basal cultural deposits at San Roque are dated approximately 1100–1000 B.C., notably later than the nearby Achugao Site where the same pottery type was found inter-mixed with the presumably older dentate-stamped and zone-filled styles that Butler (1994, 1995) named as the "Achugao Type". The same chronological order is attested at sites that exhibit strong stratigraphic separation of pottery-bearing layers, as at House of Taga in Tinian, where the curvilinear garland decorations certainly are not the earliest but rather first appear at some point between 1300 and 1100 B.C. and then continue for another few centuries. Additional excavations at San Roque have confirmed the same eponymous pottery decorative type in direct association with another, characterised by bold-line incisions and circle stamping (Ray et al. 1996), very different from the older expression of finely dentate-stamped designs that were apparently absent at San Roque.

The new type of diagnostic bold-line incised pottery was thoroughly described from collections at the Ypao Site at the far western edge of Tumon Bay in Guam, and hence it has been known as the "Ypao Type" (Leidemann 1980). Most of this pottery was red-slipped, but some examples displayed grey non-slipped surfaces. Dating of the basal cultural layer at Ypao suggests that the associated pottery was produced as early as 1023–768 B.C. (Olmo and Goodman 1994). In stratified sites where this type of pottery very certainly was not the earliest in a series, its associated layers have been dated as early as 1100 B.C. and continuing as late as 200 B.C., for example at Ritidian, House of Taga, and Unai Bapot. Within this broad time span, the frequency of Ypao Type pottery declined after 500–400 B.C. The same pottery type has been found in several other sites, apparently consistent with the proposed long date range, although confirmatory radiocarbon dating has been vague in many of these cases.

Overall, the pottery of this time period followed a trend of coarser and less refined production, perhaps interpreted as "bolder", as seen in vessel thickness, size, and decorative patterns. Although lacking the fine details of earlier pottery, the new forms displayed much the same general effects of older pre-existing traditions of red-slipped surfaces, and the decorated varieties displayed the same highlighting by white lime infill for a white-on-red contrast. These end results generally continued from the preceding time period, but the specific expressions certainly were coarser or bolder.

After 1100 B.C., the amount of labour for each pot had lessened, so that more items could be produced within the same devotion of time. This trend is difficult to interpret, but at least two hypothetical possibilities can be considered. First, the speedier production may have been a response to an increased demand by a growing population. Second, this trend may have been a way for the individual potters to have more time for other tasks, noting that people must have been concerned with meeting their basic economic subsistence after the evident change in coastal ecology following 1100 B.C. Potentially both of these scenarios are accurate, and both are congruent with a context of people adapting to the challenges of a changing natural–cultural landscape.

Similar to the trend noticed in pottery production, shell beads and ornaments overall became coarser and less refined after 1100 B.C. The older type of polished *Cypraea* sp. shell bead was no longer manufactured at all, while other shell goods lessened in their overall frequencies. Disc-shaped pendants and small ring pendants of *Conus* spp. shells greatly declined in popularity after 1100 B.C. and were no longer produced after about 700 B.C. Meanwhile, polished tiny *Conus* spp. shell beads continued in production through about 500 B.C., at which time they were replaced by larger and less polished or unpolished *Conus* sp. shell beads. Only the larger shell items, like *Conus* spp. shell bangles, were produced consistently throughout the Marianas chronological sequence.

Adzes and flaked tools of this period appear to be continuations of pre-existing traditions. These general-utility tools exhibit only very little chronological change, such as a decline in the usage of chert. After 1100 B.C., adzes mostly were made of cut and polished shells of giant clams (*Tridacna* sp.), while chert became used only very rarely for adzes. Chert continued as the most common material for small flaked tools with sharp edges. Adzes were necessary for forest clearing and wood-working, while flaked tools were utilised for numerous tasks of cutting and slicing in diverse contexts.

Fishing gear reveals no major change during this period, with the possible exception of trolling lure pieces attested for the first time. The few known fishing-related artefacts resemble the same forms of small rotating hooks made of shells as in earlier centuries, continuing the tradition of working primarily with *Isognomon* sp. shells and occasionally with *Turbo* spp. shells. Two possible broken pieces of trolling lure shanks have been dated to this period, but neither occurrence is entirely convincing as a mere fragment lacking its full diagnostic shape. One was reported at Mangilao in Guam (Dilli et al. 1998). Another was found at Ritidian in a layer containing thicker red-slipped pottery, covered by a layer of branch coral debris dated 1360–1242 B.C. and followed by a continuation of the same cultural habitation layer dated 1161–796 B.C. If trolling lure shanks can be verified of this age, then the practice of pelagic fishing can be assigned to a much older context than previously has been known from Marianas fishing gear. Large-sized fish vertebrae imply deep-sea fishing, even without finding preserved mouth parts of fish for indisputable taxonomic identifications, and deep-sea fishing skills may be expected among some of the people who lived in the remote Mariana Islands. Nevertheless, the archaeo-faunal assemblages and artefacts point to strong emphasis on nearshore resources that were most readily accessible close to the shoreline-oriented habitations.

Remnants of housing structures and other features exhibit a continuation of the same post-raised houses amidst various hearths and pits, but the total amount of on-ground activity increased with the newly available coastal landforms. Stilt-supported houses continued to be inhabited at shorelines, most efficient for above-water and in-water activities, but the lowering sea level resulted in increasingly more stable masses of sands and silts in stranded positions for on-ground activities (see Fig. 9.2). After 1100 B.C., instead of using small scattered patches of sands and periodic low-tide exposures, now people habitually were using stable dry landforms within their preferred coastal zones on a continual daily basis. Meanwhile, the surrounding shoreline ecological niches were transforming.

Regional Context

During the centuries 1100–700 B.C., the Mariana Islands no longer were the solitary inhabited islands of Remote Oceania (see Fig. 1.1). People began living in another large island group in western Micronesia, Palau, by 1000 B.C. (Fitzpatrick 2003), while numerous islands were being colonised for the first time throughout Southern Melanesia and West Polynesia almost entirely in the range of 1100–800 B.C. (Kirch 1997; Spriggs 1997). While the Marianas landscape continued to evolve as an established natural–cultural system that already had been functioning for some centuries, the other newly inhabited island landscapes were undergoing initial stages of human–environment interactions.

Although clearly in a later dating and separate context from the Marianas, first settlement in Palau by 1000 B.C. resulted from a similar long-distance migration of people coming from Island Southeast Asia. The language history in Palau reveals a similar early detachment from an archaic Malayo-Polynesian linguistic ancestry as was the case for Chamorro language in the Mariana Islands, but the two languages

(Palauan and Chamorro) descended from different subsets of the generalised Malayo-Polynesian setting (Blust 2013). Both languages can be traced to different sources within Island Southeast Asia where archaic forms of Malayo-Polynesian once were spoken, distinctly differentiated from the Oceanic-speaking language groups that developed everywhere else throughout Remote Oceania (see Figs. 5.5 and 5.6).

Outside the Marianas and Palau, the consistently widespread Oceanic-speaking affiliation in Remote Oceania implies a tightly shared ancestry of the people who initially moved into these remote islands. Exactly this scenario is supported by the evident spread of people very quickly throughout such a broad region within just a few centuries, marked by a single shared horizon of dentate-stamped pottery known as the Lapita style (Kirch 1997). This decorated pottery tradition can be traced back to immediate roots in the Near Oceanic islands of the Bismarck Archipelago around 1350 B.C. (Denham et al. 2012; Summerhayes 2007), but its deeper roots are revealed in the pottery trail extended farther back in time to the Mariana Islands at least as early as 1500 B.C. and even farther to the northern and central Philippines around 2000 B.C. (Carson et al. 2013; Hung et al. 2011).

So far, the diagnostic red-slipped and finely dentate-stamped pottery traditions have not been found in ancient sites of Palau. Instead, the oldest Palauan site deposits have yielded plain and rather coarse earthenware in low frequencies (Clark et al. 2006; Fitzpatrick 2003). In this case, independent of the linguistic evidence, the origins of the first Palauan settlers probably differed significantly from the origins of the first Marianas settlers and also from the groups making Lapita pottery in Melanesia and Polynesia.

Direct cultural connections almost certainly did not exist between the Mariana Islanders and the groups who settled in any other part of Remote Oceania. As noted, settlement of Palau derived from a different and later context that lacked the diagnostic pottery traditions and other material markers of the Marianas. Elsewhere in Remote Oceania, the Lapita-associated settlement in Southern Melanesia and West Polynesia did in fact share the same distant ancestry of dentate-stamped pottery and other traditions as seen in the Marianas, but these traits realistically can be traced back to a source in Near Oceania and specifically in the Bismarck Archipelago. Connections may have existed previously between the Mariana Islands and the Bismarck Archipelago during the initial years of Lapita pottery-making in the Bismarcks, but these connections did not extend between the Marianas and the farther reaches of the Remote Oceanic Lapita realm in more distant locations and later-dated periods.

Despite the lack of direct cultural connection, the Remote Oceanic Lapita sites were installed in much the same shoreline-oriented niches as in the Mariana Islands. However, the preferentially targeted coastal ecosystems already were beginning to transform due to sea-level drawdown by the time people first arrived in the Remote Oceanic islands in Melanesia and Polynesia. Once again, stilt-raised houses were inhabited in inter-tidal and shallow sub-tidal zones, preferably near coral reefs and lagoons as well as near mangrove swamps (Nunn 2005, 2007). After 1100 B.C., though, regional sea level started a period of drawdown, so the newly emerged beach ridges and berms already were available for the first Lapita pottery-making settlers in Southern Melanesia and West Polynesia (Dickinson 2014).

Unlike the Marianas case, the Remote Oceanic Lapita settlements were numerous and widespread throughout a greater number of islands, where human beings invaded and colonised the Remote Oceanic environment within a rather brief time span of a few centuries. The requisite sea-crossing migrations did not exceed 900 km, as compared to the voyage in excess of 2000 km for the first Marianas settlement some centuries earlier, so the Lapita-associated movements into Remote Oceania could proceed more quickly, with ample pre-planning, and with less risk. Many of the cultural preparations for remote-distance voyaging and island survival already had developed during the several generations of pre-established Lapita occupation in the Melanesian islands of Near Oceania (Irwin 1998; Spriggs 1997). The results of a broadly sweeping Remote Oceanic Lapita invasion were described in Chap. 3 as a "siege of ecological imperialism" when people imposed a foreign landscape system on the natural environment (Crosby 1986). The effects of an artificially transported landscape were magnified in the fragile island ecosystems of Remote Oceania that had evolved for several centuries without any human presence (Kirch 2000). This widespread event is attested in massive depopulations and outright extinctions of birds and other animals (Steadman 1995), along with replacement of native forests by freshly imported trees and tuber-root crops at massive scales throughout the region (Kirch 2010).

Likely due to their similar ecological settings as seen in the Marianas, the first settlement sites in other parts of Remote Oceania appear to be entirely residential and lacking mortuary features, with two notable exceptions of burial sites in settings that apparently were outside the primary residential zones. These two cases offer potential insights into mortuary practice that conceivably may have occurred in the Mariana Islands around the same time, although currently no burial features of this age are yet known in the Marianas or even in Island Southeast Asia. One case was inside the rockshelter of Chelechol ra Orrak in a tiny limestone rock island off the coast of the main land mass of Babeldaob in Palau, dated about 1000 B.C. (Fitzpatrick 2003; Fitzpatrick and Nelson 2011). Another was a burial ground at Teouma in Vanuatu, dated about 1000 B.C. and apparently separate from the central habitation area (Bedford et al. 2006; Valentin et al. 2010). In both of these sites, human skulls or fragments of skulls were removed and re-positioned, suggestive of a deeply shared mortuary tradition despite the lack of any direct cultural connection between Palau and Vanuatu. This interpretation deserves more consideration, but so far it cannot be tested against any known burials in either the Mariana Islands or in Island Southeast Asia pre-dating about 700–500 B.C. Nonetheless, the use of human bone for creating pointed artefacts in the Marianas has been attested at this early age, and reverence of curated ancestral skulls has been mentioned in later historical contexts.

Comparisons between the Marianas and elsewhere in Remote Oceania must be understood as not at all implying direct migrations of people from the Marianas to these other remote island locations. Rather, the comparisons offer clues about how people adapted to similar circumstances in different parts of the Remote Oceanic region as a whole. Any apparent similarities may be attributed to parallel choices by people facing much the same conditions independently, but to a certain extent these outcomes derived from a shared deeper ancestry of how people perceived their environments and proceeded to make their new homes in these landscapes. For instance, the apparent preference for living in certain shoreline niches can be traced back to a much older tradition in Island Southeast Asia, coincidentally also the homeland region for other traits such as betelnut chewing and producing red-slipped pottery with finely dentate-stamped and white lime-infilled decorations. A similar ancestry may be considered for cultural perceptions of the dead, special treatment of the skulls of deceased individuals, and by extension perhaps a set of religious beliefs and views of the world that are not yet decipherable from the scant archaeological clues.

While no direct linkage can be claimed between the Marianas and the first settlements in other parts of Remote Oceania, a different form of connectivity existed between the Marianas and Island Southeast Asia. At least in a few points, the chronology of pottery types in the Marianas paralleled the chronology in the Philippines and parts of Indonesia. In particular, shifts toward bolder and coarser designs are attested in each of these areas, and approximately the same temporal pace has been outlined (Bellwood 1997; Hung 2008; Simanjuntak 2008). Nevertheless, these points of similarity are few in total, and the overall patterns exhibit an ongoing independence of the Marianas material culture and landscape system from the sequences of events and processes that unfolded in Island Southeast Asia.

The parallel pottery chronologies in the Marianas and Island Southeast Asia indicate a small but recognisable degree of cultural contact and continuity across this broad and diverse region. This outcome negates the notion of a completely isolated population in the Mariana Islands, but rather people in the Marianas must have experienced at least a few contacts directly or indirectly with the changing contexts and new developments in Island Southeast Asia. The same can be said of the diverse communities scattered throughout the Philippines and Indonesia at this time, sharing many of the same new trends in pottery decorative styles over an extended period. Although cross-community contacts certainly occurred in order to account for the parallel pottery chronologies, they did not entail large-scale sharing in every aspect of material culture. Evidently people still did not transport any domesticated animals to the Marianas despite their significant cultural roles in Island Southeast Asia. Furthermore, the language histories of different groups continued to diverge independently.

The apparent cross-community and cross-regional contacts conceivably may be attributed to periodic trading partnerships, development of guest–host relationships, and maintenance of connections with ancestral homelands. Each of these contexts may have co-occurred, and possibly they mutually encouraged one another in concert. These kinds of patterns tend to emerge in places where immigrant groups sustain communications with their homeland regions, as discussed in general terms by David W. Anthony (1990) and elaborated in his treatment of the development of Bronze Age Indo-European societies (Anthony 2007). In the Marianas case, contacts with an Island Southeast Asian homeland fluctuated in frequency and intensity over time, possibly with indirect down-the-line chains of connections, but overall any contacts must have been rare due to the extremely long distance between the Mariana Islands and any possible homeland.

References

Anthony, D. W. (1990). Migration in archaeology: The baby and the bathwater. *American Anthropologist, 92*, 23–42.
Anthony, D. W. (2007). *The horse, the wheel, and language: How Bronze-age riders from the Eurasian steppes shaped the modern world*. Princeton, NJ: Princeton University Press.
Athens, J. S., & Ward, J. V. (2004). Holocene vegetation, savannah origins and human settlement of Guam. In V. Attenbrow & R. Fullagar (Eds), *A Pacific Odyssey: Archaeology and anthropology in the Western Pacific: Papers in Honour of Jim Specht*, Records of the Australian Museum, Supplement 29 (pp. 15–30). Sydney: Australian Museum.
Athens, J. S., & Ward, J. V. (2006). *Holocene Paleoenvironment of Saipan: Analysis of a core from Lake Susupe* (Micronesian Archaeological Survey Rep. No. 35). Saipan: Commonwealth of the Northern Mariana Islands Division of Historic Preservation.
Bedford, S., Spriggs, M., & Regenvanu, R. (2006). The Teouma Lapita site and the early human settlement of the Pacific Islands. *Antiquity, 80*, 812–828.
Bellwood, P. (1997). *Prehistory of the Indo-Malaysian Archipelago* (Rev. ed.). Honolulu: University of Hawaii Press.
Blust, R. (2013). *The Austronesian languages* (Rev. ed.) (Asia-Pacific Linguistics Open Access Monographs 008). Canberra: Research School of Pacific and Asian Studies, Australian National University.
Butler, B. M. (1994). Early prehistoric settlement in the Mariana Islands: New evidence from Saipan. *Man and Culture in Oceania, 10*, 15–38.
Butler, B. M. (Ed.). (1995). *Archaeological investigations in the Achugao and Matansa areas of Saipan, Mariana Islands* (Micronesian Archaeological Survey Rep. No. 30). Saipan: Commonwealth of the Northern Mariana Islands Division of Historic Preservation.
Carson, M. T. (2011). Plaeohabitat of first settlement sites 1500–1000 B.C. in Guam, Mariana Islands, Western Pacific. *Journal of Archeological Science, 38*, 2207–2221.
Carson, M. T. (2012). Evolution of an Austronesian landscape: The Ritidian site in Guam. *Journal of Austronesian Studies, 3*, 55–86.
Carson, M. T. (2014a). Paleoterrain research: Finding the first settlement sites of Remote Oceania. *Geoarchaeology, 29*, 268–275.
Carson, M. T. (2014b). *First settlement of Remote Oceania: Earliest sites in the Mariana Islands*. New York: Springer.
Carson, M. T., Hung, H.-C., Summerhayes, G., & Bellwood, P. (2013). The pottery trail from Southeast Asia to Remote Oceania. *Journal of Island and Coastal Archaeology, 8*, 17–36.
Clark, G., Anderson, A., & Wright, D. (2006). Human colonization of the Palau Islands, Western Micronesia. *Journal of Island and Coastal Archaeology, 1*, 215–232.
Cleghorn, P. L., & McIntosh J. (2002). *Second phase of archaeological investigations at Chalan Kanoa Elementary School*, Chalan Kanoa, Saipan, Commonwealth of the Northern Mariana Islands (Report prepared for Wil Chee Planning). Kailua, HI: Pacific Legacy.
Crosby, A. W. (1986). *Ecological imperialism: The biological expansion of Europe, 900–1900*. Cambridge, England: Cambridge University Press.
Denham, T. P., Ramsey, C. B., & Specht, J. (2012). Dating the appearance of Lapita pottery in the Bismarck Archipelago and its dispersal to Remote Oceania. *Archaeology in Oceania, 47*, 39–46.
Dickinson, W. R. (2014). Beach ridges as favored locales for human settlement on Pacific islands. *Geoarchaeology, 29*, 249–267.
Dilli, B. J., Haun, A. E., Goodfellow, S. T., & Deroo, B. (1998). *Archaeological Mitigation Program, Mangilao Golf Course Project Area, Mangilao Municipality, Territory of Guam, Vol. 1, Introduction, research design, and data recovery results* (Report prepared for Jetan Sahni). Hilo, HI: Paul H. Rosendahl, Ph.D.
Fitzpatrick, S. M. (2003). Early human burials in the Western Pacific: Evidence for AC 3000-year-old occupation on Palau. *Antiquity, 77*, 719–731.

Fitzpatrick, S. M., & Nelson, G. C. (2011). Purposeful commingling of adult and child cranial elements from the Chelechol ra Orrak Cemetery, Palau. *International Journal of Osteoarchaeology, 21*, 360–366.
Hung, H.-C. (2008). *Migration and cultural interaction in southern coastal China, Taiwan and the northern Philippines, 3000 BC to AD 100: The early history of Austronesian-speaking populations*. Ph.D. Dissertation, Australian National University, Canberra.
Hung, H.-C., Carson, M. T., Bellwood, P., Campos, F., Piper, P. J., Dizon, E., et al. (2011). The first settlement of Remote Oceania: The Philippines to the Marianas. *Antiquity, 85*, 909–926.
Irwin, G. J. (1998). The colonisation of the Pacific Plate: Chronological, navigational and social issues. *Journal of the Polynesian Society, 107*, 111–143.
Kirch, P. V. (1997). *The Lapita peoples: Ancestors of the Oceanic world*. Cambridge, England: Blackwell.
Kirch, P. V. (2000). *On the road of the winds: An archaeological history of the Pacific Islands before European contact*. Berkeley, CA: University of California Press.
Kirch, P. V. (2010). Peopling of the Pacific: A holistic anthropological perspective. *Annual Review of Anthropology, 39*, 131–148.
Leidemann, H. H. (1980). *Intra-site variation at Ypao Beach, Guam: A preliminary assessment*. M.A. Thesis, University of Guam, Mangilao.
Nunn, P. D. (2005). Reconstructing tropical shorelines using archaeological data: Examples from the Fiji Archipelago, Southwest Pacific (Special Issue). *Journal of Coastal Research, 42*, 15–25.
Nunn, P. D. (2007). Space and place in an ocean of islands: Thoughts on the attitudes of the Lapita people towards islands and their colonization. *South Pacific Studies, 27*, 26–35.
Olmo, R. K., & Goodman, W. L. (1994). *Archaeological investigations for Ypao Beach Park ground penetrating radar survey, Guam* (Report prepared for Guam Department of Parks and Recreation). Honolulu: International Archaeological Research Institute.
Ray, E., Fortini, W. R., Jr., & Babauta, J. L. (1996). *Archaeological data recovery at Akitsu Shoji's residence in San Roque, Saipan, Commonwealth of the Northern Mariana Islands* (Report prepared for Division of Historic Preservation, Department of Community and Cultural Affairs, Commonwealth of the Northern Mariana Islands). Saipan: A. B. Business Management and Consulting Services.
Simanjuntak, T. (Ed.). (2008). *Austronesian in Sulawesi*. Yogyakarta, Indonesia: Center for Prehistoric and Austronesian Studies.
Spoehr, A. (1957). *Marianas prehistory: Archaeological survey and excavations on Saipan, Tinian and Rota. Fieldiana: Anthropology, Vol. 48*. Chicago, IL: Chicago Natural History Museum.
Spriggs, M. (1997). *The Island Melanesians*. Cambridge, England: Blackwell.
Steadman, D. W. (1995). Prehistoric extinctions of Pacific Island birds: Biodiversity meets zooarchaeology. *Science, 267*, 1123–1131.
Summerhayes, G. R. (2007). The rise and transformations of Lapita in the Bismarck Archipelago. In S. Chiu & C. Sand (Eds.), *From Southeast Asia to the Pacific: Archaeological perspectives on the Austronesian expansion and the Lapita cultural complex* (pp. 141–184). Taipei: Center for Archaeological Studies, Research Center for Humanities and Social Sciences, Academia Sinica.
Swift, M. K., Harper, R. A., & Fleming, M. A. (2009). *Archaeological assessment: Kalabera Cave development project, Marpi, Saipan* (Report prepared for Herman B. Cabrera and Associates). Saipan: Swift and Harper Archaeological Research Consulting.
Valentin, F., Buckley, H. R., Herrscher, E., Kinaston, R., Bedford, S., Spriggs, M., et al. (2010). Lapita subsistence strategies and food consumption patterns in the community of Teouma (Efate, Vanuatu). *Journal of Archaeological Science, 37*, 1820–1829.

Chapter 10
700 B.C.–A.D. 1, Broadened Horizons

After 700 B.C., people shifted away from shoreline-oriented settlements and instead lived slightly farther from the seashore in stable backbeach zones and other coastal terrain. The prior strong emphasis on narrowly defined nearshore niches was no longer sustainable, due to a continually lowering sea level, changing ecological structure, human population growth, and new social developments. Mariana Islanders of course maintained their coastal lifestyles as much as was practical under the changing conditions, but now landward areas and land-based resources gained noticeably more prominence while many of the nearshore niches had become unpredictable and unreliable. These new patterns persisted until approximately A.D. 1 or slightly later, when a temporarily stable sea level prompted another set of systemic changes in the Marianas natural–cultural landscape.

Site Inventory and Dating

The period 700 B.C.–A.D. 1 is defined by a shift in the Marianas landscape system, with lessened emphasis on the nearshore ecosystem and increased reliance on land-based resource zones. Looking only at pottery types and a few other artefact categories, this period may be sub-divided in segments before and after 500–400 B.C., but the longer period as proposed here refers to a larger view of systemic operation of the natural–cultural landscape. In this view, the Marianas landscape as a unified system underwent a major transition around 700 B.C., and another such transition occurred after approximately A.D. 1.

As in the preceding centuries, sites of this time range have been found only in the larger southern-arc islands (Fig. 10.1). By this time, at least two sites in Rota can be added to the known site inventory, at Mochong dated at least as early as 500 B.C. (Takayama and Intoh 1976; Ward and Craib 1983) and Teteto-Guata dated at least as early as 400 B.C. (Hunter-Anderson and Butler 1995:30), although both sites could be a few centuries older. In addition to a shift in settlement slightly farther from the

M.T. Carson, *Archaeological Landscape Evolution*,
DOI 10.1007/978-3-319-31400-6_10

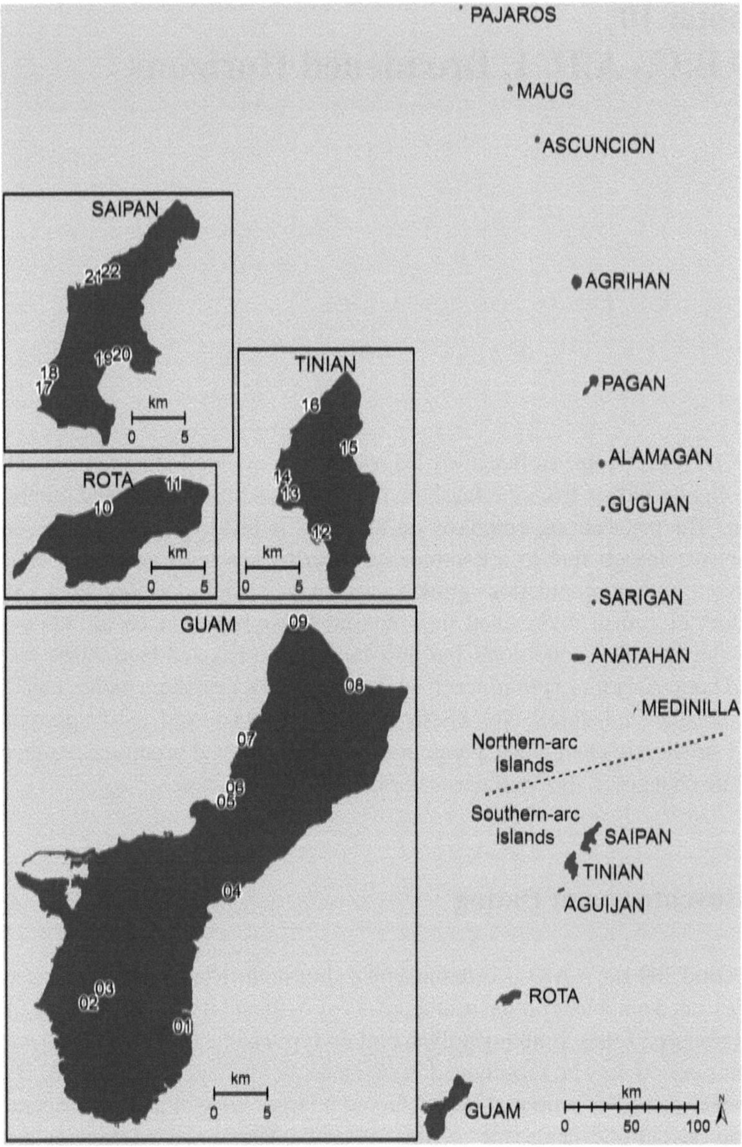

Fig. 10.1 Major site locations, 700 B.C.–A.D. 1. 01: Ipan. 02: Almagosa. 03: Fena. 04: Mangilao. 05: Tumon. 06: Naton. 07: Ague. 08: Tarague. 09: Ritidian. 10: Teteto-Guata. 11: Mochong. 12: House of Taga. 13: Kahet. 14: Atgidon. 15: Dangkulo. 16: Unai Chulu. 17: Chalan Piao. 18: Chalan Kanoa. 19: Laulau Rockshelter. 20: Unai Bapot. 21: Achugao. 22: San Roque

active beach fronts and seashores, sites of this period are found in a few farther inland locales. Habitation areas shifted slightly and occupied a few more places in the islands of Guam, Tinian, and Saipan. At least a few open campsites, small caves, and rockshelters bear evidence of cultural use along rough limestone coasts as well as in farther inland terrain of Guam and Tinian (Dixon and Schaefer 2014), for example at sites in the vicinities of Almagosa, Fena, Ague, Kahet, and Atgidon.

In most of the locations where people already had been living at the ancient shoreline zone prior to 700 B.C., cultural layers in the range of 700 B.C.–A.D. 1 are poorly represented. People continued to use these areas for short-term work, limited resource collection, and other assorted non-residential activities, while dense primary habitations had shifted slightly elsewhere. In the settings of peripheral non-residential activities, rather diffuse cultural deposits contain a mixture of pottery types and other artefacts over this extended time span. At House of Taga and at Unai Bapot, for instance, artefacts and midden still accumulated over these ancient housing zones within gradual build-up of newer sediments, but no clear living surfaces or intact residential structural remains are found again in exactly these same locations until much later. While the sea level was lowering and the coastlines were changing, the precise locations of former shoreline zones no longer were used directly as habitation sites. Rather, people installed their habitations in slightly different locations.

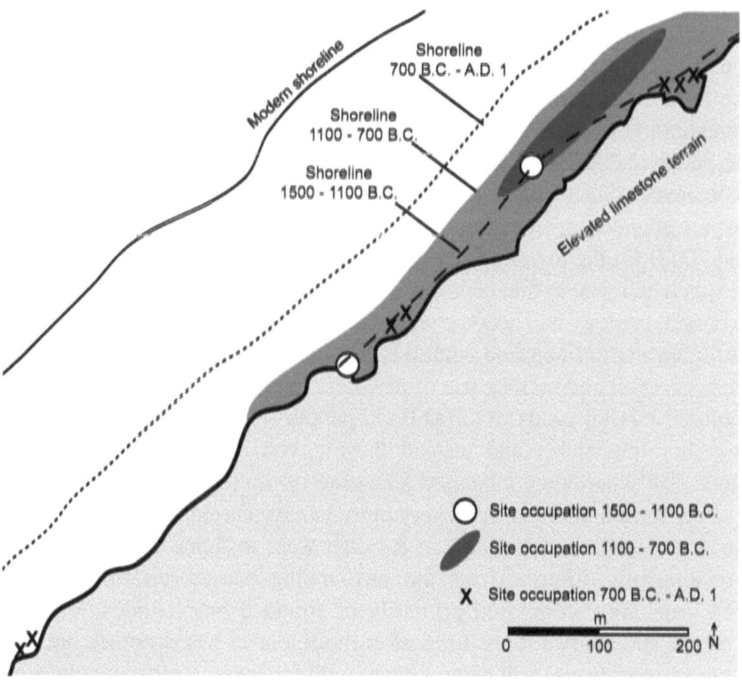

Fig. 10.2 Coastal habitat change after 700 B.C. at Ritidian, Guam. Based on data from Carson (2011, 2012, 2014a)

The clearest archaeological documentation of this period comes from places where people moved into newly stable landforms for the first time, for example at Ritidian in northern Guam (Fig. 10.2). Around 700 B.C. at Ritidian, the former habitation of a singular elongate beach ridge ended, and instead people shifted landward to occupy a number of scattered small pockets of slightly elevated terrain along the outer base of the island's limestone cliff, often very close to caves in the cliff face. The next major change in these cliff-base habitations occurred after A.D. 1, when the footprint size of each habitation grew substantially. Also after A.D. 1, people began using inland terrain more intensively, including a number of formal residential sites in areas of elevated limestone plateau and volcanic hills, marking another transition in the Marianas landscape system.

Landforms

As noted, the major characteristics of the Marianas landscape during 700 B.C.–A.D. 1 related to steady decline of sea level and reconfiguration of coastal terrain, further contributing to shifts in natural habitat structure and cultural use of these newly modified ecologies. The most visible transformations occurred in coastal zones with the greatest substantive geomorphic change combined with archaeological records of cultural activities, while the material traces in inland terrain provided sparser evidence at this time. Nonetheless, a definite shift occurred towards increasing the frequency and intensity of cultural usage of landward and inland zones, leading to modification of the island-wide landscapes in terms of physical terrain, plant and animal populations, and cultural perception of the environment.

The geoarchaeological records at Ritidian are particularly illustrative of how people increased their landward and land-based activities, while lessening but of course not abandoning their close relations with the coastal and marine zones. Around 700 B.C., the lowering sea level had retreated to a point perhaps 1 m lower than it was when people first occupied the Ritidian area at 1500 B.C., and the associated coastal ecology had become significantly different for the new generations of people living there. For some centuries already, people had been adapting to the lowering sea level and making use of new beach ridges that formed on the continually exposed coastal sands. By 700 B.C., people stopped following after the prograding shoreline niche, and instead they moved landward to the base of the limestone cliff where they inhabited a number of scattered patches of slightly elevated sandy terrain, in most cases very close to cave entrances.

The new cliff-base habitations at Ritidian were installed in small patches that were qualitatively different from the surrounding coastal terrain. These slightly elevated spots were composed primarily of stranded beach sands, left behind at these higher and landward positions after the sea level had dropped, but now they started to support in situ soil development with terrigenous silts and clays over the stable patches of land with local forest growth. In these stable settings with new soil development, trees and other vegetation began to grow more strongly than was possible in the larger surroundings of pure beach sands. In situ forest growth is noted by the presence of snail shells, including *Pythia* sp. and the now rare tree snail *Partula* sp., previously absent in the contexts of alkaline calcareous beach sands without terrigenous soil development at Ritidian. Elsewhere in the surrounding beach sands at this time, only limited vegetation could grow in the shifting sandy surfaces, whereas the cliff-base localities offered the most stable conditions for soil development and vegetation growth.

At Ritidian, the cliff-base habitations were situated close to caves that likely offered precious resources of fresh water. These new cliff-base habitations were in addition to the pre-existing tradition of using light zone overhang openings for specialised ritual purposes near the entrances of Ritidian Beach Cave and Ritidian Star Cave. Water could be collected when dripping through the limestone after rains, or it could be gathered from shallow pools that were becoming increasingly shallower as the level of the island's natural aquifer lowered over the dropping base of sea level. Curiously, these particular caves near the cliff-base habitations contain no cultural deposits in their interior spaces until much later, although handprints and other pictographs on the cave walls so far have not been dated.

At the same time of these new developments at Ritidian, other coastal environs began to change with new types of land surfaces, for example at Ipan in southeast Guam (see Fig. 3.15). Prior to 500 B.C. at Ipan, slope-eroded rocky silt and clay had accumulated over the low-lying coastal terrain. These sediments washed from the inland hillslopes, settling in re-deposited beds at the base of the slopes but slightly landward from the prograding shoreline of that time. Within this widespread sedimentary layer, its overall stable conditions supported vegetation growth and soil formation, and eventually a cultural habitation layer was emplaced at the top of the soil horizon. Within the habitation layer, a *Gafrarium* sp. shell was dated 808–496 B.C., and thick-walled red-slipped pottery with incised designs appeared congruent with this age range. The terminal point of this buried soil horizon at Ipan was dated by preserved organics in the soil at 112 B.C.–A.D. 74, referring to the last time when this unit of soil had supported organic growth, directly overlain by thick storm-surge debris from at least two distinct episodes before habitation continued again at this location some centuries later.

While people were adjusting to the new coastal conditions, cultural use of inland terrain was increasing. Beginning at least as early as 500 B.C., limited inland habitation is evidenced in low frequencies of broken pottery, burned coral pebbles and cobbles, and charcoal at scattered small sites across inland zones, including both elevated limestone plateau and volcanic hills. Additionally, slope erosion continued to produce thick layers of re-deposited slope-eroded materials in the lake-bottom and swamp-bottom archives of the larger southern-arc islands (Athens and Ward 2004, 2006), thus implying sustained human-caused forest clearance and disturbance of the inland and upland habitats where the slope erosion must have originated. Some of the slope-eroded sediments filled over valley floors and low-lying coastal terrain, where the resulting new land surfaces supported stronger vegetation growth than previously was possible.

Resource Zones

An increasing emphasis on land-based resources should not be misunderstood as implying a total abandonment of coastal zones after 700 B.C. in the Marianas. In fact, coastal habitats consistently were essential parts of living landscapes throughout the archaeological sequence, but fluctuations definitely occurred in the physical forms and cultural usage of both coastal and inland areas. One of these fluctuations contributed to a substantive change in the Marianas landscape after 700 B.C., when coastlines no longer supported the same habitats as they did previously.

By 700 B.C. in the Marianas, people no longer could live in the older targeted shoreline niches where their ancestors had installed the region's first settlements. These kinds of habitats no longer existed in the same places that now were stranded higher and inland from a lowered sea level, and the newly forming lower shorelines did not support the same kinds of ecosystems as had existed previously. In particular, the once abundant populations of *Anadara* sp. shellfish in shallow-water habitats by now were found in scarce amounts, and they would become virtually non-existent in the Marianas by A.D. 1 except in just a few locations of surviving mangrove swamps.

After 700 B.C., the inter-tidal and shallow sub-tidal zones consisted of fresh sandy deposits, now covered by pulverised storm-surge sands, and coral reefs were adjusting to new conditions of a continually lowering sea level. The older surfaces of inner reefs and lagoons already had been stranded in landward positions, and by now the middle to outer reef surfaces were dying as well. New coral reef growth proceeded at the edges of the dead and dying reef margins, at a lower absolute elevation following in accordance with the sea-level drawdown.

Along with the sea-level drawdown, the new coastal zones continued to change slowly but steadily throughout this period. Some shellfish taxa, like *Anadara* sp., were quite vulnerable to the disruption of their natural habitat, while others like *Gafrarium* sp. more easily could tolerate the changing conditions. People adjusted by collecting different types of shellfish and other items, but overall the nearshore habitats could not supply the specific combinations of natural resources that people had harvested previously.

As the nearshore habitat zones continued to adjust naturally, people could not sustain their older traditions of shoreline-oriented settlements as described in Chaps 8 and 9, and instead they began to rely increasingly on the comparatively more stable land-based resource zones. As has been noted, this shift did not entail abandonment of coastal living, but rather it involved lessened reliance on nearshore resources while people invested more time in managing the inland forests and growing various land-dependent plant foods. This shift in emphasis most likely was driven by a dietary need for basic subsistence foods, but it also involved new developing relations between people and the inland terrain zones and habitats.

During this period, a larger population size very likely enhanced the rate and magnitude of landscape change. More people made larger volume and diversity of demands on the landscape, and meanwhile the frequency and intimacy of human–environment interactions must have increased. This scenario is commensurate with the greater number of sites in more diverse settings of the islands, including stable landward portions of coasts, elevated limestone plateau terrain, and volcanic hills. Although each new habitation was rather small individually, the total amount of residential space did increase, and very importantly it also diversified after 700 B.C.

Population growth presumably was steady throughout the Marianas cultural sequence in the inherently constrained island setting, but the growth after 700 B.C. was challenged by a change in the coastal resource supplies. This combination of factors probably could not have been foreseen, and in any case the results forced a transformation in the Marianas landscape system. People began to use the available resource zones in new ways, and they necessarily entered qualitatively different relations with each other and with their inhabited environments. These trends thereafter continued in escalating proportions.

Material Culture

The key trends in artefact forms and styles continued from prior centuries, coupled with new residential lifestyles in stable land-based settings no longer directly at the active shorefronts. Pottery and shell ornaments were made with lessening attention to their finer details, likely supplying more people but with less labour per item in production. Chert greatly diminished as a raw material for manufacturing adzes and flaked tools, while volcanic stone became more popular yet still secondary behind the use of giant clam *Tridacna* sp. shells for making adzes.

During the centuries 700 B.C.–A.D. 1, earthenware pottery underwent at least a few transitions in vessel forms and decorative styles (Fig. 10.3). Overall, people made thicker and coarser vessels, occasionally decorated with bolder or broader incised patterns. The outcurved and carinated vessel profiles of earlier periods had rapidly declined after 700 B.C., while straight-sided vessel profiles by default became most popular. Red slip and white lime-infilled decorations of the Ypao Type declined significantly after 500–400 B.C. but lingered until about 200 B.C. in at least a few sites. Within the Ypao Type pottery collections, grey-coloured non-slipped surfaces were increasingly common. These declining decorative traits after 500 B.C. were replaced by emergence of assorted simple incised patterns over plain, non-slipped surfaces. At the same time, another form of vessel began to gain popularity as a large, flat-bottomed pan that would become more popular in later centuries continuing through A.D. 500.

The ending of Marianas red-slipped earthenware was a gradual process, involving steady decline over a few centuries roughly 500 through 200 B.C. In a larger perspective, this decline started even earlier, about 1100 B.C., when production transitioned from very thin and finely decorated red-slipped pottery to thicker and coarser vessels with bold-line incisions of the Ypao Type. This later type began to decline in popularity around 500–400 B.C., and it ceased production approximately around 200 B.C.

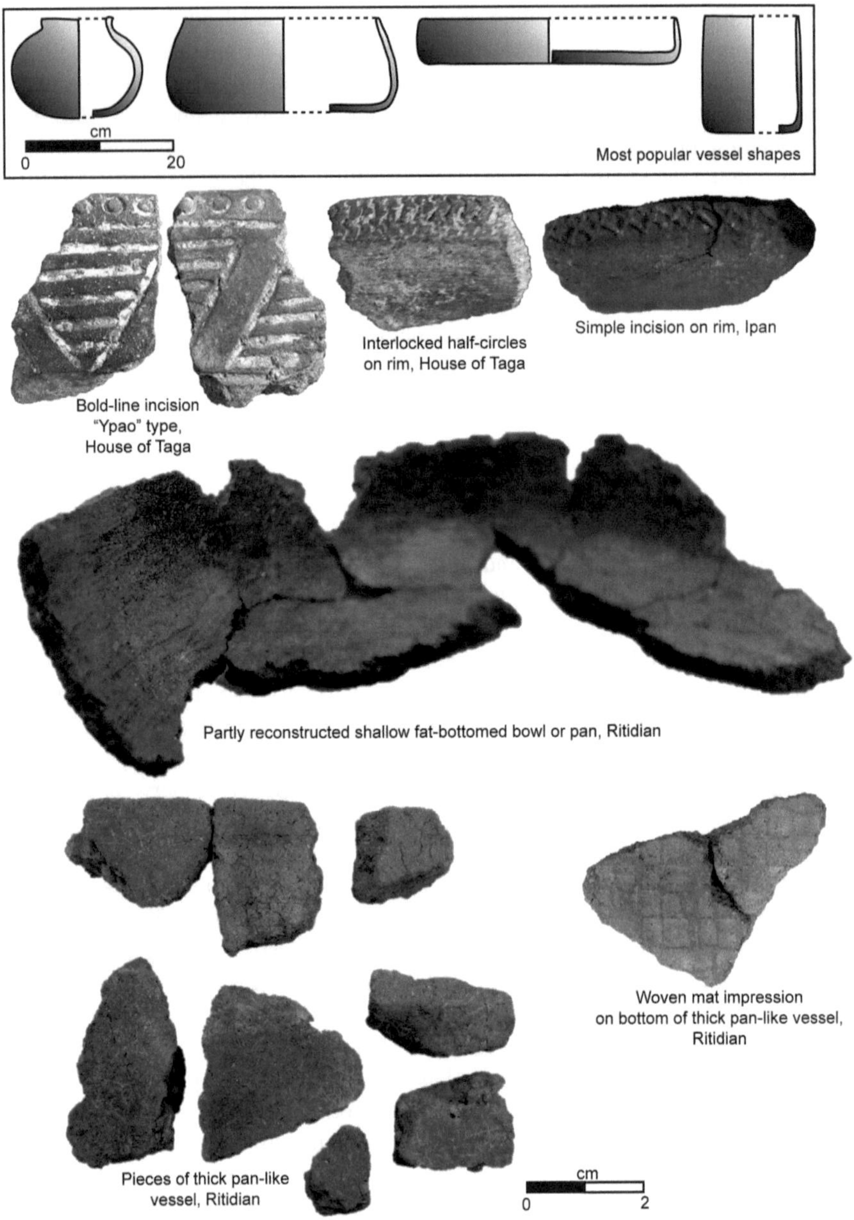

Fig. 10.3 Pottery types, 700 B.C.–A.D. 1

During the few centuries of declining production of Ypao Type pottery, non-slipped vessels are found in variable frequencies. Some but certainly not all of these objects exhibited a number of different decorative patterns, made by thick-line incisions noticeably bolder or broader than the fine-line incisions of older periods. These particular vessels with the occasional decorations mostly were simple bowls and jars.

By the time when the Ypao Type terminated, an entirely different pottery type was being produced, characterised as a large and flat-bottomed pan (Moore and Hunter-Anderson 1999). Overall production was quite low beginning possibly as early as 500 B.C., more common after 200 B.C., and most popular after A.D. 1 as will be presented in Chap. 11. The clay paste was coarse, made with medium to coarse beach-sand temper inclusions, and the surfaces were not treated with any slip or decoration. Although probably not visible most of the time when in use, some of the bottoms retained impressions of woven mats, where the wet clay once had rested on plaited matting while drying before firing. The interior surfaces often were coated with black residue, apparently from starchy and sugary foods that had been grilled or simmered in the shallow pans. This proposed usage may be implied further by the thick base portions, typically more than 1 cm, as compared to the much thinner side walls surrounding the shallow but large-sized interior up to 40 cm in diameter.

By A.D. 1, the Marianas pottery traditions had been greatly transformed, in general becoming less labour intensive for manufacture, more serviceable for larger numbers of end-users, and likely related to preparation of different types of meals. Traits of red slip and white lime infill had disappeared, replaced by non-slipped vessels with simple incised decorations or no decorations at all. Simple bowls and jars continued to be produced in variable sizes for single-serving and group-serving contexts, plus new products of flat-bottom pans were gaining popularity apparently along with trends in local food preparation.

Shell ornaments mirrored the overall trends in pottery production, in terms of trending towards less labour-intensive manufacture and less refined output over time (Fig. 10.4). The tiny and polished *Conus* spp. shell beads, about 1–2 mm diameter, had ceased production by 500 B.C., instead replaced by larger *Conus* spp. beads with little or no polished edges up to 1 cm diameter. As with the pottery, the trend in shell beads began much earlier, around 1100 B.C., when a specific type of sliced *Cypraea* sp. bead and a wafer-thin shell disc both had ceased production entirely. Other shell ornaments were produced in later periods, using various shell taxa and fashioned into diverse shapes. Despite these trends in the finely detailed shell beads and pendants, larger *Conus* spp. shell bangles continued to be produced without any obvious major change in form and style.

Noticeable change occurred in the relative amounts of chert, volcanic stone, and *Tridacna* sp. shells used for adzes and other tools (see Fig. 10.4). Continuing the trend from prior centuries, chert declined in its usage for any kind of artefact, almost never made into an adze after 700 B.C. but continuing in lower frequencies as flaked-edge tools, flaking cores, and debitage. Volcanic stone meanwhile increased in its usage especially for making polished adzes, and some of these were ovoid in section, rather than the invariably rectangular shapes of adzes made of chert or giant clam *Tridacna* sp. shells. Although volcanic stone was used with escalating frequency, it consistently remained secondary behind the popularity of *Tridacna* sp. shells for adze production.

In the new residential settings of elevated landforms inland from prograding shorelines, house structures no longer needed to be raised as tall as had been the case for stilt-raised houses very close to sea level and within range of high-tide

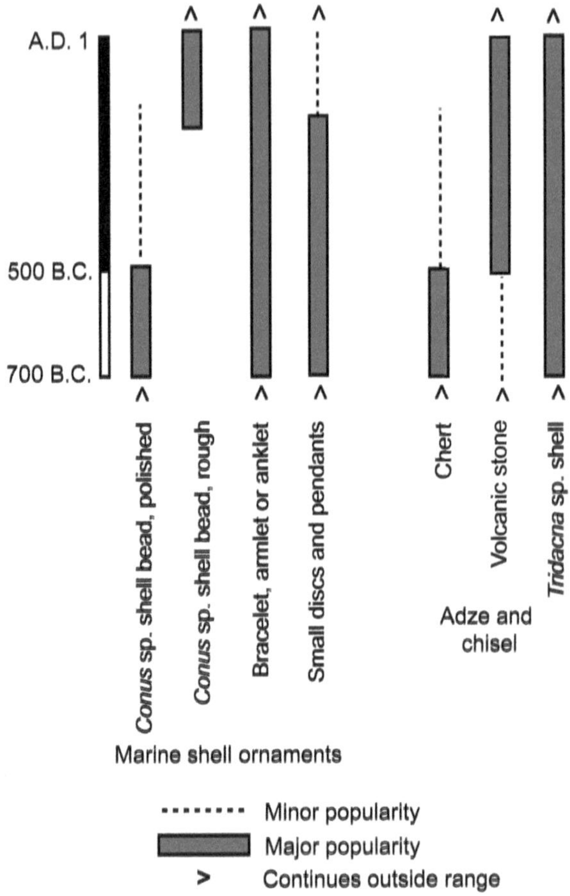

Fig. 10.4 Major trends is shell ornaments and adze materials, 700 B.C.–A.D. 1

events and frequent storm-surge washing. The new houses still were constructed on poles, as seen in arrangements of post moulds, but they did not need to be elevated as high as during the preceding traditions near the ancient seashore. Raised floors could provide advantages of height above insects, crabs, and lizards as well as perhaps exposure to cooling effects of airflow above the otherwise wind-blocking trees and foliage near ground level. The newer post moulds were mostly 15–20 cm in diameter, smaller than the older post moulds of 20–40 cm diameter, thereby suggesting less total supporting strength for smaller, lighter, or lower structures.

This period of more stable coastal terrain and increasing land-based activities has produced the oldest known formal burial graves of the Mariana Islands, preserved in a slightly elevated backbeach at Naton in Tumon of Guam (DeFant 2008). Several dozens of individuals were buried in a space dedicated specifically as a cemetery without immediate residential occupation. Prior to this time, very little available terrain was suitable for formal burial pits in sufficiently deep and sandy sediments, and these spaces already were used for residential housing. By the time of the Naton Site burials, enough stable land was available for use as a cemetery at least in this one setting and probably in others that have not yet been discovered.

At the Naton Burial Site, reportedly as many as 200 burial features here pre-date the many others belonging to *latte* period, and therefore they pre-date approximately A.D. 1000 (DeFant 2008). The Ypao Type of thick and red-slipped pottery suggests an age prior to 200 B.C. for at least some of these burials. Very small and thoroughly polished *Conus* spp. shell beads have been found with some of the individuals, suggesting a date prior to 500 B.C. Four of these beads produced radiocarbon dates approximately in the range of 800–500 B.C. This date range may yet be refined, because *Conus* spp. shells so far have yielded variable and unreliable radiocarbon dating in the Marianas region (Carson 2014b:30).

In the Naton Site burials, individuals were interred in pits slightly larger than the size of their bodies, and occasionally two people were buried together in the same grave (DeFant 2008). Grave goods were sparse, but they consisted of small earthenware bowls and occasionally stone adzes, found on the internal edges of some of the burial pits. The positions of the *Conus* spp. shell beads indicate that they had been strung as necklaces over the chest area of individuals.

In earlier periods, people may have been buried in places apart from the habitation zones that have been excavated, or perhaps people may have followed a different mortuary practice that did not result in preserved grave interments. Interments were known only very rarely elsewhere in the Remote Pacific Islands as early as 1000 B.C., as noted in Palau (Fitzpatrick and Nelson 2011) and in Vanuatu (Valentin et al. 2010), but those two cases both involved burial in places separate from the residential occupations. Similar early burials in the Marianas, if they were spatially distinct from the habitation sites, either have not been preserved or else have not yet been discovered. The practice of burial beneath residential houses would become popular after another several centuries, during the *latte* period after A.D. 1000, when family lineages developed intimate ties with specific land areas, as will be discussed in Chap. 13.

The known habitation sites of 700 B.C.–A.D. 1 were more numerous than in previous years, but they individually covered smaller spaces of scattered land patches supporting fewer houses per site. Overall, greater numbers of people were living in spatially distinctive areas. The dynamics of family relations, trading networks, and other partnerships cannot be known precisely, but a degree of independence may be inferred by the spatial separations. On the other hand, the material culture traits remained widely shared without significant spatial diversity from one site to another, therefore suggesting that people generally maintained commonalities of their cultural practices through inter-group communication and cooperation.

During this period, the inhabited landscape of the Marianas evolved with stronger relations between people and land-based resources. People never stopped using the coastal areas, but their experiences with coasts were changing, because the shorelines and nearshore ecosystems in themselves were transforming. Concurrent with the fluctuating coastal zones, the land-based components of the landscape system were being developed more strongly and interfacing more intimately with a growing human population.

Regional Context

While the Marianas landscape system was shifting towards greater emphasis on land-based resources and adjusting to significantly transformed nearshore ecosystems, parallel processes occurred in other inhabited parts of Remote Oceania. The same sea-level drawdown as in the Marianas affected the island communities making Lapita-style pottery in Southern Melanesia and West Polynesia. In these other islands around 500 B.C., people stopped producing the finely dentate-stamped Lapita pottery, lessened their reliance on nearshore ecological niches, and enhanced their usage of land-based resource zones (Nunn 2013).

The mirrored trends in the Mariana Islands and the Remote Oceanic Lapita region suggest that people responded to the effects of sea-level drawdown in much the same way and around the same time, as documented in several separate cases across the region (Nunn and Carson 2015). Despite no direct cultural connections between the Marianas and these other Remote Oceanic communities, the same outcomes are evidenced in environmental change, loss of finely decorated pottery, and shifting ecological resource emphasis from nearshore to land-based zones. In each of these separate island groups, the same trends proceeded in terms of people developing more intensive land-based traditions and closer ties with terrestrial settings.

Coastal life always would remain essential in small and remote islands, but at a certain point attachments with the land strengthened and contributed to a very different sense of inhabited landscape than had operated when people were living directly at selected types of shorelines. Hiro Kurashina (1991) described this transition in terms of settlement patterns in the Marianas Islands, as a shift away from a narrow spectrum of specific coastal niches and instead towards a broader spectrum of encompassing coastal and inland habitats, each with various sub-category niches. The transition gained visibility in the Marianas material record after 700 B.C., as compared to a date closer to 500 B.C. in the Remote Oceanic Lapita region. In each island community, these trends later would persist over time but proceed in variable ways.

While people strengthened their broad-spectrum land-use systems, no new provinces of Remote Oceania were settled by long-distance sea-crossing migrations, except possibly in Yap (see Fig. 1.1). The oldest known formal sites with definite artefacts in Yap date not much earlier than A.D. 100 (Intoh 1997; Intoh and Leach 1985), but the native vegetation may have been altered by a human-caused disturbance around 1000 B.C. (Dodson and Intoh 1999). Most archaeological work has focused on surface-visible village ruins in Yap, so the oldest buried land surfaces probably have not been discovered. The Yapese language hints at a different ancestry than all other Oceanic-speaking communities within the Micronesian branch of Oceanic language grouping (Ross 1996), so a slightly older settlement date and context may be expected.

A reduction in overseas colonising migrations coincided with a period of people spending more time in stable landward terrain. The Remote Oceanic Lapita colonising period had been effectively complete since 800 B.C., followed by adjustments to the changing coastal ecologies and development of enlarged broad-spectrum land-use strategies. This context implies increasingly closer relations between people and their island terrain, likely involving escalating population levels as well as growing independence of the numerous communities.

Beyond the islands where people already had settled by 800 B.C., very few additional islands still existed in the western Pacific that could have supported human settlement. The low-lying atolls of Central and Eastern Micronesia had not yet emerged above sea level, and they would not become stable and inhabitable until about A.D. 100 (Dickinson 2003). Other than Yap in western Micronesia, the only other possibly inhabitable islands of the Pacific were in the higher land masses such as Kosrae and Pohnpei, as well as much farther eastward in the regions of Central and Eastern Polynesia, where human settlement did not occur until some centuries later with significant new developments in sailing technology and navigational knowledge after A.D. 1000 (see Fig. 1.1).

Although the overall trend during the period 700 B.C.–A.D. 1 was towards more and more self-reliance and independence of separate groups, a number of long-distance sea-crossing contacts did still occur between established communities outside the Marianas. These contacts are evident in findings of volcanic glass and basalt tools in sites of Melanesia and Polynesia, traded between island groups of distinctive geological sources (Spriggs et al. 2011). These kinds of materials so far have not been found in the Mariana Islands, so the people of the Marianas did not engage in the same inter-island networks. Along with traded materials, presumably people exchanged various types of information, stories of recent events, and new ideas. Importantly, these findings demonstrate connections among established populations but not the movement of people into completely new territories.

Overseas trading contacts and communication networks potentially could enable groups in any one island to overcome periodic hardships, such as unusually fierce storms, crop failure, blankets of volcanic ash fall, or other calamities affecting specific locations. When these kinds of localised events occur, people can move elsewhere or rely on external contacts and partnerships to help in a recovery. This strategy was ineffective, however, when a crisis affected all groups cross-regionally, for example when people found dwindling supplies of their preferred resources in nearshore zones affected by sea-level drawdown.

Inter-island contacts became increasingly plausible after 700 B.C., when several island communities were well established, but very little long-distance contacts related to the Mariana Islands at this time. In the Marianas, external trade contacts are not at all evidenced with other parts of Remote Oceania, but some parts of the pottery sequence continue to be parallel with findings in the Philippines and Indonesia in Island Southeast Asia. Distinctive incised designs on non-slipped pottery are found in each of these areas around 500 B.C., in particular referring to rows of interfacing or interlocked half-circles (see Fig. 10.3).

The contacts between the Marianas and Island Southeast Asia were limited, possibly in contexts of just a few people moving between the distant regions or linked by down-the-line networks. These contacts notably did not involve imports of items that ordinarily might be expected as valued commodities in the Marianas. Examples of non-traded cultural items were domesticated pigs, dogs, and chickens that played key roles in developing land-based food-production systems in Island Southeast Asia and most parts of Remote Oceania, yet they were excluded from the Marianas (Wickler 2004). Other curious exclusions were stone or baked clay items of spindle whorls and bark cloth beaters for making clothing.

The few communications between the Marianas and Island Southeast Asia can be contrasted against a scene of lively trading partnerships across the South China Sea. After 500 B.C., several communities in Island Southeast Asia were linked with counterparts in Mainland Southeast Asia, trading iron, bronze, gold, and glass that never reached as far as the Marianas (Hung et al. 2013). Also across the South China Sea, a shared pottery style was evident in vessels with scroll-like designs, named Kalanay in the Philippines and Sa Huynh in Vietnam, now understood as the diagnostic marker of a Sa Huynh-Kalanay Interaction Sphere (Solheim 1957). These pervasive trading networks across the South China Sea clearly did not extend in the direction towards the Mariana Islands, where instead the only sea-crossing material link is seen in a different type of simple incised pottery.

Regardless of any overseas contacts and networks, communities in the Mariana Islands interacted with each other, maintaining friendships, kinships, and other relations in a functioning society. Increasing self-reliance is evident among the separate habitation sites after 700 B.C. in terms of accessing individual resource catchments, but a definite unity in material culture signifies that people must have shared most of the same cultural and social traditions. The types and styles of artefacts, composition of dietary foods, and housing structures were consistent from one site to another, and they followed the same chronological trends.

Despite the stress of increasing population size and changing landscape use after 700 B.C., the Mariana Islands retained large uninhabited areas and whole uninhabited islands. Although people shifted the focus of their habitations and resource usage patterns, they were not forced to live in the most marginal areas at this time. Growing numbers of people continued living only in the larger southern-arc islands, while the smaller northern-arc islands did not yet support permanent habitations. With only very slight if any impacts from human interference, these northern islands likely supplied birds, fruit bats, crabs, turtles, and other natural resources for periodic visitors. On the other hand, these islands lacked accessible sources of fresh water, and the active volcanoes at Anatahan, Pagan, and Pajaros may have caused further discouragement against permanent settlement. By comparison, the southern-arc islands still offered better opportunities.

References

Athens, J.S., & Ward, J.V. (2004). Holocene vegetation, savannah origins and human settlement of Guam. In V. Attenbrow & R, Fullager (Eds.), *A Pacific odyssey: Archaeology and anthropology in the Western Pacific: Papers in honour of Jim Specht* (pp. 15–30). Records of the Australian Museum, Supplement 29. Sydney: Australian Museum.

Athens, J.S., & Ward, J.V. (2006). *Holocene paleoenvironment of Saipan: Analysis of a core from Lake Susupe.* Micronesian Archaeological Survey Report Number 35. Saipan: Commonwealth of the Northern Mariana Islands Division of Historic Preservation.

Carson, M. T. (2011) Palaeohabitat of first settlement sites 1500–1000 B.C. in Guam, Mariana Islands, Western Pacific. *Journal of Archaeological Science, 38*, 2207–2221.

Carson, M. T. (2012) Evolution of an Austronesian landscape: the Ritidian Site in Guam. *Journal of Austronesian Studies*, 3, 55–86.

Carson, M. T. (2014a) Paleoterrain research: finding the first settlement sites of Remote Oceania. *Geoarchaeology, 29*, 268–275.

Carson, M. T. (2014b). *First settlement of Remote Oceania: Earliest sites in the Mariana Islands*. New York: Springer.

DeFant, D. G. (2008). Early human burials from the Naton Beach Site, Tumon Bay, Island of Guam, Mariana Islands. *Journal of Island and Coastal Archaeology, 3*, 149–153.

Dickinson, W. R. (2003). Impact of mid-holocene hydro-isostatic highstand in regional sea level on habitability of islands in Pacific Oceania. *Journal of Coastal Research, 19*, 489–502.

Dixon, B., & Schaefer, R. (2014). Archaeological investigation of caves and rock shelters in Guam and Tinian: a synthesis of their use through time. *Journal of Pacific Archeology, 5*, 52–74.

Dodson, J., & Intoh, M. (1999). Prehistory and palaeoecology of Yap, Federated States of Micronesia. *Quaternary International, 59*, 17–26.

Fitzpatrick, S. M., & Nelson, G. C. (2011). Purposeful commingling of adult and child cranial elements from the Chelechol ra Orrak Cemetery, Palau. *International Journal of Osteoarchaeology, 21*, 360–366.

Hung, H.-C., Nguyen, K. D., Bellwood, P., & Carson, M. T. (2013). Coastal connectivity: Long-term trading networks across the South China Sea. *Journal of Island and Coastal Archaeology, 8*, 384–404.

Hunter-Anderson, R. L., & Butler, B. M. (1995). *An overview of Northern Marianas prehistory*. (Micronesian Archaeological Survey Rep. No. 31). Saipan: Commonwealth of the Northern Mariana Islands Division of Historic Preservation.

Intoh, M. (1997). Human dispersals into Micronesia. *Anthropological Science, 105*, 15–28.

Intoh, M., & Leach, F. (1985). *Archaeological investigations in the Yap Islands, Micronesia: First Millennium BC to the present day*. British Archaeological Reports International Series, S277. Oxford: Archaeopress.

Kurashina, H. (1991). Prehistoric settlement patterns on Guam. *Journal of the Pacific Society, 51*, 141–158.

Moore, D. R., & Hunter-Anderson, R. L. (1999). Pots and pans in the intermediate pre-latte (2500–1600 bp) Mariana Islands, Micronesia. In J.-C. Galipaud & I. Lilley (Eds.), *Le Pacifique de 5000 à 2000 avant le Présent: Suppléments à l'Histoire d'une Colonisation* (pp. 487–503). Paris: Institut de Recherche pour la Développement.

Nunn, P. D. (2013). The end of the Pacific? Effects of sea level rise on Pacific Island livelihoods. *Singapore Journal of Tropical Geography, 34*, 143–171.

Nunn, P. D., & Carson, M. T. (2015). Sea-level fall implicated in profound societal change around 2570 cal yr BP (620 BC) in western Pacific Island groups. *Geo: Geography and Environment, 2*, 17–32.

Ross, M. D. (1996). Is Yapese Oceanic? In B. Nothofer (Ed.), *Reconstruction, classification, description: Festschrift in Honor of Isidore Dyen* (pp. 121–166). Hamburg: Abera.

Solheim, W. G., II. (1957). The Kalanay pottery complex in the Philippines. *Artibus Asiae, 20*, 279–288.

Spriggs, M., Reepmeyer, C., Angraenni, P. L., Neri, L., Ronquillo, W. P., Simanjuntak, T., et al. (2011). Obsidian sources and distribution systems in Island Southeast Asia: A review of previous research. *Journal of Archaeological Science, 38*, 2871–2881.

Takayama, J., & Intoh, M. (1976). *Archaeological excavation of Latte Site M-13, Rota, in the Marianas*. (Reports in Pacific Archaeological Survey, No. 4). Tokyo: Tokai University.

Valentin, F., Buckley, H. R., Herrscher, E., Kinaston, R., Bedford, S., Spriggs, M., et al. (2010). Lapita subsistence strategies and food consumption patterns in the community of Teouma (Efate, Vanuatu). *Journal of Archaeological Science, 37*, 1820–1829.

Ward, G., & Craib, J. L. (1983). *Mochong archaeological research. Document on file at Historic Preservation Office*. Saipan: Commonwealth of the Northern Mariana Islands.

Wickler, S. K. (2004). Modelling colonisation and migration in Micronesia from a zooarchaeological perspective. In M. Mondini, S. Munoz, & S. K. Wickler (Eds.), *Colonisation, migration and marginal areas: A zooarchaeological approach* (pp. 28–40). Oxford, England: Oxbow Books.

Chapter 11
A.D. 1–500, Temporary Stability

The Marianas landscape system developed with overall stability during the years A.D. 1–500, in both coastal and inland areas. Sea level continued lowering but at a very slow or even negligible rate, and it was effectively stable for at least a short time while coral reefs and other natural habitats strengthened in coastal areas. Much of the same circumstances that improved coastal living in the Marianas made the small and low-lying atolls of Central and Eastern Micronesia habitable for the first time. In the Mariana Islands, though, people maintained a pre-existing pattern of mixed coastal and land-based subsistence and habitation, while they were quick to make use of opportunities in stabilised beaches and productive coastal resource zones that had been unsteady in the preceding centuries.

Site Inventory and Dating

Archaeological sites of this period for the most part expanded from pre-existing habitations, but a few others were established for the first time in nearby areas (Fig. 11.1). The prior pattern of cliff-base habitations continued around the edges of elevated limestone plateau in the southern-arc islands, for example at Ritidian in Guam, but it also appeared in additional similar settings that previously lacked habitation sites. On the sandy shores of Tumon Bay in Guam, temporarily stable beach surfaces did not necessarily allow new residential occupations, but they did enable expanded peripheral support zones of campsites and workshops. Inland habitations and temporary campsites grew in their number and intensity of use at this time, mostly in patches of slightly higher ground portions of limestone plateau and ridgetops of volcanic hills.

M.T. Carson, *Archaeological Landscape Evolution*,
DOI 10.1007/978-3-319-31400-6_11

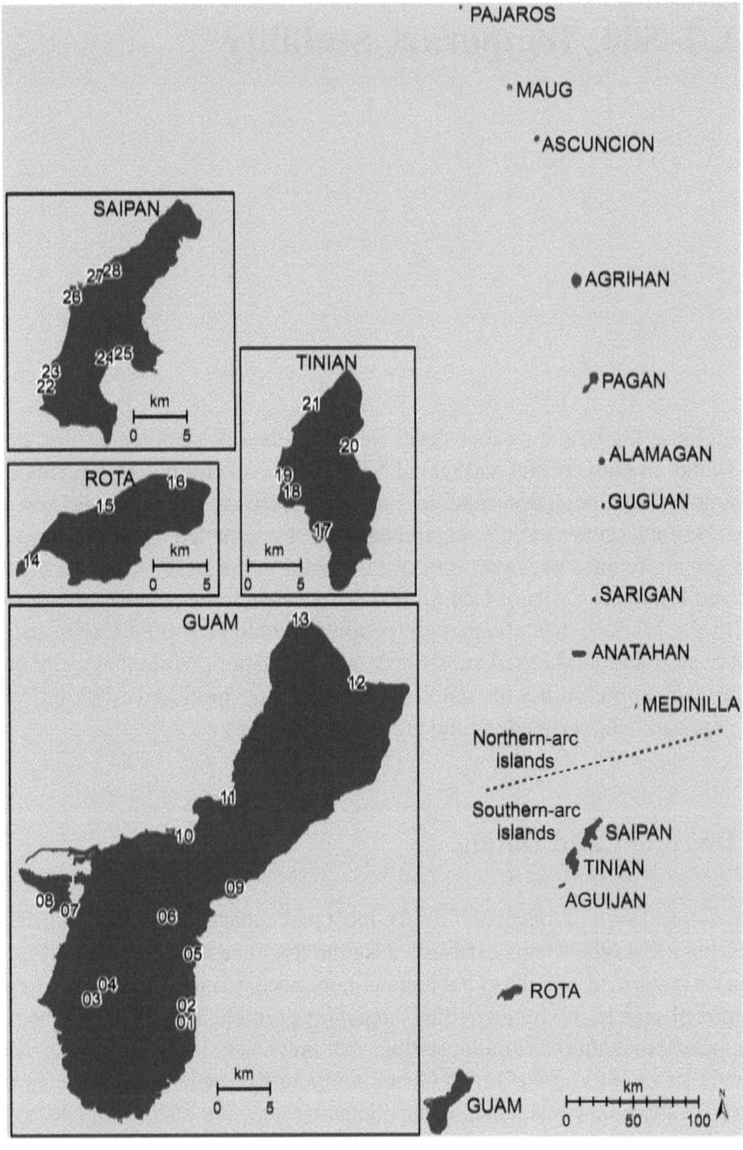

Fig. 11.1 Major site locations, A.D. 1–500. 01: Ipan. 02: Talofofo. 03: Almagosa. 04: Fena. 05: Ylig. 06: Manenggon. 07: Apra. 08: Tipalao. 09: Mangilao. 10: Hagatna. 11: Tumon. 12: Tarague. 13: Ritidian. 14: Songsong. 15: Teteto-Guata. 16: Mochong. 17: House of Taga. 18: Kahet. 19: Atgidon. 20: Dangkulo. 21: Unai Chulu. 22: Chalan Piao. 23: Chalan Kanoa. 24: Laulau Rockshelter. 25: Unai Bapot. 26: Garapan. 27: Achugao. 28: San Roque

In the kinds of sites known from this period in stable landform settings, charcoal generally has been well preserved and has constituted the preferred material for radiocarbon dating. The cultural layers, however, often have been disturbed at least partially by later cultural use of the same sites, thus creating some ambiguities when assigning specific date ranges to collections of artefacts and midden within the inter-mixed deposits. Rare exceptions for dating this period are the temporarily stable beach surfaces, as in Tumon, where people intensively used these land surfaces just for a few centuries or less, resulting in stratigraphic isolation of the deposits as short-lived records of the first few centuries A.D.

Beginning about A.D. 1, a considerably slowed rate of declining sea level created overall stable conditions for development of coastal resource zones keeping pace with the changing sea level. These conditions are most strongly evidenced in a tight range of approximately A.D. 100–200. Before and after this brief window in time, coastal ecosystems sustained overall stability during very little regional change in sea level, but frequent storm-surge episodes created a different kind of instability especially in the most vulnerable low-elevation coastal landforms.

Following A.D. 500, people began installing formal residential sites in the broad coastal landforms at a larger scale than had been the case during the preceding few centuries. This transition serves as a convenient bracket for defining the period A.D. 1–500. An exact timing of this transition likely varied from one place to another, but overall it can be estimated about A.D. 500.

Landforms

The noted temporary stability of coastal zones allowed new cultural use patterns of the prograding sandy beaches. These enlarged beaches were composed primarily of storm-surge sands, overlaying deeper and much older beds of ancient lagoon facies. The storm-surge layers are characterised by large-scale build-up of pulverised clasts of sands, corals, and shells mixed with broken tips of branch corals. These deposits apparently accumulated during successive episodes of storm-surge overwash, effectively blending into massive sand units.

Within the profiles of massive storm-surge beach sand units, occasionally a number of thin lenses are visible as relicts of the surfaces of singular storm-surge episodes. In some instances, a dark organic staining indicates an incipient organic horizon of 1–2 cm thickness where a beach surface once supported a temporarily stable habitat that ceased when it was covered suddenly by storm-surge debris. In other cases, bioclasts of *Halimeda* sp. algal growth from a shallow-water lagoon were thrown onto the beach during high-tide or storm-surge events, found at the former highest tide mark and usually in small patches or lenses less than 1 cm in thickness.

A clear example of a preserved incipient organic horizon was found inside the profile of storm-surge build-up along the shore of Tumon Bay in Guam (Fig. 11.2; see also Fig. 3.11). This example is especially informative for preserving the shape

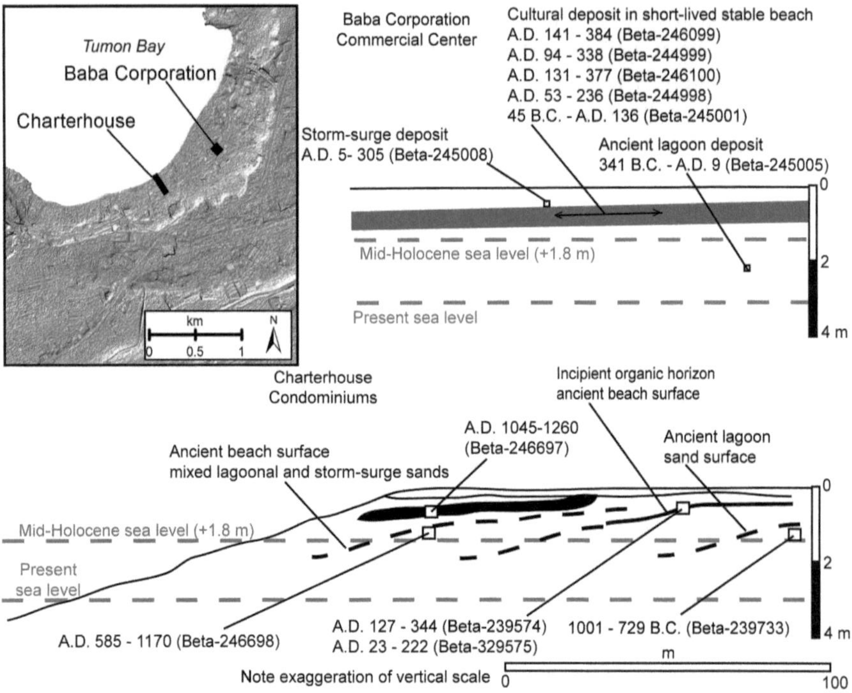

Fig. 11.2 Palaeo-beach in Tumon, Guam. Example from the Baba Corporation project is based on data from Carson and Peterson (2010). Example from Charterhouse project is based on data from Carson and Peterson (2009). Both examples are updated with information from Carson (2011, 2014)

of the slightly sloping ancient land surface, dipping slightly downward towards the former shoreline. Two radiocarbon dates of charcoal particles in this thin layer overlapped in the range of A.D. 127–222 (Carson and Peterson 2009). In other portions of the fully extended beach profile, additional ancient land surfaces of different ages were identified by sudden shifts in beach sand compositions of intact lagoon facies, pulverised storm-surge debris, and other indicators.

At approximately the same absolute elevation of this old beach surface in Tumon, intensive cultural activity was preserved about 700–800 m farther northeast along the edge of the embayment (see Fig. 11.2). At this location, a short-lived temporary land surface of approximately A.D. 100–200 was captured between an underlying bed of lagoon facies material and an overlaying deposit of storm-surge debris (Carson and Peterson 2010). The lagoon habitat existed here at some point during the range of 341 B.C.–A.D. 9. The cultural use of hearths was securely dated by charcoal directly from these features, including two dates from one hearth overlapping at A.D. 141–338, another two from a separate hearth overlapping at A.D. 53–136, and finally one other hearth charcoal sample at A.D. 131–377. The storm-surge overlaying the cultural deposit was dated at some point in the range of A.D. 5–305, most probably near the middle to end of this range in order to accommodate the multiple dates of the hearths that must have pre-dated the storm-surge event.

While the coastal zones were gaining a new stability, no major change occurred in island interior landforms, other than continued slope erosion with ongoing forest clearance and inland habitation. Eroded sediments were re-deposited at lower elevations, especially noticeable at the bases of volcanic hillslopes in southern Guam and in the valley floors of large basins such as Hagatna and Pago in central Guam. The re-deposited sedimentary build-up at this time did not yet support the large habitable flat lands at the coastal-fronting mouths of Hagatna and Pago, where instead marine or brackish swampy habitats persisted until approximately A.D. 1000.

Resource Zones

The new coastal stability allowed coral reefs to grow and expand more than had been possible during the earlier conditions of more rapid sea-level drawdown. These developing reef and lagoon ecosystems around A.D. 1–500 formed the basis of the near-shore ecosystems seen today around the Mariana Islands. During these centuries, *Anadara* sp. shellfish were harvested in extremely small numbers or often not at all, as their natural habitats of swampy conditions no longer existed except in a few rare spots. Instead, gastropods such as *Strombus gibberulus* inhabited the freshly formed clean beach sand substrates. Meanwhile, various shellfish and swimming fish were attracted to the new healthy coral reefs.

The lowered sea level by this time made some of the freshwater sources inaccessible, for example in pools at the bottoms of certain caves, although other sources still were available in much lower elevation cave floors. At Ritidian in northern Guam, a number of cliff-base habitations continued outside caves where these water sources must have been accessible when sea level was 1.8 m higher than today's level, but the lowered sea level by A.D. 1–500 would have caused the floating freshwater lens to drop beneath these particular cave floors. Curiously, the sediments inside these same caves began to accumulate charcoal, ash, and discarded shells during the first few centuries A.D., when the freshwater sources no longer existed to be polluted or protected. Other deeper caves meanwhile retained freshwater pools in places like Sasajayan and Pagat in eastern Guam, and dense habitation sites later would develop around these locations especially after A.D. 1000.

Other than in basal cave pools, fresh water could be accessed in natural seeps at low elevations near the coast. In this regard, the thicker and broader sandy beaches after A.D. 1 allowed more opportunities for people to find these seeps in clean lenses that discharged just above the brackish or salty water. These conditions would develop over the next several centuries with continued coastal progradation.

The enlarged coastal landforms of the first few centuries A.D. occurred region-wide, and fringes of coastal plains began to emerge around the small northern-arc islands of the Marianas. This event can be inferred from the beach profile chronologies as presented for Guam. It further is inferred by the emergence of the small atolls of Central and Eastern Micronesia above sea level all around the same time about A.D.

1–200 (Dickinson 2003). So far no habitation sites have been found in the northern-arc islands prior to A.D. 1000, but the availability of new resource zones must have been noticed by periodic visitors.

Material Culture

So far the best documentation of material culture from the period A.D. 1–500 has been at the Baba Corporation Commercial Site in the central portion of Tumon in Guam, previously noted for its preservation of a short-lived beach surface dated about A.D. 100–200 (see Fig. 11.2). Excavation uncovered 60 m² of the ancient site surface (Fig. 11.3), at its time the largest of any hand-excavated single exposure in the Mariana Islands, although now it is ranked second behind the 90 m² excavated at House of Taga in Tinian. Additionally informative excavations of this period were at the cliff-base habitations at Ritidian, as well as a number of caves and rockshelters.

The material culture of A.D. 1–500 is characterised primarily by its pottery forms. By A.D. 1, traditions of red-slipped surfaces and lime-infilled decorations had ceased. The two most popular pottery vessel shapes were flat-bottomed large pans and narrow jars or flasks (Fig. 11.4). The flat-bottomed pans first appeared as early as 500 B.C., but they became increasingly popular after 200 B.C. and most common in the range of A.D. 1–500. Small cups were produced in almost all known time periods in the Marianas, but those of the period A.D. 1–500 were straight-sided (flasks) or occasionally with slightly restricted tops (jars).

The flat-bottomed pans were made of coarse clay and using large-grained beach sand temper, resulting in friable clay fabrics. When the wet clays were drying before firing, they apparently were rested on woven mats, thus creating impressions of plaited

Fig. 11.3 Excavation of an ancient living surface with hearth features by the author at the Baba Corporation project in Tumon. Low-altitude kite-suspended digital imagery performed through remote-controlled system Ty Tarantino and Asa Peterson

Fig. 11.4 Pottery types, A.D. 1–500

strips of palm fronds on their bases. The pans ranged in size from 20 to 40 cm diameter, mostly circular but occasionally oblong in shape, while the short side walls of 3–5 cm height contained rather shallow interior space. These interior surfaces often retained blackened residues, containing phytoliths of banana leaf and assorted starchy foods (Horrocks et al. 2015).

Fig. 11.5 Shell and stone artefacts, A.D. 1–500

The small jars or flasks were made of finer clay paste and using small- to medium-grained beach-sand temper, resulting in hard clay fabrics noticeably different from those of the flat-bottomed pans. These objects resembled single-serving cups, manufactured with remarkable consistency all about 6–8 cm diameter and 8–10 cm tall. Very rarely were any of these items decorated with simple incisions.

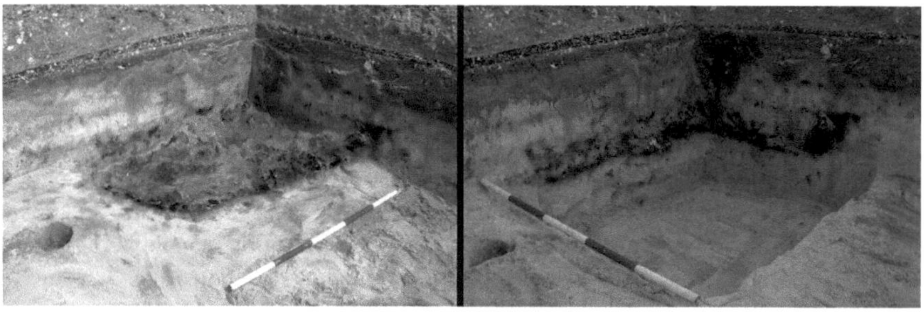

Fig. 11.6 Example of a hearth feature before (*left*) and after (*right*) excavation at the Baba Corporation project, Tumon. Scale bar is in 20-cm increments

Stone and shell artefacts do not appear especially distinctive of this time period. The collections from the largest site excavation of this period at the Baba Corporation Site in Tumon included a few broken shell fishing hooks, worked shell by-products, two abrading stones, a worked piece of crystallised limestone, two *Tridacna* sp. shell adze blades, and a few bits of *Tridacna* sp. shell flakes (Fig. 11.5). Other sites of this age so far have yielded smaller numbers of the same inventory of materials, plus pieces of flaked volcanic stone and chert.

The Baba Corporation Site revealed 28 hearths within the excavated area of 60 m^2. The hearths were oval in plan view, mostly 1–2 m long and 30–50 cm deep (Fig. 11.6). They contained burned heating stones, charcoal remains, variable amounts of discarded shells and fish bones, and broken pottery. Other artefacts were concentrated almost entirely within 1 m around the hearth perimeters, clearly the focal points of cultural activities.

Although it certainly was an intensively utilised surface at A.D. 100–200, the Baba Corporation Site did not include a residential habitation in precisely the location of the archaeological excavation. No post moulds or other housing structural remains were found in this lower cultural layer, although several intrusive post moulds had originated in an upper and later-aged layer. In this case, the intensive hearth use on the beach appears to have been an outlying activity in support of a residential zone probably slightly farther landward or perhaps in another more stabilised portion of the coastal terrain at Tumon.

In the known residential deposits of this time period, such as at the cliff-base habitations at Ritidian in northern Guam, the findings include dense charcoal and ash, enclosing variable amounts of artefacts and midden, with a few small post moulds. The post moulds generally are 15–20 cm diameter, but so far no definite housing structure patterns have been discerned in the small test pit excavations. These habitations apparently continued from pre-existing use of the same scattered locations at Ritidian, now becoming slightly larger in each instance and presumably accommodating greater numbers of people.

Regional Context

While the Marianas landscape system continued without any formalised agricultural fields, different field complexes were developing in other Pacific Islands groups. By the first centuries A.D., mountain slopes in Samoa were modified by slope-retention features and earthen terraces (Carson 2006). The natural landforms of New Caledonia were transformed by impressively large taro-growing irrigation complexes and yam-planting mounds (Sand 2002). In the hilltops of Palau, people constructed extensive earthwork settlements at this time or perhaps slightly earlier (Liston 2009).

During this same time frame, some of the Pacific Islands began to show signs of large-scale slope erosion, caused by inland forest clearing for agricultural land use that had begun earlier at a smaller scale. The erosion-deposition patterns now creating great volumes of clays and silts re-deposited in lowland and coastal zones around the larger islands, such as seen in New Caledonia and Vanuatu (Spriggs 1997). These outcomes did not occur at this time in the Mariana Islands, perhaps in part due to the limited impacts of managing tree crops and root-tuber crops in these islands as compared to the evident terrain-altering efforts of formalised field systems in other places.

During the first few centuries A.D., stable coastlines supported more intensive and widespread cultural use of coastal landforms, not only in the Marianas but also throughout Micronesia. The low-lying small atolls of Central and Eastern Micronesia became stable above sea level and notably with an accessible freshwater lens at this time (Dickinson 2003), and indeed people began living in these areas in the Marshall Islands (Shun and Athens 1990; Weisler 1999) and Kiribati by A.D. 200 (Thomas 2009). Meanwhile, residential sites appeared for the first time in the larger island masses such as Pohnpei (Athens 1990a; Ayres 1990; Galipaud 2001) and Kosrae (Athens 1990b), as well as in Yap and Chuuk by this time if not earlier (Intoh 1997). The ability of people to inhabit the broad geographic distribution of Micronesian atolls likely encouraged settlements in the larger islands such as Pohnpei and Kosrae that previously were habitable above sea level but remotely isolated in Eastern Micronesia. Additionally, the lowered sea level would have exposed coastal plains in these larger islands, and the surrounding coastal ecosystems would have been more stable and reliable for human populations as had occurred in the Mariana Islands at this time.

Human settlement in Central and Eastern Micronesia was made possible by the lowered sea level that exposed habitable land surfaces, but at the same time people could excavate into the sandy sediments and make use of the floating freshwater lens. In the atolls of the Marshall Islands, excavated pits for growing swamp taro have been dated around A.D. 1–200 (Weisler 1999), dug into layers of sands that had been stranded above the lowering sea level. These sands contained foraminifera from marine organisms, dated just prior to the ages of the excavated pits and cultural habitation layers (Yamaguchi et al. 2005; Weisler et al. 2012). The time gap therefore was minimal between the physical emergence of the atolls above sea level and the cultural settlement of these areas.

With the sudden surge of human settlement throughout Micronesia, a "sea of islands" may have become a new reality. In Hau'ofa's (1994) original formulation of the "sea of islands" concept, the ocean connected rather than isolated the numerous island groups from one another. Moreover, the ocean was perceived as a fully inhabited space. Certainly, the Mariana Islanders and all other islanders always have been aware of the world beyond their shores. The major difference beginning about A.D. 1–500, however, was that the world itself was changing significantly in its physical form and now most especially in the ability of people to inhabit it.

Human settlement across much of Central and Eastern Micronesia signalled one of the major episodes of human migration in Pacific Oceania (see Fig. 1.1). A suddenly larger expanse of the region was inhabited within a brief time span by tightly related Oceanic-speaking groups who must have come from source areas in Melanesia or West Polynesia where the ancestral Oceanic languages were spoken (Bender et al. 2003a, b). This event or series of events was very much like a repeat of the Lapita-associated settlement of Southern Melanesia and West Polynesia some centuries previously, except that now people probably were much better prepared for the long-distance sailing and conditions of remote-island living. The resulting human occupation of Central and Eastern Micronesia therefore represents one of the major stages in a more or less continuous human exploration and settlement of the Pacific.

While widespread Micronesian settlement was re-defining the inhabited space of Pacific Oceania, cross-regional commerce had intensified farther to the west in Island Southeast Asia, but these networks did not extend as far as the Marianas or any other Pacific Islands. One such network already had operated since at least 500 B.C., connecting communities across the South China Sea, evidenced in the exchange of specialised ear rings, pendants, and other products among groups in Taiwan, the Philippines, and Vietnam (Hung et al. 2013). By A.D. 100–200, another such large network involved traders from India, engaged with groups in Indonesia (Ardika and Bellwood 1991). Iron and bronze objects, glass beads, specialised forms of personal ornaments, and unique pottery designs all bear witness to these trade networks, but no such materials entered the Marianas or other parts of Remote Oceania at this time.

References

Ardika, I. W., & Bellwood, P. (1991). Sembiran: The beginning of Indian contact with Bali. *Antiquity, 65*, 221–232.

Athens, J. S. (1990a). Non Madol pottery, Pohnpei. In R. L. Hunter-Anderson (Ed.), *Recent advances in Micronesian archaeology*, Micronesica Supplement 2 (pp. 17–32). Mangilao: University of Guam Press.

Athens, J. S. (1990b). Kosrae pottery, clay and early settlement. In R. L. Hunter-Anderson (Ed.), *Recent advances in Micronesian archaeology*, Micronesica Supplement 2 (pp. 171–186). Mangilao: University of Guam Press.

Ayres, W. S. (1990). Pohnpei's position in Eastern Micronesian prehistory. In R. L. Hunter-Anderson (Ed.), *Recent advances in Micronesian archaeology*, Micronesica Supplement 2 (pp. 187–212). Mangilao: University of Guam Press.

Bender, B. W., Goodenough, W. H., Jackson, F. H., Marck, J., Rehg, K. L., Sohn, H.-M., et al. (2003a). Proto-micronesian reconstructions—1. *Oceanic Linguistics, 42*, 1–110.

Bender, B. W., Goodenough, W. H., Jackson, F. H., Marck, J., Rehg, K. L., Sohn, H.-M., et al. (2003b). Proto-micronesian reconstructions—2. *Oceanic Linguistics, 42*, 271–358.
Carson, M. T. (2006). Samoan cultivation practices in archaeological perspective. *People and Culture in Oceania, 22*, 1–29.
Carson, M. T. (2011). Palaeohabitat of first settlement sites 1500-1000 B.C. in Guam, Mariana Islands, Western Pacific. *Journal of Archaeological Science, 38*, 2207–2221.
Carson, M. T. (2014). Paleoterrain research: Finding the first settlement sites of Remote Oceania. *Geoarchaeology, 29*, 268–275.
Carson, M. T., & Peterson, J. A. (2009). *Archaeological research at the new Charterhouse Condominiums Project, location of former Tropicana Hotel in Tumon, Tamuning Municipality, Guam* (Report prepared for Architects Laguana and Cristobal). Mangilao: Archaeology Office, Micronesian Area Research Center, University of Guam.
Carson, M. T., & Peterson, J. A. (2010). *Archaeological research at the Baba Corporation Commercial Center in Lot 5093, Tumon, Tamuning Municipality, Guam* (Report prepared for Architects Laguana and Cristobal). Mangilao: Archaeology Office, Micronesian Area Research Center, University of Guam.
Dickinson, W. R. (2003). Impact of mid-holocene hydro-isostatic highstand in regional sea level on habitability of islands in Pacific Oceania. *Journal of Coastal Research, 19*, 489–502.
Galipaud, J.-C. (2001). Le peuplement initial de Pohnpei. *Journal de la Société des Océanistes, 112*, 49–60.
Hau'ofa, E. (1994). Our sea of islands. *Contemporary Pacific, 6*, 147–161.
Horrocks, M., Peterson, J. A., & Carson, M. T. (2015). Pollen, starch and biosilicate analysis of archaeological deposits on Guam and Saipan, Mariana Islands, northwest Pacific: Evidence for Chamorro subsistence crops and marine resources. *Journal of Island and Coastal Archaeology, 10*, 97–110.
Hung, H.-C., Nguyen, K. D., Bellwood, P., & Carson, M. T. (2013). Coastal connectivity: Long-term trading networks across the South China Sea. *Journal of Island and Coastal Archaeology, 8*, 384–404.
Intoh, M. (1997). Human dispersals into Micronesia. *Anthropological Science, 105*, 15–28.
Liston, J. (2009). Cultural chronology of earthworks in Palau, Western Micronesia. *Archaeology in Oceania, 44*, 56–73.
Sand, C. (2002). Creations and transformations of prehistoric landscapes in New Caledonia, the southernmost Melanesian islands. In T. N. Ladefoged & M. W. Graves (Eds.), *Pacific landscapes: Archaeological approaches* (pp. 11–34). Los Osos, CA: Bearsville Press.
Shun, K., & Athens, J. S. (1990). Archaeological investigations at Kwajalein Atoll, Marshall Islands, Micronesia. In R. L. Hunter-Anderson (Ed.), *Recent advances in Micronesian archaeology*, Micronesica Supplement 2 (pp. 231–240). Mangilao: University of Guam Press.
Spriggs, M. (1997). Landscape catastrophe and landscape enhancement: Are either or both true in the Pacific? In P. V. Kirch & T. L. Hunt (Eds.), *Historical ecology in the Pacific Islands: Prehistoric environmental and landscape change* (pp. 80–104). New Haven, CT: Yale University Press.
Thomas, F. R. (2009). Historical ecology in Kiribati: Linking past with present. *Pacific Science, 63*, 567–600.
Weisler, M. I. (1999). The antiquity of aroid pit agriculture and significance of buried A horizons on Pacific atolls. *Geoarchaeology, 14*, 621–654.
Weisler, M. I., Yamano, H., & Hua, Q. (2012). A multidisciplinary approach for dating human colonization of Pacific atolls. *Journal of Island and Coastal Archaeology, 7*, 102–125.
Yamaguchi, T., Kayanne, H., Yamano, H., Najima, Y., Chikamori, M., & Yokoki, H. (2005). Excavation of pit-agriculture landscape on Majuro Atoll, Marshall Islands, and its implications. *Global Environmental Research, 9*, 27–36.

Chapter 12
A.D. 500–1000, Sustained Use of Coastal and Inland Zones

The period A.D. 500–1000 is the least known archaeologically of any part of the Marianas chronological sequence, but it is by no means a "dark age" in the archaeological record. This period was characterised by increasing numbers of settlements in coastal and inland areas, where cultural deposits contained mostly plain and simple pottery made for technical functionality. A rising population base may be inferred, magnifying in density in the southern-arc islands without yet expanding at a large scale into the northern-arc islands.

Site Inventory and Dating

Archaeological findings of this period so far are known entirely from subsurface cultural deposits, most often found beneath surface-visible sites of the megalithic *latte*-building period but also inside caves and rockshelters (Fig. 12.1). The cultural deposits almost always have been disturbed to some degree by super-imposed residential occupations, human burial features, crop growth activities, and cave use post-dating A.D. 1000. In a few rare cases of rapid sedimentary build-up, the later occupation layers did not intrude so much into the lower pre-existing cultural deposits of A.D. 500–1000, for example as seen in the gradual material transitions preserved at the Mochong Site in Rota (Takayama and Intoh 1976; Ward and Craib 1983). In most cases, though, the surviving intact remnants are only a few cm thick beneath disturbed or truncated horizons, as in the middle–upper layers at the Baba Corporation Site in Tumon of Guam (Carson and Peterson 2010).

Given the typically disturbed or truncated contents of the cultural layers, their time range is indicated primarily by pottery types, attached with only very limited radiocarbon dating, but it is most easily defined in contrast to the immediately earlier and later periods. The earlier traditions of large flat-bottomed pans apparently ended by A.D. 500, and the earlier preference for using beach sand temper began to diminish in pottery-making. The later traditions of thickened-rim pottery gained

M.T. Carson, *Archaeological Landscape Evolution*,
DOI 10.1007/978-3-319-31400-6_12

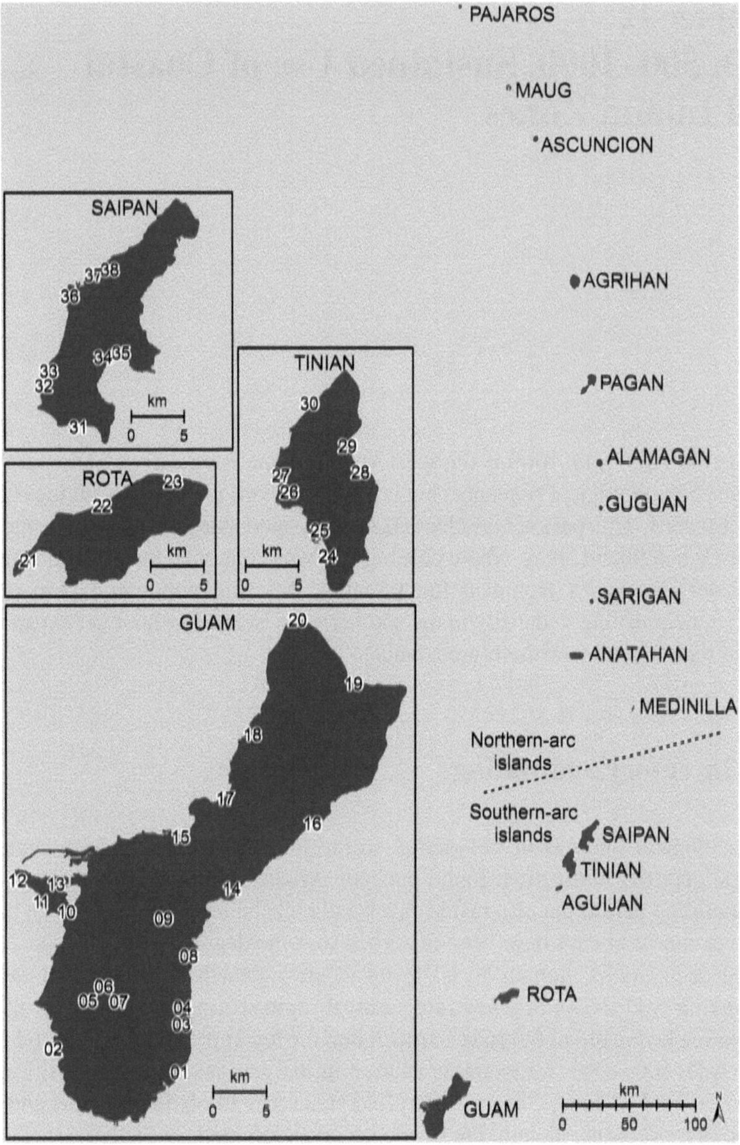

Fig. 12.1 Major site locations, A.D. 500–1000. 01: Nomna. 02: Fouha. 03: Ipan. 04: Talofofo. 05: Almagosa. 06: Fena. 07: Lost River. 08: Ylig. 09: Manenggon. 10: Apra. 11: Tipalao. 12: Orote. 13: Sumay. 14: Mangilao. 15: Hagatna. 16: Pagat. 17: Tumon. 18: Finegayan. 19: Tarague. 20: Ritidian. 21: Songsong. 22: Teteto-Guata. 23: Mochong. 24: Tachungnya. 25: House of Taga. 26: Kahet. 27: Atgidon. 28: Masalok. 29: Dangkulo. 30: Unai Chulu. 31: Obyan. 32: Chalan Piao. 33: Chalan Kanoa. 34: Laulau Rockshelter. 35: Unai Bapot. 36: Garapan. 37: Achugao. 38: San Roque

their greatest popularity after A.D. 1000, associated with megalithic *latte* ruins and several other obvious material indicators of this later period that will be discussed in Chap. 13.

The period A.D. 500–1000 may be characterised as a time of increasing population density along with more intensity of human–environment relations in the Marianas landscape. By A.D. 500, coastal plains supported larger scale residential housing than had occurred in the preceding centuries, while inland residential sites continued to grow as well. This mode of land use eventually would reach a critical threshold and shift into a new system of formalised village layouts in the same pre-established sites plus numerous others throughout the Mariana Islands after A.D. 1000. The overall pattern of A.D. 500–1000 indicates that more and more people lived in much the same habitat configuration as had been the case for some centuries already in the southern-arc islands. At least a few habitations may be expected of this age in the northern-arc islands, although none yet have been verified.

Landforms

After A.D. 500, coastlines approached the shape of today's conditions, but beach progradation continued with a slow pace of sea-level drawdown and periodic storm-surge deposits. People began to make more sustained use of these new coastal areas as habitation zones, unlike the preceding periods of only short-lived beach use for non-residential activities while housing zones had been situated farther landward. The pattern after A.D. 500 brought more residential housing sites closer to the shores in safely stable coastal landforms.

While coastal sites became larger in their available stabilised landforms, the inland terrain sustained the pre-existing patterns of forest clearance and habitation activities. The inland habitation sites were constrained to areas of level or roughly level land, such as the high-ground patches of limestone plateau terrain in each of the larger southern-arc islands, as well as a few ridgetops and tablelands in the volcanic hills of southern Guam. These sites mostly were the same as in prior centuries, but more were added by this time. The surrounding terrain likely supported various forms of horticulture, arboriculture, and forest management.

Prior to A.D. 1000, the Marianas landscape system in general emphasised coastal plains and limestone plateau habitats, while cultural use of volcanic hilly terrain still was in the minority overall. Although their lifestyle may not have gained much popularity at this time, people in southern Guam certainly were adept at living on hill ridges and tablelands while tending to crops and forests on the adjacent slopes and along stream drainages. The use of hilly terrain would increase at a large scale after A.D. 1000, not only in southern Guam but also in the smaller northern-arc islands, and some of this activity likely started at a small scale not yet clearly documented during the centuries of A.D. 500–1000.

Resource Zones

By A.D. 500, coastal habitats had been stabilised long enough for their biomass to increase and diversify, and this trend would continue for another several centuries. Coastal ecosystems supported healthy coral reefs and lagoons in several places around the islands, although some of course were larger and supported more biomass than others. Most of these were immediately adjacent to broad sandy beaches and coastal plains. Others bordered coastal benches of natural limestone terraces close to sea level.

Swamps, mangroves, and saltwater wetlands by now had become rarities in the Marianas, found in very few locations such as the Hagatna Basin in Guam. Saltwater wetland habitat persisted in Hagatna until about A.D. 1000 (Carson 2011), but the wetland in Susupe of Saipan transitioned from saltwater to freshwater conditions around A.D. 500 (Athens and Ward 2006). Without these supporting natural habitats, shellfish like *Anadara* sp. no longer could feed large populations, but instead people were harvesting the growing supplies of other shellfish living in the new coral reefs and sandy-bottom clear lagoons.

Plant growth by now was strong over the enlarged coastal plains, but here people had been managing the types of plants that grew throughout the geological lifetime of these landforms. By comparison, the forests in much older island interior terrain retained several native species despite the consecutive centuries of cultural mediation. Overall, the plant communities were becoming increasingly shaped by human interventions throughout the long-inhabited southern-arc islands, but such likely was not yet the case in the northern-arc islands that did not support large-scale residential sites prior to A.D. 1000.

Other than rainwater collection, sources of fresh water by this time were found mostly in natural seeps in low-lying coastal terrain, as well as in stream drainages of southern Guam and southeast Saipan. Accessible freshwater pools by this time were scarce inside caves, except for a few large and deep caverns. Additional groundwater sources could be found in places like Susupe in Saipan, where saltwater marshes were transitioning into freshwater wetlands.

Material Culture

Material culture of this period is known from only small samples, but the most diagnostic material so far includes earthenware pottery. Simple bowls and jars were made in variable sizes (Fig. 12.2). Rim profiles mostly were straight-sided as seen in the immediately preceding centuries, while a few vessels were made with slightly incurving rims that would gain considerably more popularity in later centuries. Decorations were extremely uncommon, except for occasional examples of roughly brushed exteriors and rare instances of simple incisions or impressions on and near rims.

Fig. 12.2 Pottery types, A.D. 500–1000

Pottery temper inclusions by this time showed a declining use of calcareous beach sands, while volcanic tempers and quartz grains were becoming more popular. Increasing use of volcanic and quartz tempers may have related to natural inclusions in clays obtained from volcanic hills and stream drainages, but from a technological standpoint these materials would allow higher firing temperature than was possible with only calcareous beach sand tempers. After A.D. 500, most potsherds were made with mixed calcareous and other temper types, but a few were made exclusively with volcanic or quartz sands. By A.D. 1000, most potsherds were made with coarse volcanic tempers and occasionally some quartz grains, but coarse calcareous materials were used in a few coastal settlements.

The vessel forms and material compositions both suggest an emphasis on producing technically serviceable pottery without much investment in artistic output or labour-intensive manufacture. The major change after A.D. 500 was an increasing use of volcanic and quartz tempers that could allow higher firing temperature and thus more durable products of larger size. A trend towards lessening artistic decoration had been progressing for several centuries, and by now pottery was almost entirely undecorated. By A.D. 500, pottery may have been valued more for fulfilling technical functioning of storage, cooking, and serving rather than for artistic expressions.

The Mochong Site in Rota offers an instructive example of a cultural deposit pre-dating A.D. 1000 directly beneath a surface-visible *latte*-associated layer (Takayama and Intoh 1976; Ward and Craib 1983). The carved limestone pieces of *latte* house posts were based in a cultural deposit containing diagnostic thickened-rim pottery and post-dating A.D. 1000, directly overlaying sediments bearing straight-rim or narrowed-rim pottery. The oldest cultural layer here began to form as early as 500 B.C. (Takayama and Intoh 1976). The portion of the sedimentary deposit inferred to date to A.D. 500–1000 contained broken pieces of simple earthenware bowls, made with mixed tempers of beach sands and volcanic sands with rare quartz inclusions.

In the full stratigraphic sequence at Mochong, a gradual transition was noted in the clay paste and tempers inclusions in the pottery. The oldest layer about 500 B.C. contained friable pottery made with calcareous beach sand and quartz sand tempers, and the surface-associated layer after A.D. 1000 contained more durable pottery made with volcanic sand tempers (Leach et al. 2008). The pottery inferred to come from the period of A.D. 500–1000 exhibits a mid-range transition of declining use of beach sands, almost no more quartz sands at this point, and rising popularity of volcanic sands. Most of these potsherds contained mixed recipes of beach sands and volcanic sands within the clay paste.

Unlike the gradual cultural transition preserved at the Mochong Site, most other cultural layers of this period have been disturbed and truncated by later occupation layers, as seen at the Baba Corporation Site in Tumon (see Fig. 11.2). Formal residential use occurred here after a storm-surge event had covered a lower cultural layer containing numerous hearths but no definite habitation structural remains, as described in Chap. 10. The later residential occupation zone was used over an extended period of roughly A.D. 500–1000, but the surviving cultural layer had been severely truncated with less than 20 cm remaining and containing only very few artefacts or midden.

The residential occupation layer at the Baba Corporation Site contained 19 post moulds within the 60 m^2 excavation area (Fig. 12.3). The post moulds consistently were 15–20 cm in diameter, except for two instances with larger diameters of 30–40 cm. A few were lined with bracing stones of small cobbles. Three of the post moulds retained large chunks of charcoal, presumably from the burned remnants of the original wooden posts, and they produced radiocarbon dates of A.D. 351–546, A.D. 565–669, and A.D. 1042–1255. The other post moulds retained dark-stained sediments but no large pieces of charred wood.

Although somewhat lacking in artefact collections, the known cultural layers of this period have been found in greater numbers than those of the prior several centuries. In the larger southern-arc islands of the Marianas, population sizes were increasing, and communities were becoming densely crowded within the inherently limited inhabitable space of these islands. Precise numbers of people cannot be estimated at this time, but the overall trend implies mounting pressure on the natural resources.

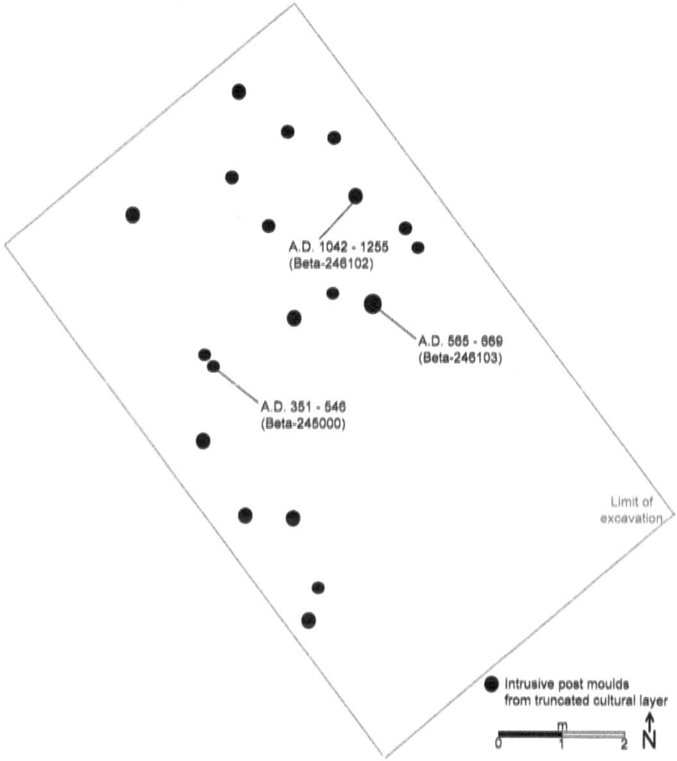

Fig. 12.3 Post moulds at the Baba Corporation project in Tumon, Guam, noting radiocarbon dating results. Based on data from Carson and Peterson (2010)

Regional Context

During the years A.D. 500–1000, the Mariana Islands were at a remote far edge of an extensively inhabited seascape. Connections with Island Southeast Asia by now had diminished, in particular noting total absence of the iron, bronze, glass, and other durable materials otherwise traded among communities in the Philippines and Indonesia. Decorated pottery styles in the Marianas no longer were shared with any parts of Island Southeast Asia or anywhere else, and in fact the Marianas pottery by now was being produced without much attention at all for decorative output. Meanwhile, the Mariana Islands were just one of several inhabited places in a "sea of islands" that supporting well-established and often inter-connected populations throughout Micronesia, Melanesia, and West Polynesia (see Fig. 1.1).

By A.D. 500, the only major region of the Pacific not yet inhabited was farther to the east in Central and East Polynesia. Limited exploratory incursions into these islands likely occurred prior to A.D. 1000, in advance of formal colonising migrations

(Irwin 1992:80–81). Such an early scouting activity may be hinted by disturbance of native vegetation in Mangaia of the Cook Islands as early as A.D. 500 (Ellison 1994; Kirch and Ellison 1994), although cultural habitation deposits with definitive artefacts and midden have been dated there no earlier than A.D. 1000 (Kirch et al. 1995). People lived in formal residential sites in the Marquesas Islands of East Polynesia after A.D. 1000 (Rolett 1998; Rolett and Conte 1995), perhaps made possible by slightly earlier voyages of discovery.

Long-distance ocean-crossing contacts and exchange probably operated at concurrent local, regional, and inter-regional scales among island communities of Pacific Oceania engaged in multiple contexts of networking (Fitzpatrick 2008), but the material archaeological evidence so far has been lacking of such networks involving the Mariana Islands. Historically known inter-connectivity has been a fundamentally definitive cultural trait of the societies of Micronesia as described by Glenn Petersen (2009) and by Paul Rainbird (1994), and very likely these practices were active by A.D. 500 when the many islands across Micronesia were inhabited by growing populations. Nonetheless, these traditions do not indicate a role for the Mariana Islands except as a place where people from Yap, Chuuk, and other Micronesian islands may have stayed for occasional short periods.

Especially for people living in the vulnerable and marginal small atolls of Micronesia, long-distance contacts and networks of social alliances may have been an essential part of life for sustaining access to food and other resources during times of devastating storms and crop shortages. In the Mariana Islands, however, these outcomes may have been achievable among the spatially separated communities in different islands, and perhaps the most extreme cataclysms could be overcome by drawing on supplemental resources from the northern-arc islands. Moreover, the lack of formalised agricultural fields in the Marianas in a sense avoided the dangers of over-reliance on subsistence food-producing systems that very easily could be destroyed by the region's frequent typhoons or any other sudden calamities.

After A.D. 500, pottery-making in the Marianas reflected overall independence of local traditions, not shared with external sources. A set of largely utilitarian pots in the Marianas can be contrasted against an evident decline of pottery altogether in other parts of the Pacific at this time. By A.D. 1000, pottery-making was a dying or dead art in Central and Eastern Micronesia (Athens 1990a, b; Ayres 1990), and well as in West Polynesia (Burley and Clark 2003; Carson 2014), although it continued in the Mariana Islands, Palau, and Melanesia. In this context, the persistence of pottery traditions in the Mariana Islands could suggest a degree of independence apart from inter-island contacts that may have been more active farther to the southeast in Micronesia and Polynesia.

References

Athens, J. S. (1990a). Non Madol pottery, Pohnpei. In R. L. Hunter-Anderson (Ed.), *Recent advances in Micronesian archaeology* (Micronesica Supplement 2), pp. 17–32. Mangilao: University of Guam Press.

Athens, J. S. (1990b). Kosrae pottery, clay and early settlement. In R. L. Hunter-Anderson (Ed.), *Recent advances in Micronesian archaeology*, Micronesica Supplement 2 (pp. 171–186). Mangilao: University of Guam Press.

Athens, J. S., & Ward, J. V. (2006). *Holocene paleoenvironment of Saipan: Analysis of a core from Lake Susupe* (Micronesian Archaeological Survey Rep. No. 35). Saipan: Commonwealth of the Northern Mariana Islands Division of Historic Preservation.

Ayres, W. S. (1990). Pohnpei's position in Eastern Micronesian prehistory. In R. L. Hunter-Anderson (Ed.), *Recent advances in Micronesian archaeology*, Micronesica Supplement 2 (pp. 187–212). Mangilao: University of Guam Press.

Burley, D. V., & Clark, J. T. (2003). The archaeology of Fiji/Western Polynesia in the post-Lapita era. In C. Sand (Ed.), *Pacific archaeology: Assessments and prospects*, Cahiers de l'Archéologie en Nouvelle-Calédonie 15 (pp. 235–254). Nouméa: Département d'Archéologie, Services des Musées et du Patrimoine de Nouvelle-Calédonie.

Carson, M. T. (2011). Palaeohabitat of first settlement sites 1500–1000 B.C. in Guam, Mariana Islands, Western Pacific. *Journal of Archaeological Science, 38*, 2207–2221.

Carson, M. T. (2014). De-coding the archaeological landscape of Samoa: Austronesian origins and Polynesian culture. *Journal of Austronesian Studies, 5*, 1–41.

Carson, M. T., & Peterson, J. A. (2010). *Archaeological research at the Baba Corporation Commercial Center in Lot 5093, Tumon, Tamuning Municipality, Guam* (Report prepared for Architects Laguana and Cristobal). Mangilao: Archaeology Office, Micronesian Area Research Center, University of Guam.

Ellison, J. C. (1994). Paleo-lake and swamp stratigraphic records of Holocene vegetation and sea-level changes, Mangaia, Cook Islands. *Pacific Science, 48*, 1–15.

Fitzpatrick, S. M. (2008). Maritime interregional interaction in Micronesia: Deciphering multi-group contacts and exchange systems through time. *Journal of Anthropological Archaeology, 27*, 131–147.

Irwin, G. J. (1992). *The prehistoric exploration and colonisation of the Pacific*. Cambridge, England: Cambridge University Press.

Kirch, P. V., & Ellison, J. C. (1994). Paleoenvironmental evidence for human colonization of remote Oceanic islands. *Antiquity, 68*, 310–321.

Kirch, P. V., Steadman, D. W., Butler, V. L., Hather, J., & Weisler, M. I. (1995). Prehistory and human ecology in Eastern Polynesia: Excavations at Tangatatau Rockshelter, Mangaia, Cook Islands. *Archaeology in Oceania, 30*, 47–65.

Leach F., Davidson, J., Claridge, G., Ward, G., & Craib, J. (2008). The physical and mineralogical characteristics of pottery from Mochong, Rota, Mariana Islands. In G. Clark, F. Leach & S. O'Connor (Eds.), *Islands of inquiry: Colonisation, seafaring and the archaeology of maritime landscapes*, Terra Australis29 (pp. 435–452). Canberra: Australian National University E Press.

Petersen, G. (2009). In *Traditional Micronesian societies: Adaptation, integration, and political organization in the Central Pacific*. Honolulu: University of Hawaii Press.

Rainbird, P. (1994). Prehistory of the north-west tropical Pacific: The Caroline, Mariana, and Marshall islands. *Journal of World Prehistory, 8*, 293–349.

Rolett, B. V. (1998). *Hanamiai: Prehistoric colonization and cultural change in the Marquesas islands (French Polynesia)*. New Haven, CT: Yale University Press. Yale University Publications in Anthropology.

Rolett, B. V., & Conte, E. (1995). Renewed investigation of the Ha'atuatua Dune (Nukuhiva, Marquesas Islands): A key site in Polynesian prehistory. *Journal of the Polynesian Society, 104*, 195–228.

Takayama, J., & Intoh, M. (1976). *Archaeological excavation of Latte Site M-13, Rota, in the Marianas* (Reports in Pacific Archaeological Survey, No. 4). Tokyo: Tokai University.

Ward, G., & Craib, J. L. (1983). *Mochong archaeological research. Document on file at Historic Preservation Office*. Saipan: Commonwealth of the Northern Mariana Islands.

Chapter 13
A.D. 1000–1700, A Sea of Islands and Monuments

The vast majority of Marianas archaeology has related to the years A.D. 1000–1700, locally known as the *latte* period when people lived in houses raised on stone pillars, prior to intensive foreign-imposed change pervading through nearly all aspects of social life and landscape. As the last tangible link with a pre-Spanish native Chamorro culture and heritage, *latte* have become ideologically charged icons in the Mariana Islands, and by extension the *latte* period has dominated perceptions of the past. *Latte* megalithic ruins, broken pottery, stone-grinding basins, and other materials of this period are found throughout the islands in great abundance in surface-visible and near-surface contexts, studied by archaeologists but also by historians, cultural anthropologists, educators, political activists, religious practitioners, artists, tourism developers, and others interested in the native Chamorro past for numerous reasons.

In one estimation (Carson 2012b:3), 'more than 90% of Marianas archaeology has involved the *latte* period', so this chapter necessarily provides only a summary as relevant for understanding the evolution of the Marianas natural–cultural landscape system. This chapter situates the *latte* period within the full chronological sequence of the Marianas, as just one period that happens to be copiously documented. Despite the impressive volume of information, several key questions persist about how and why the apparent widespread and intensive land use patterns of the *latte* period actually developed in relation to the preceding several centuries of an inhabited landscape.

Site Inventory and Dating

By A.D. 1700, almost every inhabitable part of the Mariana Islands supported residential complexes, temporary campsites, simple concentrations of broken pottery, or at least some material sign of cultural use of the landscape. No map can depict all of the sites of this period adequately, but rather this chapter considers the overall patterns and the sites that so far have yielded the most useful information (Fig. 13.1).

M.T. Carson, *Archaeological Landscape Evolution*,
DOI 10.1007/978-3-319-31400-6_13

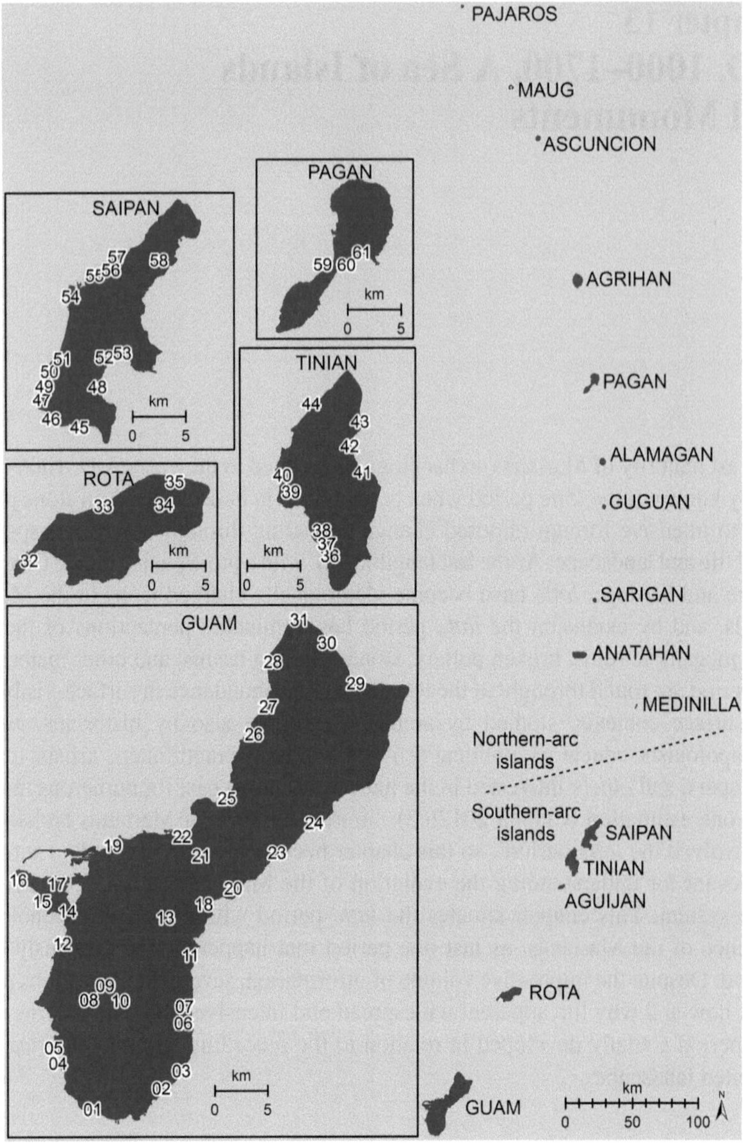

Fig. 13.1 Major site locations, A.D. 1000–1700. 01: Malesso. 02: Inarajan. 03: Nomna. 04: Umatac. 05: Fouha. 06: Ipan. 07: Talofofo. 08: Almagosa. 09: Fena. 10: Lost River. 11: Ylig. 12: Agat. 13: Manenggon. 14: Apra. 15: Tipalao. 16: Orote. 17: Sumay. 18: Pago. 19: Asan. 20: Mangilao. 21: Barrigada. 22: Hagatna. 23: Sasajayan. 24: Pagat. 25: Tumon. 26: Finegayan. 27: Haputo. 28: Urunao. 29: Tarague. 30: Jinapsan. 31: Ritidian. 32: Songsong. 33: Teteto-Guata. 34: As Nieves. 35: Mochong. 36: Tachungnya. 37: Taga Quarry. 38: House of Taga. 39: Kahet. 40: Atgidon. 41: Masalok. 42: Dangkulo. 43: Chiget. 44: Unai Chulu. 45: Obyan. 46: Agingan. 47: Afetna. 48: Dandan. 49: Chalan Piao. 50: Chalan Kanoa. 51: Chalan Kiya. 52: Laulau Rockshelter. 53: Unai Bapot. 54: Garapan. 55: Achugao. 56: San Roque. 57: Paopao. 58: Kalabera Cave. 59: Apansanme'na. 60: Apansantatte. 61: Regusa

Erik Reed's (1952) update of Hans Hornbostel's (1925) map of *latte* settlements in Guam fairly reflects the extent of these archaeological remnants (Fig. 13.2), and a similar scene may be expected in the other islands. At the time of written records in the late 1600s, people lived in several villages throughout the southern-arc and northern-arc islands (Fig. 13.3). In each of these islands, even today after centuries of imperialistic regimes, traditional place names still are known, and the ruins of *latte* sites testify to formal residential occupations (Marche 1889, 1982; Yawata 1945).

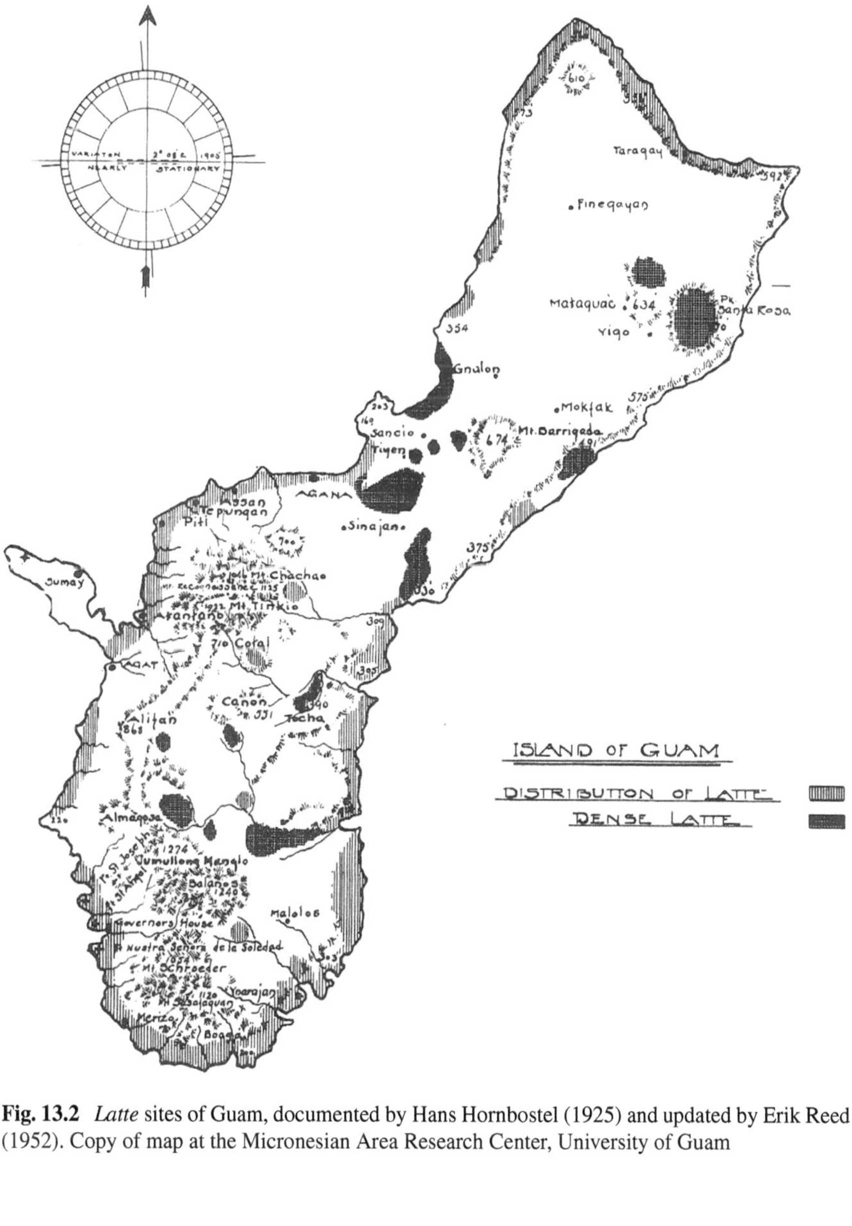

Fig. 13.2 *Latte* sites of Guam, documented by Hans Hornbostel (1925) and updated by Erik Reed (1952). Copy of map at the Micronesian Area Research Center, University of Guam

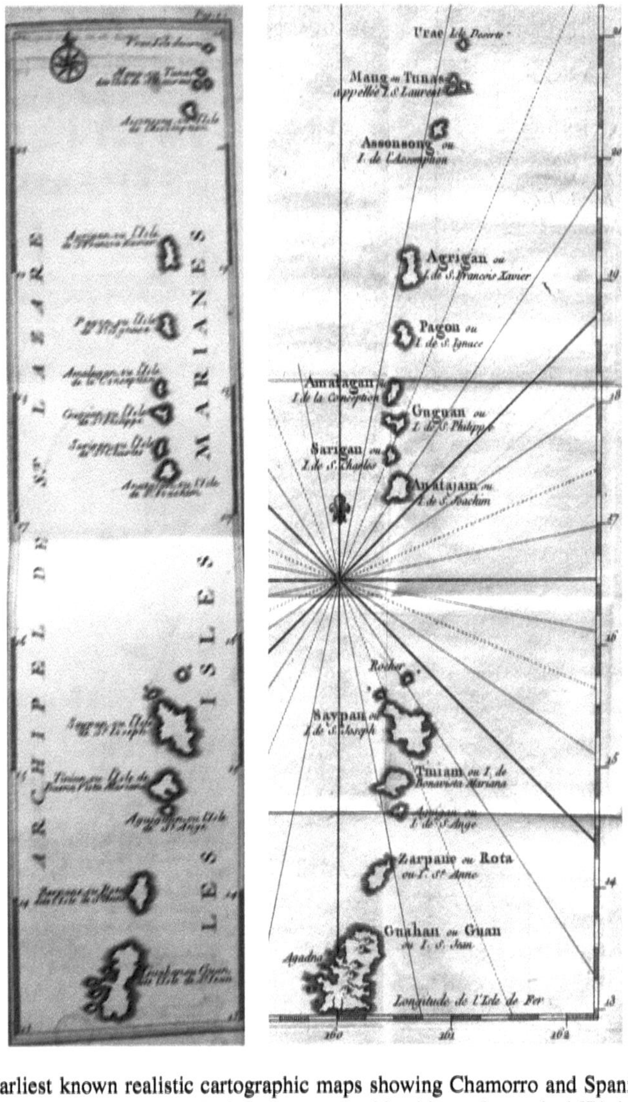

Fig. 13.3 Earliest known realistic cartographic maps showing Chamorro and Spanish names of the Mariana Islands. The image at the left was prepared by Alonso Lopez in 1671, here shown in a photograph of the original faded parchment that was folded and pressed inside the curved book pages published by le Gobien (1700), curated at University of Hawaii Library, Pacific Rare Books Collections. The image at the right is the more popularly known map that has survived in good condition in multiple copies, prepared by Bellin in 1762, based on observations of the original chart by Lopez and the memories of Pedro Morales, here shown in a photograph of a copy at the Micronesian Area Research Center, University of Guam

Fig. 13.4 Architectural model of *latte*, prepared by John Aguon and reprinted by permission from Laguana et al. (2012)

The conventionally understood *latte* period is named after the megalithic ruins of former houses. Individual *latte* sets consisted of paired rows of house pillars or columns of stone known as *haligi*, each topped by hemispherical capital stones known as *tasa* (see Fig. 4.6). The original wooden superstructures disintegrated long ago, and typically the stone pillars and capitals now are found in fallen and often broken condition. Artistic renderings have been proposed (Fig. 13.4), based on principles of architectural engineering and knowledge of post-raised houses in Island Southeast Asia.

A generic house-supporting function of *latte* is known from historically recorded observations and local traditions (Laguana et al. 2012; Thompson 1940). Specific houses presumably were used as residences, workshops, storage sheds, meeting houses, and for other purposes with considerable variation in their size, shape, and material contents. Surface collections and excavations have verified distinctive spatial patterns of artefacts and midden at different *latte* sets, thereby suggesting that people used different areas for specific kinds of work and other contexts (Bayman et al. 2012a, b; Craib 1986; Dixon et al. 2006). Interpretations have been varied about how the *latte* structures and their associated materials potentially reflect past economic, social, and political organisation at different scales of households, villages, whole islands, or even larger units (Cordy 1983, 1985; Cunningham 1992; Goddard 1995; Graves 1986; Peterson 2012; Russell 1998).

The extended period of A.D. 1000–1700 may be categorised overall as the *latte* period, but many the surviving known megalithic *latte* ruins date probably to the 1600s. The surface-visible *latte* ruins refer to the last time when people lived at these sites, prior to the Spanish-Chamorro wars and forced re-location of the surviving native Chamorro populations during the Spanish-imposed programme of *reducción* in the late 1600s. The *latte* villages effectively were abandoned by A.D. 1700, except for probably a few sites where stalwart stragglers may have resisted the Spanish occupation. Most of the associated cultural deposits contain a few pieces of European or Asian iron, glass, and porcelain dated to the late 1600s.

Regardless of association with *latte* structures, deposits containing fragments of thickened-rim earthenware bowls mostly date no earlier than A.D. 1000 (Fig. 13.5), so this age range by convention refers to the beginning of the *latte* period in the

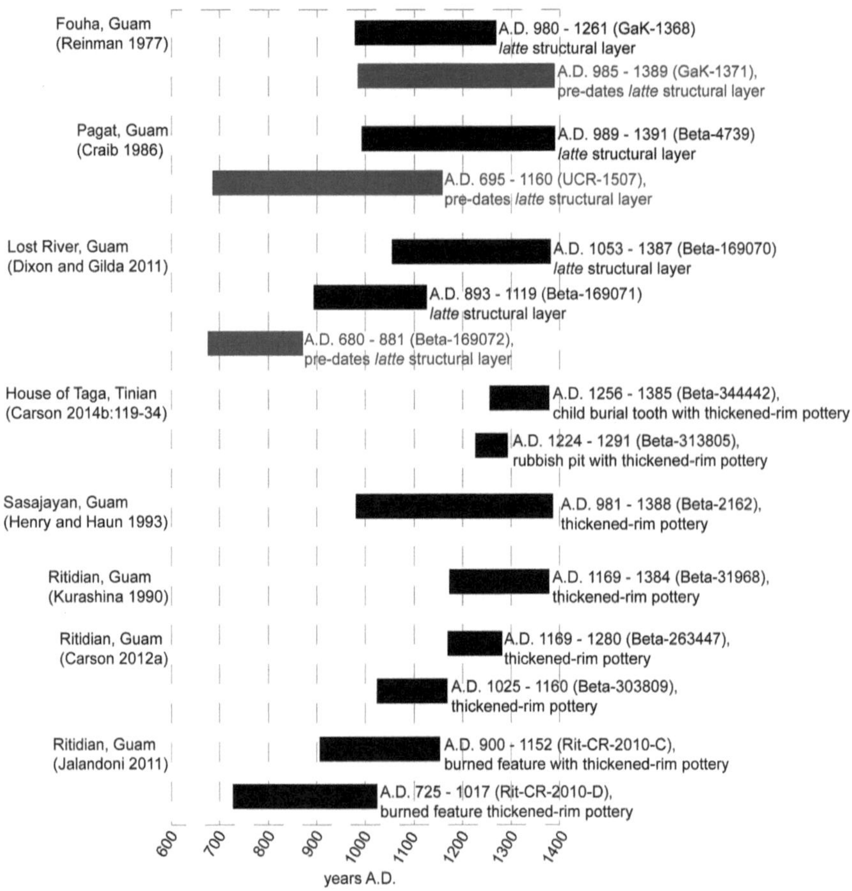

Fig. 13.5 Radiocarbon dating of archaeological deposits in relation to thickened-rim pottery and *latte* structural remains

Mariana Islands. Although stratigraphic associations do not always allow convincing dating of specific *latte* structures at this early age, the uniformity of cultural deposits over several centuries nonetheless suggests the likelihood that *latte* structures were made and used throughout this period. A precise starting date can be argued at any point in broader range of perhaps A.D. 800–1200, but here it is proposed around A.D. 1000 based on the radiocarbon dating as presented. Very likely, at least some of the traditions of the *latte* period emerged through gradual transitions over a few human generations, but the resulting archaeological record resembles an overall shift in material culture, settlement pattern, and land use system.

In cultural deposits containing the footing bases of *latte* pillars, radiocarbon dates of charcoal so far suggest an oldest possible age around A.D. 1000, but a precise age is debatable when considering the margins of error in dating calibrations. From the Lost River *latte* site in southern Guam, Boyd Dixon and Laura Gilda (2011) reported ages of A.D. 893–1119 and A.D. 1053–1387 for the base of a cultural deposit associated with *latte* ruins, overlaying an older deposit pre-dating the *latte* structure that was dated A.D. 680–881. Similarly at the base of a *latte*-associated layer at the Pagat Site in eastern Guam, John Craib (1986:134–135) reported a dating of A.D. 898–1393, overlaying another dating of A.D. 695–1160 for the immediately lower layer that pre-dating the *latte* occupation. At the Fouha Site in southern Guam, Fred Reinman (1977:43–48) reported a date of A.D. 980–1261 around the base of the surface-associated layer with *latte* ruins, once again overlaying a deeper cultural deposit but in this case with a similar dating of A.D. 985–1389.

The cultural materials of A.D. 1000–1700 are characterised most productively by fragments of thickened-rim earthenware bowls. In some cases, these potsherds are the only material evidence, found in concentrations directly on the surface and with no subsurface component. In other cases, the subsurface findings extend mostly 20–40 cm but occasionally deeper, with little or no stratigraphic differentiation in the pottery or other materials.

Some of oldest radiocarbon dates of the *latte* period have come from cultural deposits containing the diagnostic thickened-rim pottery, although they were spatially separated from any actual *latte* structures. At the Ritidian Site in northern Guam, directly underlying an early Spanish missionary centre, carbonised coconut nutshells yielded two radiocarbon dates of A.D. 725–1017 and A.D. 900–1152 for the ages of charcoal-rich features containing thickened-rim pottery (Jalandoni 2011:105). In three other portions of the widespread deposit bearing the thickened-rim pottery at Ritidian, dating has been as early as A.D. 1025–1160, A.D. 1169–1280 (Carson 2012a, 2014a), and A.D. 1169–1384 (Kurashina 1990:188). The nearby *latte* sets at Ritidian, however, include foreign-introduced materials such as iron, glass, and porcelain within the surface-visible cultural deposits, suggesting dates of the late 1600s just prior to abandonment of the area (Bayman et al. 2012a, b), and the most intact *latte* village structures bear deposits with dates of the A.D. 1500s through 1600s. The Sasajayan Site in Eastern Guam reportedly lost its *latte* structures during massive land-clearing, but surviving remnants of truncated cultural deposits with dense concentrations of thickened-rim pottery have been dated as old as A.D. 981–1388 (Henry and Haun 1993).

Many sites are undated by radiocarbon, yet a general range of A.D. 1000–1700 can be assigned according to the presence of *latte* ruins or thickened-rim pottery. In the northern-arc islands, for instance, several of these sites are known, but so far just two sites have yielded radiocarbon dates and only in the island of Pagan. Three dating results from the Resuga Site are calibrated at A.D. 1224–1444, A.D. 1284–1663, and A.D. 1437–Modern (Egami and Saito 1973:213–214). Four dating results from the Apansanme'na Site are calibrated at A.D. 1281–1400, A.D. 1450–1636, A.D. 1446–1635, and A.D. 1514–Modern (Athens 2011:325). At both sites, settlement certainly occurred by the A.D. 1400s but may have started as early as the A.D. 1200s.

Large-scale settlement of the northern-arc islands overall occurred after A.D. 1000 as part of the *latte* period landscape system in the Marianas. The exact timing may have been slightly later than the emergence of *latte* architecture and thickened-rim pottery in the southern-arc islands, perhaps after A.D. 1200 as suggested by the limited radiocarbon dating currently available. Future research may yet discover earlier sites, but so far no evidence has been found of older pottery types or deeper cultural deposits in these islands.

The period A.D. 1000–1700 is considered here in terms of an apparently cohesive landscape system in the Mariana Islands, although it must have incorporated variation both geographically and chronologically. The period can be subdivided in a number of different ways that do not necessarily agree with one another, for example according to change in climate, population growth, Chamorro–Spanish interactions, relative positions of subsurface versus surface-visible contexts, or probability distributions of radiocarbon dates. In any case, the landscape system of this period can be studied through a greater amount of information and from more diverse viewpoints than has been possible so far regarding any other period in the Marianas sequence.

Landforms

The overall landform structure during the period A.D. 1000–1700 was similar to the modern setting seen today, but people interfaced with the Marianas landforms quite differently during this period as compared to any other time before or since then. Very little of the physical landform structure has changed since this period, other than minor modifications of coastal progradation, storm-surge build-up, and ongoing slope erosion-deposition patterns. The major distinguishing characteristics at this time were the cultural activities that took place throughout nearly all landforms and thereby intensified human–environment interactions at the scale of the landscape.

By A.D. 1000, stable landforms had developed in coastal and low-lying areas that previously had been unstable or uninhabitable. At the mouth of the Pago River in Guam, a stable coastal plain developed for the first time. Peat deposits began to form in the bottom of an enclosed freshwater wetland habitat in the Hagatna Basin of Guam, previously a saltwater swamp connected with the ocean. In the far northern-arc island of Pagan, a sandy berm stabilised along the northwest coast and trapped a brackish and slightly sulphurous lake that potentially could be used as a water supply or as a therapeutic treatment. Overall by A.D. 1000, people could make use of much more stable coastal and lowland terrain than ever had been the case previously throughout the Mariana Islands.

Island interior landforms did not undergo any major alteration in morphology, but these areas were utilised extensively after A.D. 1000. Residential sites were occupied in several parts of limestone plateau terrain and over nearly all volcanic hill ridges and tablelands. In addition to formal residential complexes, campsites and ambiguous activity areas are evident in the innumerable surface-visible and near-surface concentrations of diagnostic thickened-rim potsherds of this period.

Resource Zones

The Marianas landscape during the period A.D. 1000–1700 reveals an abundant material record of people using nearly every inhabitable space and resource zone, but this impressive result evolved from long-term land use practices in a changing environment beyond the temporal scope of the *latte* period itself. Dye and Cleghorn (1990) noted that the apparent land use patterns on the surface today potentially have skewed archaeological attention too heavily in favour of the few centuries close to A.D. 1700, during the last time when people were making and using the ubiquitous thickened-rim pottery diagnostic of the time period. Moreover, the archaeological record of the *latte* period has been magnified by durable stone objects instead of a prior use probably of perishable wooden materials for house posts and grinding basins, as well as extremely large amounts of pottery that created larger and longer-lasting archaeological footprints.

During this period overall, the Marianas archaeological record shows an even distribution of cultural usage of a broad spectrum of resource zones in coastal settings, limestone plateau terrain, hilltop ridges, tablelands, and valley floors. All of these resource zones contributed significantly to the Marianas landscape system, whereas previously some had been emphasised more than others. During the *latte* period, the archaeological record shows abundant evidence of Kurashina's (1991) model of a transition to broad-spectrum use of resources, but importantly this transition had been underway in different aspects over the several preceding centuries. Coastal zones in particular had gone through a series of changing natural conditions and cultural use patterns, but coastal zones by A.D. 1000 were reliably stable and productive throughout the Mariana Islands. Meanwhile, cultural use overall had been steadily increasing in land-based island interior resources, and these zones by A.D. 1000 were strongly established within a broad-spectrum land use strategy.

By A.D. 1000, coral reefs, lagoons, and other components of coastal ecosystems had been stable for at least a few centuries, despite minor fluctuations in sea level and short-lived interruptions by storm-surge events. By this time, coastal zones provided ample and reliable supplies of shellfish, seaweeds, and swimming fish for large resident populations, unlike during the more rapid sea level drawdown of much earlier centuries. Shells of the gastropod *Strombus gibberulus* overwhelmingly dominated food-refuse middens after A.D. 1000 (see Fig. 5.10), likely because these gastropods lived in great abundance in the sandy substrates of the new beaches around the Mariana Island (Amesbury 1999). When exposed during low-tide events, these sandy substrates and some portions of nearby coral reefs could provide hundreds of shellfish, easily collectible even by children with very little training and minimal adult supervision.

Plant-based food production at this time, as in prior periods, relied on informal plantings of various tree and root-tuber crops, plus versatile use of culturally managed forests. The absence of formally constructed agricultural fields in one view may have limited the potential annual yields of crop harvests, but in another view it avoided the potential dangers of catastrophic loss due to overemphasis on any particular narrow range of subsistence foods. In this non-formalised broad-spectrum approach, almost any landform setting could support effective plant growth in the humid tropical environment, and people certainly were aware of the varied productivity of soils and terrain types for growing different crops with deeper or shallower roots, requirements for more or less water retention, degree of sunlight exposure, and so on.

Widespread land use after A.D. 1000 can be linked with an increasing role of food production, through at least three material observations. First, the surging numbers of sites required a larger food supply, particularly of starchy plant foods to support basic human nutrition. Second, the placement of more sites in inland zones required greater emphasis on land-based activities overall, most likely including cultivation of root-tuber and tree crops. Third, grinding basins (*lusong*) were manufactured in stone for the first time in the Marianas, indicative of increased demand for food processing tasks and more central role of formalising these tasks in the society.

Resource zones after A.D. 1000 were increasingly modified by human actions, both directly and indirectly. Deliberate human interventions resulted in the expansion of culturally managed communities of plants, replacing the native forests with trees and root-tuber crops useful for basic subsistence, supplemental flavouring, medicinal and narcotic effects, construction materials, and artworks. Secondarily, these human-caused disturbances resulted in declining numbers of native plants now outnumbered by the culturally managed taxa, as well as more productive growth of grasses and weeds that tend to flourish in disturbed and open ground. Coincidentally, the oldest rat bones in the Marianas appear to date around A.D. 1000 (Pregill and Steadman 2009; Wickler 2004), thus indicting another impact on the local environment and possibly causing further decline of native forests and birds.

Despite the widespread character of land use after A.D. 1000, cultural use of the resource zones was not necessarily intensified, but rather the food-producing land use patterns were expanded geographically. Expanding the geographic range of food production is fundamentally different from intensifying the magnitude of production within a given limited area (Leach 1999; Morrison 1994). Intensification involves a change in the use of a resource, for example resulting in greater volume of edible foods within a given limited land area through implementing new technologies of water management, higher-yielding crops, or decreasing fallow periods between crop cycles (Boserup 1965; Brookfield and Hart 1971). In most cases globally, larger numbers of people first will encourage expansions into new territories, and eventually an increasing density of populations will force an investment in new intensification of land use and food production after no options are available for expansion (Boserup 1965, 1981).

In the Mariana Islands prior to Spanish influences, people did not invest in irrigation systems or other formalised food production complexes that potentially could maximise the crop yields within each unit of land area. Instead, people expanded to inhabit more land areas, and each of these areas produced foods and other resources through informal horticulture and arboriculture. The daily tasks of horticulture and arboriculture did not appear to change substantially in terms of their technical operation, but rather people employed much the same techniques of prior periods applied over larger and more diverse land areas.

Patterns of resource usage after A.D. 1000 may not have been intensified per se, but people very likely developed more intimate relations with the Marianas landscape. Due to inhabiting larger numbers of residential sites in more places, people must have used place names to refer to more and more refined areas, subareas, and specific locations within the landscape. Traditions of historical and mythical events became attached to each place. Sedentary habitations throughout the Mariana Islands now created prolonged intimate relations between people and the inhabited landscape, significantly with a new character of sustained interactions with the landscape by a large population in every place of formalised settlement.

According to the broad geographic distribution of residential sites after A.D. 1000, communities of people lived in diverse coastal and inland resource zones in each of the southern-arc and northern-arc islands. Opinions vary with regard to what these patterns might imply about economic, social, political, or other distinctions among the groups living in one area versus another. If people mostly were self-sufficient and relied largely on the resources immediately around their residential sites, then perhaps groups sustained traditions of independence from one another. On the other hand, inter-community relations somehow must have allowed for the geographically separate residential sites to share much the same inventory of pottery forms and styles, shell and stone tools, housing structures, and food-refuse compositions. The evident datasets potentially could have resulted from a variety of hypothetical scenarios of inter-community trade, family management of resources within and between occupied territories, political hierarchies of re-distribution of resources, or any combination of these possibilities.

After A.D. 1000, the apparently widespread resource-use patterns very likely were facilitated by continuously stable conditions of the Little Climatic Optimum (LCO). For approximately three centuries about A.D. 1000–1300, most settings around the world experienced warmer temperatures and greater rainfall (Lamb 1965), but most importantly these conditions were steady and predictable. People could rely on the same familiar seasonal patterns and shifts in temperature and rainfall, year after year, continuously over a few centuries with very little inter-annual variation.

Conditions of the LCO allowed several generations of people to maximise the productivity of their inhabited environments, especially important in regions where seasonal variations influenced the growth of subsistence crops. In many cases worldwide, population levels increased with the reliable productivity of crop growth during this period. In the Mariana Islands, the archaeological record shows the emergence of formalised villages and land use patterns throughout the islands after A.D. 1000. Similar population growth and expansion has been noted throughout much of the Pacific Islands at this same time (Nunn et al. 2007).

About A.D. 1300, global climate shifted into overall cooler and drier conditions, with increased punctuated storminess and overall unpredictable weather events. This period has been described as a Little Ice Age (LIA) that lasted until about A.D. 1850 (Fagan 2000). In many parts of the world, the conditions of the LIA potentially were disastrous for increased numbers of people who had grown to depend on the maximal output of agricultural productivity during the LCO at A.D. 1000 through 1300. In many of the Pacific Islands, archaeological records indicate disruption of social order, shifting of residential locations, and other upheaval after A.D. 1300–1400 (Nunn 2000; Nunn et al. 2007), arguably due to conflicts among increased numbers of people relying on limited food supplies and natural resources.

During the *latte* period of A.D. 1000–1700 in the Mariana Islands, both the LCO and the LIA occurred, but the shifting climate may or may not have caused the same outcomes here as had happened in other parts of the world. The overall stable warmer and wetter conditions of the LCO likely encouraged population growth and expansion, but the unpredictable conditions of the slightly cooler and drier LIA probably did not cause too much alarm in the humid tropical environment of the Marianas. All of the tree and root-tuber crops still could grow rather exuberantly, even with slight reduction in temperature and rainfall.

The informal food production and broad-based resource usage in the Marianas allowed considerable flexibility to overcome periodic instability in climate. These traditions very likely developed as part of coping with the frequent typhoons that afflict the Mariana Islands. Without overspecialising in certain subsistence crops or overinvesting in a set of intensive agricultural fields, people could avoid the catastrophic loss of the population's food supply due to a typhoon or due to drought, flood, warfare, or other event. These traditions may have enabled people in the Mariana Islands to sustain through the LIA just as well as through any other period of instability.

Regardless of the long-term benefits of diversified resource management, human populations in the Mariana Islands needed access to sources of fresh water. Natural seeps could be accessed in the larger coastal plain landforms, while streams and freshwater wetland habitats provided other important sources of groundwater in the larger southern-arc islands. The smaller northern-arc islands, however, offered only very limited opportunities for finding natural seeps in coastal plains, and the only standing body of water was a brackish and slightly sulphurous lake in the island of Pagan.

Collection and storage of rainwater may have been essential for the survival of many groups, especially noting the large number of residential sites in diverse settings after A.D. 1000. Rainwater collection was of the most urgency in the northern-arc islands with limited or no supplies of accessible groundwater, and it was of variable importance elsewhere. In this regard, at least some of the large-sized bowls after A.D. 1000 probably were used for collecting rainwater.

Material Culture

The material culture of the period A.D. 1000–1700 most popularly is known through the iconic ruins of *latte* house structures, but several artefacts and other remains offer more abundant and diverse characterisation of this period. In addition to the stone pillar and capital components of *latte* ruins, other large stone products include grinding basins (*lusong*) and pestles (*lummok*). Thickened-rim pottery is especially abundant throughout the islands, along with variable amounts of assorted shell, stone, and bone artefacts as well as food-refuse middens, hearths, mounds of ash and charcoal, and rock art.

The largest standing *latte* set in the Mariana Islands was the famous House of Taga in Tinian (Fig. 13.6). The combined height of pillars and capitals stood about 4.5 m above the ground, supporting a wooden superstructure that decayed long ago. In local legend, a strong chief named Taga lived here. Today, the site is revered as a symbol of Chamorro cultural heritage.

Archaeological investigations have verified a widespread and dense cultural deposit at and around House of Taga. Antoine-Alfred Marche (1889, 1982) recorded one of the first scientific descriptions of the megalithic ruins. Hans Hornbostel (1925) and Kotondo Hasebe (1938) reported dense concentrations of broken pottery on the surface in the 1920s and 1930s, as well as a thick cultural deposit as deep as 2 m, containing a number of burial features. After World War II, only sparse traces of pottery were visible on the surface, but subsurface excavations focused on a deep and dense cultural deposit as reported by Alexander Spoehr (1957:85–98). About 35–40 m inland from the megalithic ruins, the upper cultural layer contained diagnostic thickened-rim pottery, remnants of cobble pavings, hearths, and human burial features with radiocarbon dates of A.D. 1224–1291 for charcoal in a rubbish pit and A.D. 1256–1385 for a tooth from a child's burial.

Other *latte* structures were not nearly as tall as at House of Taga, and in fact considerable diversity is noted in size, shape, and raw material (Fig. 13.7). Variations are noted in the size ratios of pillars compared to capitals, wherein certain sites appear to have their own architectural styles. The only consistent traits are the design of paired rows of pillars and the original placement of a capital atop each pillar. The number of paired rows in the vast majority of cases was four, but the number varied for making smaller or larger overall structures.

At least some of the variability in *latte* forms can be attributed to the raw materials. In the larger southern-arc islands, quarried slabs of limestone were made into pillars, mostly about 80–100 cm high, topped by capitals of another 30–40 cm. In several sites, at least one *latte* set stood taller than its neighbours, often approaching 2 m tall. In the northern-arc islands, only volcanic stone was available, including water-rounded beach cobbles and boulders scarcely taller than 40 cm, as well as a soft volcanic tuff that could be quarried, cut, and shaped in a number of localities.

In the southern-arc islands, *latte* pillars and capitals were quarried from natural terraces of limestone. At least one major quarry known is each of these larger islands, including the Urunao Quarry in Guam, the As Nieves Quarry in Rota, the Taga Quarry

Fig. 13.6 *Latte* ruins at House of Taga

Fig. 13.7 Variations of *latte* size, form, and material. Where scale bars are shown, they are in 20-cm increments. Rick Schaefer provides a human sale in the image from Dangkulo, Tinian

Fig. 13.8 Taga Quarry, Tinian

in Tinian, and the Agingan Quarry in Saipan (April 2004). In Tinian, the Taga Quarry is situated about 1.2 km from the House of Taga itself (Fig. 13.8). In Rota, the As Nieves Quarry contains the largest *latte* pillars and capitals ever carved, even larger than those at the House of Taga, but they never were removed from the ground (Fig. 13.9).

Latte sets so far have never been found in isolation, but rather they were parts of larger settlements containing at least two *latte* structures and often many more (Fig. 13.10). In large contiguous areas of roughly level ground, several *latte* structures could be organised as was the case around House of Taga in Tinian. In other cases of smaller available land areas, as in the hilly terrain of southern Guam, the *latte* sets were organised in several clusters wherever suitable land could provide a reasonably levelled supporting base. Similarly in the rough limestone landforms as in northern Guam, *latte* sets were constructed in the available scattered patches of thin rocky sediments amidst an otherwise continual expanse of inhospitable rocky exposure.

The paired rows of the *latte* pillars for the most part were arranged with the long axis in parallel with the underlying ground elevation contours, for example parallel with a coastline or following the top of a mountain ridge. The downslope side in nearly all cases can be interpreted as the 'front' or 'face' of the house, often with the greatest concentration of habitation debris and the most likely location of burial features. Some of the 'front' or 'face' sides of *latte* were formalised by alignments of cobble-boulder slabs, resembling a patio or front porch area.

Fig. 13.9 As Nieves Quarry, Rota

Within the apparent *latte* villages, usually at least one structure was taller and longer than the others, hinting at a higher status or community-serving function. These kinds of interpretations require cross-comparison of the different *latte* structures within an inferred community area, so that the largest set in the community can be recognised as different from its counterparts. Whereas most *latte* likely served in support of household residence and assorted daily routines, the comparatively larger or more impressive *latte* in a group most likely served a different function.

At most *latte* sets, at least one grinding basin (*lusong*) was made in a semiportable boulder. The sizes and shapes of the grinding facets varied, likely reflecting differential use for processing a range of foods and possibly other materials (Fig. 13.11). A different type of *lusong* can be found in locations separate from *latte* residences, typically made in large exposures of bedrock or in immovable large boulders (Fig. 13.12). The immovable *lusong* clearly were used in non-residential contexts, often at the entrance to a cave containing pictographs or else situated near the boundary between large *latte* settlements. *Lusong* were used in combination with a hand-held stone pestle, locally called a *lummok*.

Clusters of *latte* often include a few other structural feature remains, such as the *lusong* as noted plus arrangements of cobbles and earthen mounds. Cobbles and small boulders in many cases were used as alignment markers for dividing space, for demarcating boundaries, for covering the ground as paving, and occasionally

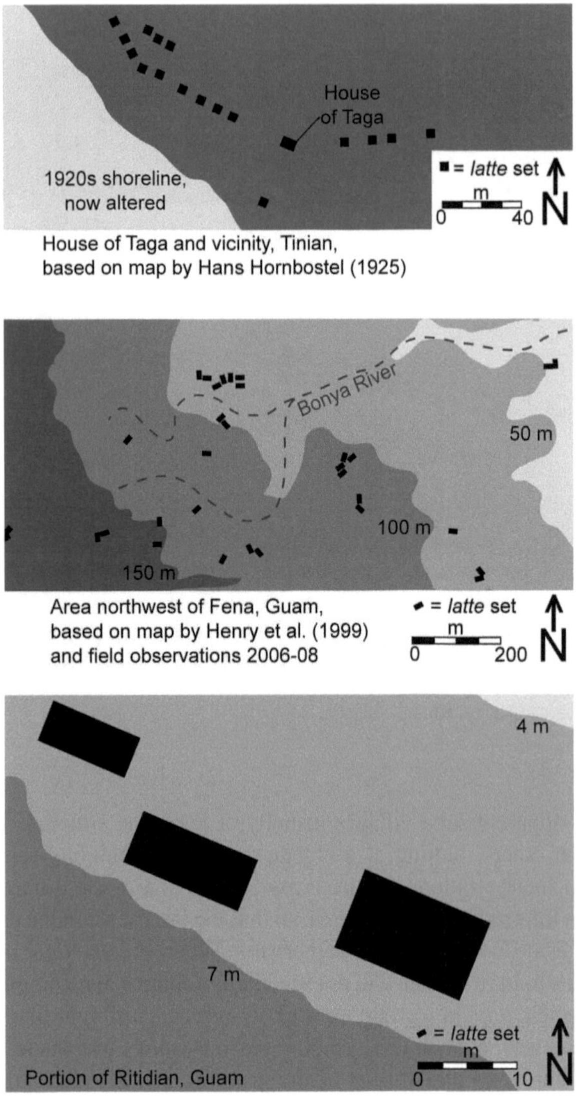

Fig. 13.10 Examples of *latte* settlement plan maps

incorporating a portable or semiportable *lusong* stone into the paving. Earthen mounds with charcoal and ash tend to be found slightly outside the immediate vicinity of *latte* clusters, of unclear purpose but perhaps multifunctional for burning of rubbish, baking pottery, or preparing slaked lime (Bulgrin 2006).

At several *latte* sites, spatial patterns of cultural activities have been ascertained through a combination of surface mapping and dispersed test pits (Fig. 13.13). In some cases, the highest concentrations of pottery and other materials are found

Broken *lusong*, showing section view, Unai Bapot, Saipan

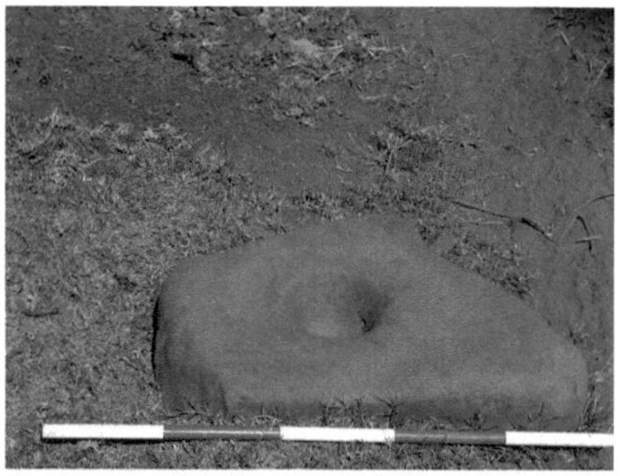

Typical *lusong*, Apansantatte, Pagan

Unusually small stone but with large *lusong* depression, Apansantatte, Pagan

Large basin-like *lusong*, Apansantatte, Pagan

Fig. 13.11 Examples of portable *lusong* grinding mortars. Scale bars are in 20-cm increments

within and immediately around the footprints of the *latte* structures. In other cases, though, the densest findings are outside the structural footprints, suggesting a toss zone of rubbish or perhaps the designation of certain areas for workshops. Various approaches have been developed for classifying the types of activity areas according to the represented materials of *latte* house remains, *lusong* grinding basins, pottery, adzes, and other items (Craib 1986; Dixon et al. 2006). Apparently different sets of household tasks were performed at two immediately adjacent *latte* sets in the Ritidian Site (Fig. 13.14), thus reflecting a probable division of labour among the residents (Bayman et al. 2012a, b).

Thickened-rim pottery has been mentioned as the most abundant artefact material at *latte* sites as well as in many other locations without *latte* ruins (Carson 2012b:42–43; Moore 2012; Spoehr 1957:108–117), but the range of forms and styles deserves more attention (Fig. 13.15). Most of these potsherds were broken from large vessels about 35–45 cm diameter, although the full range of sizes varied from 15 to 90 cm diameter. The vessel shapes mostly were large bowls and jars with bulging bodies and incurved rims, usually with rounded bases but sometimes with nearly conical bases. The clay recipes contained generally coarse-grained volcanic sand tempers

Setting of immovable *lusong*, central Pagan

Detail of immovable *lusong*, central Pagan

Immovable *lusong* outside "Upper Cave" at Ritidian, Guam

Immovable *lusong* outside small cave near boundary of Ritidian and Urunao, Guam

Fig. 13.12 Examples of immovable *lusong* grinding mortars. Scale bars are in 20-cm increments

and sometimes small amounts of quartz grains (Dickinson et al. 2001), although coarse beach sands continued to be used as tempers in some coastal sites. Wall thickness almost always was 8–10 mm, except for the rims that expanded up to 1 cm or more. Exteriors often were marked by vertical scrapings or dragging of a thick comb, edge of a bivalve shell, rope or cord, or other similar tool. Others were undecorated, but some showed signs of rough wiping or brushing. A few rare pieces have exhibited simple incisions, small punctured perforations, or thick coating of slaked lime.

Pottery production during this period provided great numbers of large-sized pots, apparently used for mixed tasks of storage, cooking, and serving. Possibly, some of the vessels were intended for specific purposes, such as for cooking of seafood versus storage of water, but so far the forms and styles appear overall uniform without any obvious functional differentiation. Exterior comb-scraping or other traits may have aided in signalling separate functional categories or perhaps ownership.

At least three traits of the thickened-rim pottery varied geographically across the Mariana Islands. First, examinations of sand tempers have shown that the different types of quartz grains can be grouped into one production centre in Guam and another in Saipan, each with a small amount of external distribution to other islands

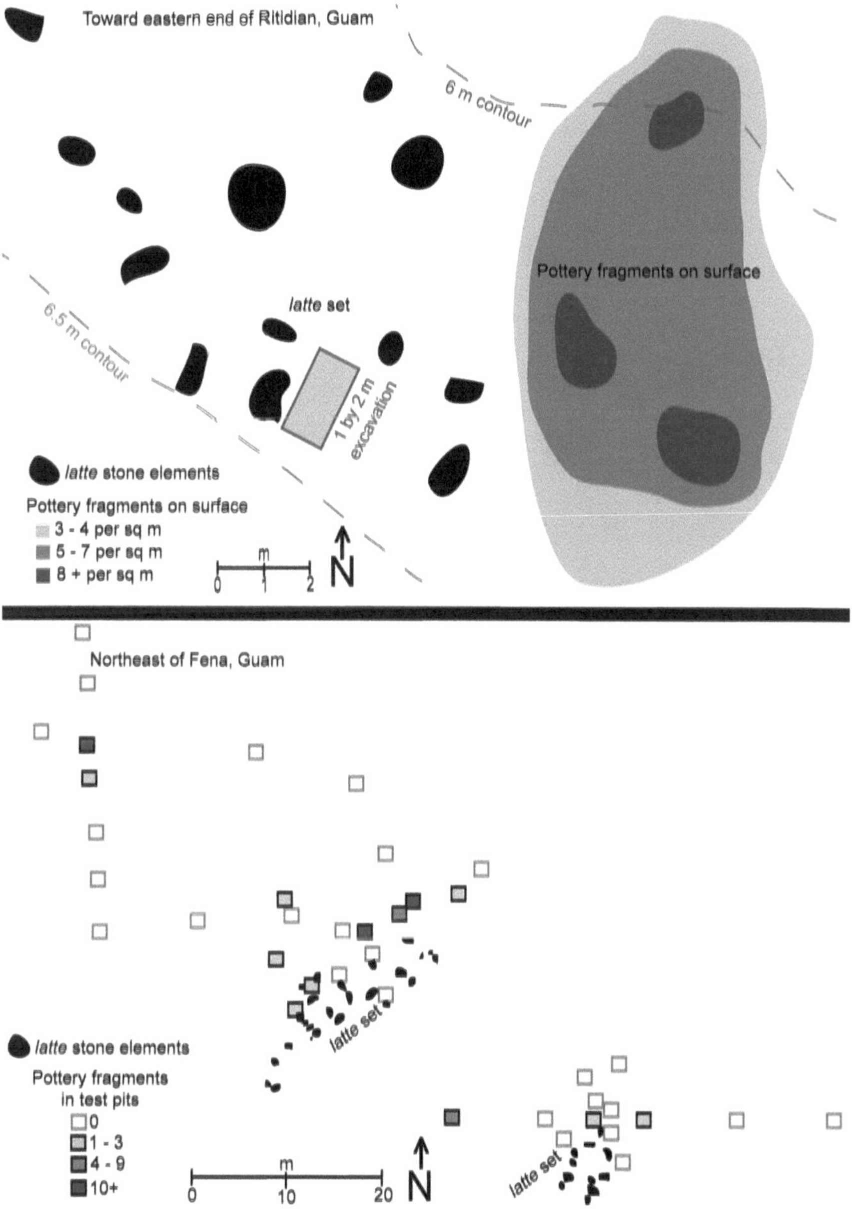

Fig. 13.13 Material densities of activity areas at *latte* sites

(Dickinson et al. 2001). Second, the rims tended to be thickest in Guam and Rota, whereas they were less pronounced farther north in the islands. Third, the external comb-scraping or cord-dragging mostly was made with broader implements and on the greatest majority of pottery in Guam and Rota, whereas mixed thinner and thicker widths were seen more commonly in the pottery collections of other islands.

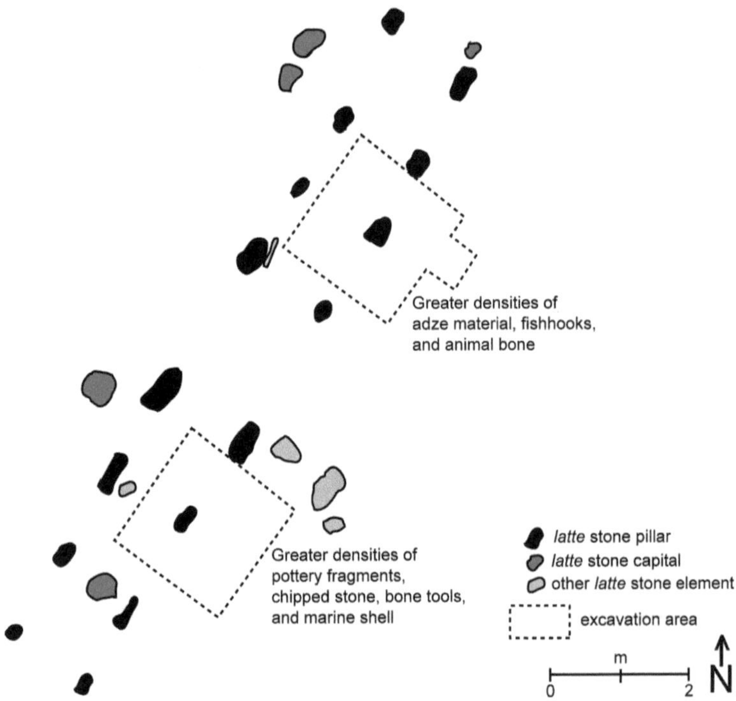

Fig. 13.14 Comparison of artefact assemblages at two adjacent *latte* sets at Ritidian. Based on data from Bayman et al. (2012a, b)

The geographic patterns may relate to at least two major groupings of pottery-making traditions or possibly exchange. One grouping included Guam and Rota, while the other involved Saipan and Tinian with limited influence in the other islands. More research will be needed for identifying differences in the pottery findings of the northern-arc islands, but so far the known materials appear to contain coarse volcanic sand tempers likely from local sources in those volcanic cone islands.

Regarding the residues adhering to pottery of this period, Darlene Moore (2012) discussed traces of taro starch, *ti* plant (*Cordyline fruticosa*), sugarcane, rice, shellfish, and bird feathers. These findings indicate definite cooking of varied foods inside at least some of the pots. Others without residues may have been used for water collection, storage, or boiling.

In addition to the sometimes overwhelming amount of potsherds, at least a few other artefacts are diagnostic of the period A.D. 1000–1700 (Fig. 13.16). Two of the most distinctive artefacts of this period are slingstones and carved bone spearpoints. Shell ornaments at this time consisted of large and only sometimes polished beads and pendants of *Conus* spp. shells, but now orange-coloured or sometimes purple-coloured *Spondylus* sp. shells gained a great popularity for fashioning into various sizes and shapes of beads, pendants, and other ornaments. Many of these items continue to be produced today by Chamorro artisans, resulting in several modern parallels, except

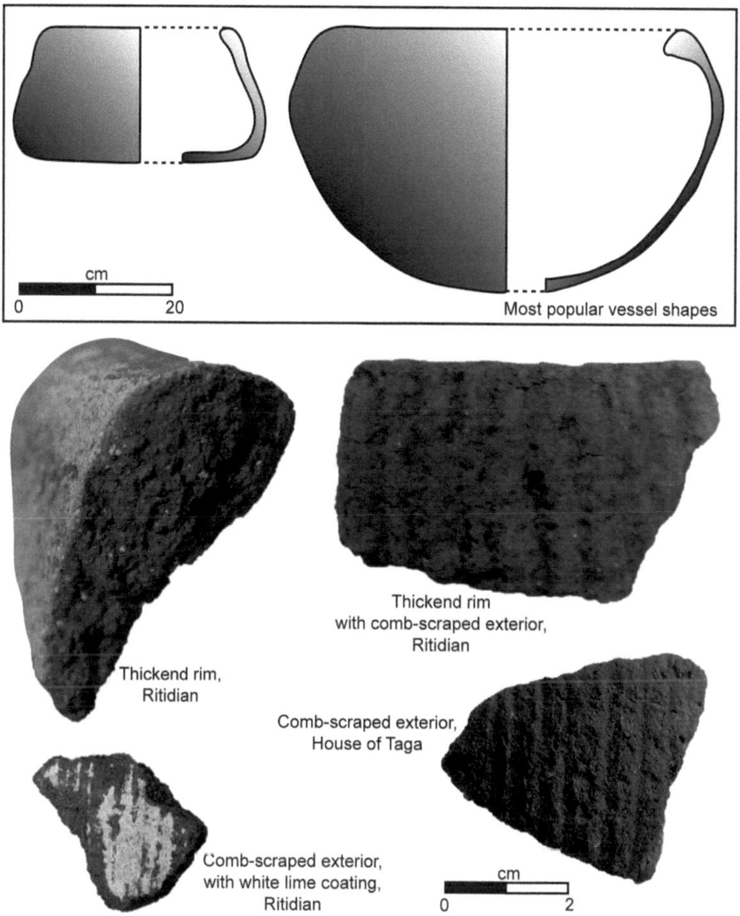

Fig. 13.15 Pottery types, A.D. 1000–1700

for the spearpoints made of human bone, but Hornbostel (1925) illustrated the most probable way of forming those spearpoints from human tibia as seems to be congruent with the archaeological findings of the objects and of burial remains (Fig. 13.17).

A crescent moon-shaped chest ornament, known as *sinahi* (literally 'new moon'), was made from *Tridacna* sp. shell, found only rarely in archaeological sites and today regarded as a sign of special rank or status. Hornbostel (1925) illustrated how the crescent-shaped pieces may have been attached as links on a cordage (Fig. 13.18), and he recounted a story that Georg Fritz (German Administrator in Saipan, 1899 through 1907) once found a complete necklace of 12 links in a burial cave. The *sinahi* could reflect a later variant of the transversely drilled pink coral links, found very rarely in the early settlement period (Chap. 8). Today, *sinahi* are worn as single central pieces on a necklace with other smaller accompanying shell beads, or sometimes they are worn in sets of three or more links along a necklace.

Fig. 13.16 Shell, stone, and bone artefacts, A.D. 1000–1700

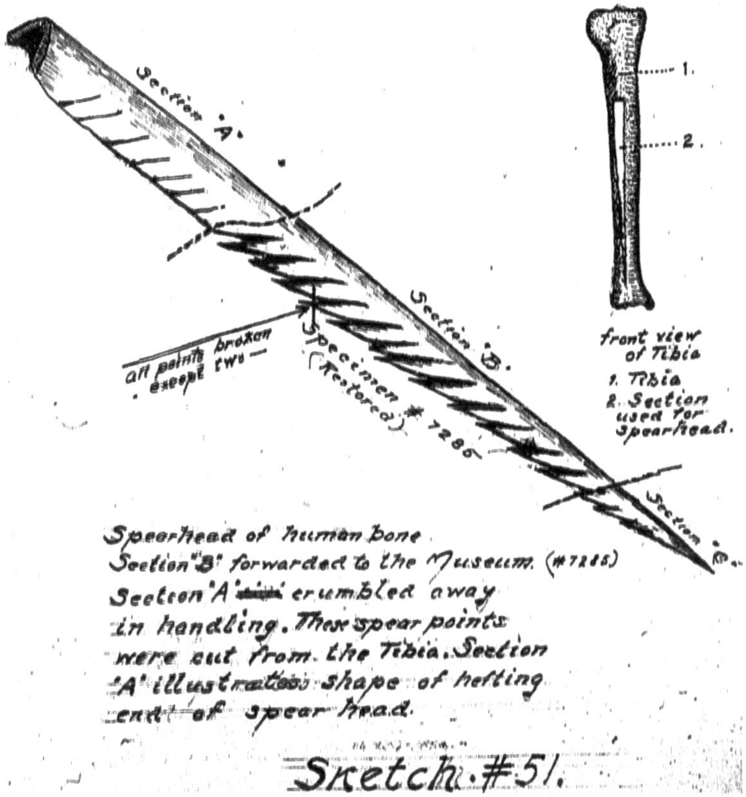

Fig. 13.17 Probable way of producing spearpoint of human bone (tibia), illustrated by Hans Hornbostel (1925)

Fishing gear is well represented during the period A.D. 1000–1700, compared to only very few items found in contexts prior to A.D. 1000. Rotating hooks, jabbing hooks, and V-shaped gorges were made of nacreous shell and rarely of bone, and the total toolkit accordingly served for capturing different types of fish. Compound objects were made of two or more pieces, including lures for catching pelagic fish and other lures for trapping octopus. Sinking weights most often were made of shaped limestone with a thin groove for a line attachment, but other forms have been reported.

A few artefacts of frequent daily use resemble more or less the same forms as seen in prior centuries. Polished adzes and chisels were made of giant clam (*Tridacna* sp.) shells, but others were made of fine-grained volcanic stone. Various shell containers for lime powder (for chewing of betelnut quids) were made expediently from shells, identified mainly by traces of white lime powder residues.

The impressive volume and geographic distribution of archaeological materials at A.D. 1000–1700 prompt questions about population size and density. Growing settlement size can be inferred for a few sites where dated stratigraphic layers show increase over time in the total footprint of residential occupation, but a region-wide picture of population growth cannot yet be discerned within the 700-year-long *latte*

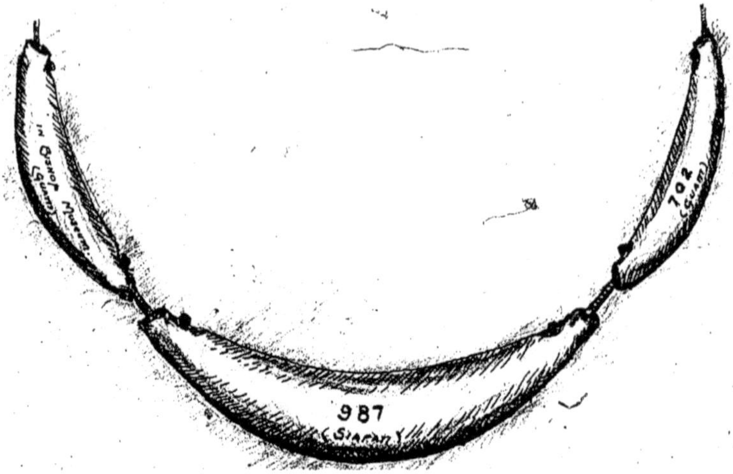

Fig. 13.18 Possible way of attaching *sinahi* links on a cordage, illustrated by Hans Hornbostel (1925)

period. Most sites contain only a single undifferentiated layer generically assigned to the *latte* period, and many of these have not been dated by radiocarbon due to lack of suitable dating material in partly disturbed surface-related contexts. As noted, nearly all of the surface-visible *latte* sites were last inhabited in the late 1600s, but those with deeper stratigraphic components have yielded earlier dates, often in the range of approximately A.D. 1400–1600 but sometimes extending as early as A.D. 1000.

Perhaps more population growth than usual occurred during a few centuries of reliably warmer and wetter climate about A.D. 1000–1300 (Nunn 2000, 2013), but then the persistence of such a large population thereafter may not have been practical. Hostile conflicts and warfare are suggested by the developments of new weaponry in the forms of slingstones and spearpoints. Inter-village rivalries were known at the time of Spanish contacts in the 1500s and 1600s, but the practice of warfare seems to have involved at least some degree of symbolic action and magic rather than outright mass killings (Farrer and Sellmann 2014).

The abundant and durable material record of A.D. 1000–1700 partly resulted from a transition of material culture output into larger amounts of longer-lasting products than ever had been the case in prior centuries. After A.D. 1000, people produced stone pillars and capitals of *latte* structures, other stone constructions of *lusong* grinding basins, and copious amounts of large-sized and thick earthenware bowls and jars. Previously, people had made wooden house posts, probably wooden boards instead of stone grinding basins, and lesser numbers of smaller-sized pottery vessels.

The shift in material output about A.D. 1000 may be described as a formalisation of material culture. Most tellingly, house structures were fixed in stone, at least in their foundational support pillars, and moreover entire villages were installed with substantial and long-lasting footprints in their underlying landforms. In the residential *latte* sites that contain some stratigraphic depth and differentiation, a long-term continuity is evident in sustained cultural use of these places.

The archaeological imprint of the *latte* period conveys a sense of permanence in the land. The villages were intended to endure indefinitely into the future, although specific house structures necessarily were rebuilt and reconfigured within the footprint of each village space. The wooden superstructures of houses could not last much longer than one person's lifespan, but the stone pillars and capitals of *latte* could be recycled and apparently supported long-term residences over perhaps a few hundred years in some cases.

The intended permanence of *latte* villages can be appreciated by considering the placement of burials beneath or around individual *latte* structures. This practice implies a link between a buried individual and the specific house of burial, and by extension a link can be proposed between a family or other lineage and its occupied village territory. Generally speaking, house burial suggests a conscious effort for the living to continue some sort of contact with the deceased as permanent residents of a place (Adams and King 2011). Connections with the deceased may have continued in other ways, for example through use of long bones (especially the tibia) for fashioning spearpoints (McNeill 2005). According to historical documents and local traditions, the heads of deceased individuals sometimes were curated for religious purposes and spiritual consultations (Farrer and Sellmann 2014).

Most often at *latte* sets in the Marianas, individuals were posed in extended position inside prepared pits, and several individuals could be buried side by side. Less common occurrences are secondary burials in pit bundles, wherein bones are disarticulated inside a small pit, sometimes holding the partial remains of more than one individual. These repeated mortuary patterns were noted by Hans Hornbostel (1925), and since then they have been confirmed at countless numbers of *latte* sites.

House burials contrasted against burials in or near caves, typically interred inside shallow pits. These settings very clearly were separated from residential sites, and therefore no connection can be proposed between the buried person and a designated lineage or village. Cave burial may have been reserved for individuals who did not meet the requirements of belonging to a particular household, family, lineage, or village territory (King 2006). Candidates for cave burial may have included people who lived unordinary lifestyles or who did not undergo rites of passage, either due to their actions during life or perhaps simply due to younger age at the time of death. Cave burials in some cases likely related to the specialised cultural use and perception of caves, notably in the caves bearing pictographs of human figures that may have depicted their associated burial rites (Cabrera and Tudela 2006).

Another type of burial at *latte* sites involved the placement of disarticulated and burned bones inside pits dense with charcoal, burned rock, often a few broken potsherds, and varied food-refuse midden. These features often are described as secondary burials, because the bones obviously were handled and re-deposited into the secondary context of a pit. These particular cases, however, suggest deliberate burning and disposal in contexts with food remains, thus leading Hornbostel (1925) to suggest the results of cannibalistic feasts. This possibility should be clarified as most importantly involving symbolic behaviour and not necessarily requiring actual consumption of human flesh (Jones et al. 2012). In any case, this proposal has not yet been tested through close examination for signs of butchery marks or human tooth impressions on bones, but these kinds of marking may not exist if the procedures were purely symbolic rather than literal.

Forms of artefacts, housing structures, village layouts, and mortuary practice collectively indicate a formalisation of residential life in the Marianas landscape after A.D. 1000. A significant shift occurred in cultural relations with the land, at this point involving deliberate acts of tying people with specific territories through permanent housing and burial. In prior periods, these patterns were not evident in the material record, but rather they emerged in formalised characters after A.D. 1000 and were repeated at a large scale throughout the Mariana Islands.

The setting after A.D. 1000 suggests a context of people for whatever reasons needing to establish clear links with land, territory, and resources that other people could understand and acknowledge. To some extent, the outcomes of a formalised village system may have minimised conflicts among people who otherwise would compete over resource zones, marriage partners, and other points of contention. In this view, population size and density must have contributed to competition and threats of hostility, probably magnified within the inherently limited resources of small and remote islands. The perceived effectiveness of the formalised village system can be ascertained through its apparently widespread adoption across the Marianas after A.D. 1000.

Formalised villages and land use patterns probably could not exist without a system of managing social and political order, especially if increasing numbers of people were involved in delineating their territories. In this regard, historical and sociological studies have proposed that Chamorro society was organised according to hierarchical groupings, apparently attested in Spanish records of the 1600s (Cordy 1983, 1985; Cunningham 1992; Russell 1998). In particular, higher and lower social ranks have been interpreted as responsible for creating larger and smaller *latte* structures, as well as for creating settlements in areas of greater or lesser resource productivity. These hypothetical scenarios may or may not be accurate, and some caution is advisable when working with information filtered through foreign observers and record-keepers in imperialistic contexts. With these cautions in mind, Scott Russell (1998:142) considered multiple historical accounts and concluded overall agreement that Chamorro society involved at least some degree of hierarchical organisation and recognised leadership roles prior to A.D. 1700, without specifying the exact social or political organisation.

As mentioned in Chap. 5, several early historical accounts from the 1500s through early 1700s are relevant for interpreting ancient society and landscape of the *latte* period. While this information is plentiful in the Marianas, it does not necessarily apply beyond the historically contingent points of reference in particular places and times. With some caution, limited historical interpretations can be proposed for the *latte* period as a whole, for example as proposed in Russell's (1998) work, at least to the extent that material culture was overall consistent from A.D. 1000 through 1700. Historical viewpoints are difficult to extend deeper into the past, noting the substantial transformations of the natural and cultural setting over time in the Marianas, but in theory at least some aspects of society were stable throughout the changing conditions.

Of the many cultural traditions accessible through histories and ethnohistories, place names hold powerful research potential for learning about ancient life and landscape, especially pertaining to contexts just prior to the Spanish *reducción* and near the end of the *latte* period prior to A.D. 1700. Even in the northern-arc islands that have been largely abandoned, traditional place names persist among periodic itinerant visitors. In the far northern island of Pagan, for instance, two of the many *latte* villages are remembered as Apansanme'na ('the dry front') and Apansantatte ('the dry back'), both visible from the central mountain ridge spine of the island at a place marked by an exceptionally large and immovable boulder with dozens of *lusong* grinding facets (see Fig. 13.12). The 'dry' trait refers to the partly emerged coral reefs that do not support productive coral growth and abundant marine life as in other parts of the islands. The 'front' refers to the leeward (west) side of the island, typically more attractive when approaching for a seacraft landing. The 'back' refers to the windward (east) side of the island, less promising for a safe landing. These places thus are linked together conceptually, with or without any obvious material archaeological traces. Although connected in a sense, the two *latte* villages exhibit distinctive usage of local stones for making the individual house-supporting pillars and capitals, such as the water-worn stones collected from the adjacent cobble-boulder beach at Apansanme'na in contrast to the quarried and shaped pieces of volcanic tuff at Apansantatte (see Fig. 13.7).

Through the naming of land units and assignment of traditions to each place, the Marianas landscape became a fully inhabited environment. As people established large-scale and long-lasting residency in new areas, each of those areas developed with its own set of place names and traditions. Many parts of the Marianas landscape thus shifted from vague or inert perceptions into active engagement with daily life. The population expansion throughout the islands after A.D. 1000 must have been a major turning point in transforming the natural terrain into a configuration of culturally named and known places. Many place names likely have much older origins, but most can be traced at least to this period. Others, however, clearly are Spanish-derived names, and a few are attributed to the Carolinian Islanders who lived in the region especially since the 1800s.

The material markers of the formalised inhabited environment may not be immediately obvious, but they are abundant. Place names and traditions are not directly visible or tangible, but they can be attached with their general vicinities that often contain remnants of *latte* villages, relicts of formerly managed forests, and prominent natural landmark features. The terrain itself is encoded with traditions about historical and mythical events that occurred there, and the same may be said of different identifiable parts of the ocean (McKinnon et al. 2014).

Prior to A.D. 1700, territories and resource zones were not defined by official rock wall fencing or other artificial constructions of persistent border markers, but very likely they could be recognised by the presence of several other characteristics of the landscape. Examples of culturally recognisable landmarks may have included *latte* structures and other houses, specifically situated *lusong* boulders at the boundaries of residential zones, forms of rock art posted at cave entrances, certain trees and rocks in important locations, and naturally occurring features such as streams and mountain ridges. These components were named and incorporated into a system of thoughts and perceptions that members of the society learned to comprehend through daily practice of inhabiting the landscape. Many of these traditions inevitably changed over time, but an apparently cohesive system operated during the *latte* period that has been retained at least partially in the place names and traditions that survive today.

Regional Context

The emergence of *latte* architecture and other formalised elements of the *latte* period coincided with a region-wide pattern emphasising stonework and especially large stone monuments throughout much of Micronesia, Melanesia, and Polynesia starting about A.D. 1000 (Carson 2013). Extensive village complexes of stone-filled house foundations replaced prior use of earthen terraces in Palau and Yap in Western Micronesia (Cordy 1987; Liston 2009; Wickler 2002), as well as in Samoa in West Polynesia (Carson 2006, 2014b). Monumental stonework complexes were built at Nan Madol and other sites of Pohnpei and at Lelu and other sites of Kosrae in Eastern Micronesia (Athens 1983; Ayres 1990; Cordy 1993). Stone-lined pathways, large pigeon-snaring mounds, and other constructions were produced in Samoa and Tonga (Burley 1996; Herdrich 1991; Herdrich and Clark 1993), along with hilltop fortifications in Samoa and Fiji (Best 1993).

Unlike the mixed residential and other functions of *latte* sites, many of the stonework complexes and other monuments in Micronesia and Polynesia were used explicitly for performance of high-status activities and religious rites, in some cases justifying political authority of the ruling elites (Athens 2007; Green 1986, 1993). The constructions may imply command of large forces of labour, because they involved several tonnes of stones, stacked and placed with dry masonry techniques, situated in places of special importance for communal gatherings. Very often, prominent upright slabs or pillars served as focal points of ceremonies, epitomised in East Polynesia as statues of human figures known as *tiki* in most places or *moai* in Easter Island (Emory 1970; Hunt and Lipo 2011).

While the *latte* village system flourished in the Marianas, the final human settlement of the farthest reaches of Pacific Oceania was underway (see Fig. 1.1). The islands of East Polynesia were inhabited beginning about A.D. 1000, and nearly every island in this broad region was colonised by A.D. 1300 (Kirch 2010). Polynesians meanwhile established small enclaves in parts of Micronesia and Melanesia, known as the Polynesian Outliers, where people maintained Polynesian languages and other traditions amidst clearly different cultural regions (Carson 2012c).

By the time of final Polynesian sea-crossing migrations, at least a few important transitions had occurred in the Polynesian material culture repertoire (Carson 2006,

2012c:29, 2014b; Kirch and Green 2001). As noted, earthen-filled terraces were replaced by stone-filled house foundations, but this shift further involved a replacement of the older Malayo-Polynesian traditions of post-raised houses by a new tradition of houses built directly at ground level on an artificially stone-covered surface. At the same time, pottery-making was no longer practiced in Polynesia, nor was it evident in Eastern Micronesia, except in very low frequencies at few sites for another few generations after A.D. 1000.

The new developments in Eastern Micronesia and Polynesia after A.D. 1000 showed a distinctively different cultural context from the *latte* period in the Mariana Islands. In the Marianas, some *latte* certainly were intended as monumental expressions, such as at House of Taga and other large constructions, but most *latte* in fact were much smaller and probably operated primarily for daily household activities. Religious and political connotations do not apply for all *latte*, and the Marianas traditions do not include the strong ideological contexts of the monuments in Eastern Micronesia and Polynesia. Furthermore, the loss of pottery-making and the loss of post-raised houses in these other regions both indicate a significantly different cultural setting than was experienced in the Mariana Islands at this time.

Now with essentially a fully inhabited 'sea of islands' across Pacific Oceania, inter-island networking necessarily expanded in ways that never before were possible. In Polynesia, stone adzes from renowned geological sources were mobilised over long distances between island communities (Best et al. 1992). In Micronesia, highly structured exchange systems operated at various scales, but the Mariana Islands may not have played a major role in these activities. Most notable was the *sawei* system of tribute, linking numerous island communities with each other and with Yap as the paramount position in the tribute system (Descantes 2005; Hunter-Anderson and Zan 1996; Sudo 1996). Large stone discs, sometimes 2 m or larger in diameter, were quarried and shaped from limestone in the small rock islands of Palau, used as 'stone money' or *rai* of special prestige overseas in Yap (Fitzpatrick 2002).

The Mariana Islands curiously are not implicated in any of the inter-island systems of trade and exchange otherwise attested in the Pacific Islands prior to Spanish presence in the region. If the *sawei* and other networks extended into the Marianas, then they did not create any clear material record. Nevertheless, hybrids of breadfruits indicate that people must have travelled between the Marianas and other distant island groups, specifically bringing the unique properties of breadfruit from the Mariana Islands to other locations across Micronesia. The indigenous seeded breadfruit of the Marianas (*dugdug* or *Artocarpus mariannensis*) at some point in time was hybridised with the more geographically widespread non-seeded species (*A. altilis*) and distributed through many of the Micronesian islands (Zerega et al. 2004, 2006). An exact timing of this breadfruit exchange is unknown, but it likely played a key role in developing broadly shared cultural traditions across Micronesia (Petersen 2006, 2009).

Curiously around A.D. 1000, rat bones appeared for the first time in archaeological sites and natural cave deposits in the Mariana Islands (Pregill and Steadman 2009; Wickler 2004), necessarily requiring an overseas contact for the arrival of these rats. Rats most likely were stowaways aboard canoes, and they may have come with any of the increasing numbers of external contacts across Micronesia or perhaps into Island Southeast Asia. In principle, they may have come from multiple possible source areas in the Asia-Pacific region, but so far no ancient rat bones from the Marianas have been analysed for possible DNA signatures of their geographic origins. The extent of the rat population is noted in Juan Pobre's first-hand account in the year 1602, mentioning an alarmingly large number of rats in the island of Rota, causing problems in harvesting the crops before they were devoured by the rats (translated in Driver 1993).

By A.D. 1000, Island Southeast Asia was an active zone of inter-regional commerce, but the Mariana Islands were not part of these activities. As mentioned in Chaps 11 and 12, several objects were traded across the South China Sea between Mainland Southeast Asia and the Philippines since 500 B.C., and traders from India had been working in Indonesia at least since A.D. 100–200. Elements of Hindu and Buddhist beliefs and practices were embraced in many areas, later with Islamic overlays in some places. In the centuries after A.D. 1000, foreign traders lived in long-established thriving colonies in Indonesia and now beginning in the Philippines. Although these groups certainly were within conceivable range of contacts with the Mariana Islands and other parts of Micronesia, apparently none of the diagnostic materials such as metals, glass, and porcelain reached the Marianas until after Spanish contact.

By the time of Ferdinand Magellan's arrival in the Mariana Islands in the year 1521, Island Southeast Asia was engaged in a globalised economy and society. Soon thereafter, Spanish colonial power in the Philippines supported a base of operations for the galleon trade from 1565 through 1815, bringing gold and silver from the Americas across the Pacific Ocean. During these years, the galleons followed a route from the Philippines around the north of the Mariana Islands on the way to the Americas, and they stopped at Guam for provisions on the way back to the Philippines. Meanwhile, the spice trade gained extraordinary momentum in Indonesia, occupying a critical role in the world trade market, but again the Mariana Islands were at the external margins of this excitement.

In their limited role related to the galleon trade, the Mariana Islands for several decades were perceived as a low priority by foreigners, until the large-scale efforts of establishing a Spanish colony there in the late 1600s. By the year 1700, nearly all aspects of life and landscape in the Marianas were indelibly changed, due to intensive Christianisation, the Spanish-Chamorro wars, and the *reducción* programme of reducing the native population and re-locating the survivors into a few easily controlled villages. The traditional *latte* villages were abandoned, local pottery production ceased, and instead people lived under Spanish rule in the officially designated villages in Guam, Rota, and Saipan.

Among the innumerable historical records pertaining to the Mariana Islands, one episode in particular captures a sense of the international setting in the 1660s, and it serves as a convenient way to bridge from this chapter to the next. The event in question involved a Chinese trader marooned in Guam. According to the chronicle of this era (Garcia 2004:190): 'Now he [the devil] raised against the Church a more dangerous persecution in the person of a *Sangley*, an idolatrous Chinese called

Choco, who arrived in these islands 20 years before the fathers, cast ashore in a storm as he was sailing from Manila to Terrenate [Ternate] in a sampan.' This single sentence reveals at least four important points:

1. The foreign record-keeper portrays local resistance against Spanish oppression in the Marianas as an act of hostility against the Christian Church.
2. The Chinese trader is described as a *Sangley*, which seems to be a version of the Hokkien Chinese word 'seng-li' (literally meaning 'business'), referring generally to Chinese traders in the Philippines.
3. At some time in the 1640s (20 years before the arrival of the Jesuit missionaries), the trader (Choco) had intended to travel between Manila (the centre of Spanish control in the Philippines) and Ternate (one of the major 'spice islands' in Indonesia), two key points of international trade at the time but controlled by different polities.
4. Choco was lost at sea somewhere between Manila and Ternate, yet he managed to arrive in Guam presumably by following dominant wind and wave patterns in his beleaguered vessel, thus leaving no doubt that a small seacraft could make this journey.

References

Adams, R. L., & King, S. M. (Eds.). (2011). *Residential burial: A multiregional exploration* (Archeological Papers of the American Anthropological Association, No. 20). Hoboken, NJ: Wiley.

Amesbury, J. R. (1999). Changes in species composition of archaeological marine shell assemblages in Guam. *Micronesica, 31*, 347–366.

April, V. N. (2004). *Latte Quarries of the Mariana Islands* (Latte: Occasional Papers in Anthropology and Historic Preservation, No. 2). Agana Heights: Guam Historic Resources Division.

Athens, J. S. (1983). The megalithic ruins of Nan Madol. *Natural History, 92*, 50–61.

Athens, J. S. (2007). The rise of the Saudeleur: Dating the Nan Madol chiefdom, Pohnpei. In A. Anderson, K. Green, & F. Leach (Eds.), *Vastly ingenious: The archaeology of Pacific material culture in Honour of Janet M. Davidson* (pp. 192–208). Dunedin: Otago University Press.

Athens, J. S. (2011). Latte period occupation on Pagan and Sarigan, northern Mariana Islands. *Journal of Island and Coastal Archaeology, 6*, 314–330.

Ayres, W. S. (1990). Pohnpei's position in Eastern Micronesian prehistory. In R. L. Hunter-Anderson (Ed.), *Recent advances in Micronesian archaeology* (Micronesica Supplement 2, pp. 187–212). Mangilao: University of Guam Press.

Bayman, J. M., Kurashina, H., Carson, M. T., Peterson, J. A., Doig, D. J., & Drengson, J. (2012a). Latte household economic organization at Ritidian, Guam National Wildlife Refuge, Mariana Islands. *Micronesica, 42*, 258–273.

Bayman, J. M., Kurashina, H., Carson, M. T., Peterson, J. A., Doig, D. J., & Drengson, J. (2012b). Household economy and gendered labor in the 17th century A.D. on Guam. *Journal of Field Archaeology, 37*, 259–269.

Best, S. (1993). At the halls of the mountain kings, Fijian and Samoan fortifications: Comparison and analysis. *Journal of the Polynesian Society, 102*, 385–447.

Best, S., Sheppard, P., Green, R., & Parker, R. (1992). Necromancing the stone: Archaeologists and adzes in Samoa. *Journal of the Polynesian Society, 101*, 45–85.

Boserup, E. (1965). *The conditions of agricultural growth: The economics of Agrarian change under population pressure*. New York: Aldine.
Boserup, E. (1981). *Population and technological change: A study of long-term trends*. Chicago: University of Chicago Press.
Brookfield, H. C., & Hart, D. (1971). *Melanesia: A geographical interpretation of an island world*. London: Methuen.
Bulgrin, L. (2006). Fina'okso antigu: Prehistoric soil mounds in the interior of Rota. *Micronesian Journal of the Humanities and Social Sciences, 5*, 31–41.
Burley, D. V. (1996). Sports, status, and field monuments in the Polynesian chiefdom of Tonga: The pigeon snaring mounds of northern Ha'apai. *Journal of Field Archaeology, 23*, 421–435.
Cabrera, G., & Tudela, H. (2006). Conversations with i man-aniti: Interpretation of discoveries of the rock art in the Northern Mariana Islands. *Micronesian Journal of the Humanities and Social Sciences, 5*, 42–52.
Carson, M. T. (2006). Samoan cultivation practices in archaeological perspective. *People and Culture in Oceania, 22*, 1–29.
Carson, M. T. (2012a). Evolution of an Austronesian landscape: The Ritidian site in Guam. *Journal of Austronesian Studies, 3*, 55–86.
Carson, M. T. (2012b). An overview of latte period archaeology. *Micronesica, 42*, 1–79.
Carson, M. T. (2012c). Recent developments in prehistory: Perspectives on settlement chronology, inter-community relations, and identity formation. In R. Feinberg & R. Scaglion (Eds.), *The Polynesian outliers: State of the art* (Ethnology Monographs No. 21, pp. 27–48). Pittsburgh, PA: University of Pittsburgh.
Carson, M. T. (2013). Austronesian migrations and developments in Micronesia. *Journal of Austronesian Studies, 4*, 25–52.
Carson, M. T. (2014a). Contexts of natural-cultural history: A 3500-year record at Ritidian in Guam. In M. T. Carson (Ed.), *Guam's Hidden Gem: Archaeological and historical studies at Ritidian, British Archaeological Reports International Series 2663* (pp. 1–43). Oxford: Archaeopress.
Carson, M. T. (2014a). De-coding the archaeological landscape of Samoa: Austronesian origins and Polynesian culture. *Journal of Austronesian Studies, 5*, 1–41.
Cordy, R. (1983). Social stratification in the Mariana Islands. *Oceania, 53*, 272–276.
Cordy, R. (1985). Settlement patterns of complex societies in the Pacific. *New Zealand Journal of Archaeology, 7*, 159–182.
Cordy, R. (1987). *Archaeological settlement pattern studies on Yap* (Micronesian Archaeological Survey Rep. No. 16). Saipan: Commonwealth of the Northern Mariana Islands Division of Historic Preservation.
Cordy, R. (1993). *The Lelu Stone Ruins (Kosrae, Micronesia): 1978–1981 historical and archaeological research, Asian and Pacific Archaeology Series 10*. Honolulu: Social Science Research Institute, University of Hawaii.
Craib, J. L. (1986). *Casas de los Antiguos: Social differentiation in protohistoric Chamorro society, Mariana Islands*. Ph.D. dissertation. University of Sydney, Sydney.
Cunningham, L. J. (1992). *Ancient Chamorro society*. Honolulu: Bess Press.
Descantes, C. (2005). Integrating archaeology and ethnohistory: The development of exchange between Yap and Ulithi, Western Caroline Islands (British Archaeological Report International Series No. 1344). Oxford: Archaeopress.
Dickinson, W. R., Butler, B. M., Moore, D. R., & Swift, M. (2001). Geological sources and geographic distribution of sand tempers in prehistoric potsherds from the Mariana Islands. *Geoarchaeology, 16*, 827–854.
Dixon, B., & Gilda, L. (2011). A comparison of an inland Latte period community to coastal settlement patterns observed on southern Guam. *People and Culture in Oceania, 27*, 65–86.
Dixon, B., Mangieri, T., McDowell, E., Paraso, K., & Rieth, T. (2006). Prehistoric Chamorro household activities and refuse disposal patterns on the Micronesian island of Tinian, Commonwealth of the Northern Mariana Islands. *Micronesica, 39*, 55–71.

Driver, M. G. (1993). *Fray Juan Pobre in the Marianas, 1602* (Miscellaneous Series No. 8). Mangilao: Micronesian Area Research Center, University of Guam.

Dye, T. S., & Cleghorn, P. L. (1990). Prehistoric use of the interior of southern Guam. In R. L. Hunter-Anderson (Ed.), *Recent advances in Micronesian archaeology* (Micronesica Supplement 2, pp. 261–274). Mangilao: University of Guam Press.

Egami, T., & Saito, F. (1973). Archaeological excavation on Pagan in the Mariana Islands. *Journal of the Anthropological Society of Nippon, 81*, 203–226.

Emory, K. P. (1970). A re-examination of East Polynesian marae: Many marae later. In R. C. Green & M. Kelly (Eds.), *Studies in oceanic culture history, Vol. 1* (Pacific Anthropological Records No. 11) (pp. 73–92). Honolulu: Department of Anthropology, Bishop Museum.

Fagan, B. (2000). *The Little Ice Age: How climate made history 1300–1850*. New York: Basic Books.

Farrer, D. S., & Sellmann, J. D. (2014). Chants of re-enchantment: Chamorro spiritual resistance to colonial dominion. *Social Analysis, 58*, 127–148.

Fitzpatrick, S. M. (2002). A radiocarbon chronology of Yapese stone money quarries in Palau. *Micronesica, 34*, 227–242.

Garcia, F. (2004). *The Life and Martyrdom of the Venerable Father Diego Luis de San Vitores of the Society of Jesus* (J. A. McDonough Ed. & M. M. Higgins, F. Plaza & J. M. H. Ledesma Trans.) (Monograph Series No. 3). Translated and edited version of original published in 1684. Mangilao: Micronesian Area Research Center, University of Guam.

Goddard, P. (1995). *Latte: The mysterious Megaliths of the Marianas*. Perth, Australia: Abrohols Publishing Party.

Graves, M. W. (1986). Organization and differentiation within late prehistoric ranked social units, Mariana Islands, western Pacific. *Journal of Field Archaeology, 13*, 139–154.

Green, R. C. (1986). Some basic components of the Ancestral Polynesian settlement system: building blocks for more complex Polynesian societies. In P. V. Kirch (Ed.), *Island societies: Archaeological approaches to evolution and transformation* (New directions in archaeology, pp. 50–54). Cambridge: Cambridge University Press.

Green, R. C. (1993). Community-level organisation, power and elites in Polynesian settlement pattern studies. In M. W. Graves & R. C. Green (Eds.), *The evolution and organisation of prehistoric society in Polynesia* (Monograph No. 19, pp. 9–12). Auckland: New Zealand Archaeological Association.

Hasebe, K. (1938). The natives of the South Seas Archipelago. *Lectures on Anthropology and Prehistory, 1*, 1–35 (in Japanese).

Henry, J. D., & Haun, A. E. (1993). *Sohbu Resort project area archaeological mitigation program findings and supplemental mitigation plans, Mangilao Municipality, Territory of Guam* (Report prepared for Sohbu Guam Development Company). Hilo, HI: Paul H. Rosendahl, Ph.D.

Henry, J. D., Haun, A. E., & Kirkendall, M. A. (1999). *Phase I archaeological survey and subsurface testing, U.S. Naval Activities Ordnance Annex, Guam* (Report prepared for United States Naval Facilities Engineering Command). Hilo, HI: Paul H. Rosendahl, Ph.D.

Herdrich, D. J. (1991). Towards an understanding of Samoan star mounds. *Journal of the Polynesian Society, 100*, 381–435.

Herdrich, D. J., & Clark, J. T. (1993). Samoan tia'ave and social structure: Methodological and theoretical considerations. In M. W. Graves & R. C. Green (Eds.), *The evolution and organisation of prehistoric society in Polynesia* (Monograph No. 19) (pp. 52–63). Auckland: New Zealand Archaeological Association.

Hornbostel, H. (1925). *Unpublished field notes, 1921–1924*. Honolulu: Record on file at Library of Bishop Museum.

Hunt, T. L., & Lipo, C. (2011). *The statues that walked: Unraveling the mystery of Easter Island*. New York: Free Press.

Hunter-Anderson, R. L., & Zan, Y. (1996). Demystifying the Sawei, a traditional interisland exchange system. *Isla: A Journal of Micronesian Studies, 4*, 1–45.

Jalandoni, A. (2011). *Casa real or not real? A Jesuit mission house in Guam*. M.A. Thesis, University of the Philippines, Diliman.

Jones, S., Walsh-Haney, H., & Quinn, R. (2012). Kana tamata or feasts of men: An interdisciplinary approach for identifying cannibalism in prehistoric Fiji. *International Journal of Osteoarchaeology, 25*(2), 127–145. doi:10.1002/oa.2269.

King, S. M. (2006). The marking of age in ancient coastal Oaxaca. In T. Ardren & S. R. Hutson (Eds.), *The social experience of childhood in ancient Mesoamerica* (pp. 169–200). Boulder: University Press of Colorado.

Kirch, P. V. (2010). Peopling of the Pacific: A holistic anthropological perspective. *Annual Review of Anthropology, 39*, 131–148.

Kirch, P. V., & Green, R. C. (2001). *Hawaiki, Ancestral Polynesia: An essay in historical anthropology*. Cambridge: Cambridge University Press.

Kurashina, H. (Ed.). (1990). *Archaeological investigations at the Naval Facility (NAVFAC) Ritidian Point, Guam, Mariana Islands* (Report prepared for United States Department of the Navy). Mangilao: Micronesian Area Research Center, University of Guam.

Kurashina, H. (1991). Prehistoric settlement patterns on Guam. *Journal of the Pacific Society, 51*, 141–158.

Laguana, A., Kurashina, H., Carson, M. T., Peterson, J. A., Bayman, J. M., Ames, T., et al. (2012). Estorian i latte: A story of latte. *Micronesica, 42*, 80–120.

Lamb, H. H. (1965). The early medieval warm epoch and its sequel. *Palaeogeography, Palaeoclimatology, Palaeoecology, 1*, 13–37.

le Gobien, C. (1700). *Histoire des Îles Mariannes Nouvellement Converties à la Réligion Chrétienne*. Paris: N. Pepie.

Leach, H. M. (1999). Intensification in the Pacific: A critique of the archaeological criteria and their application. *Current Anthropology, 40*, 311–339.

Liston, J. (2009). Cultural chronology of earthworks in Palau, western Micronesia. *Archaeology in Oceania, 44*, 56–73.

Marche, A.-A. (1889). Rapport general sur une mission aux Îles Mariannes. *Nouvelles Archives des Missions Scientifiques et Littéraires, Nouvelle Série, 1*, 241–280.

Marche, A.-A. (1982). *The Mariana Islands* (R. D. Craig Ed. & S. E. Cheng Trans.). (Publication Series No. 8). Mangilao: Micronesian Area Research Center, University of Guam.

McKinnon, J., Mushynsky, J., & Cabrera, G. (2014). A fluid sea in the Mariana Islands: Community archaeology and mapping the seascape of Saipan. *Journal of Maritime Archaeology, 9*, 59–79.

McNeill, J. R. (2005). Putting the dead to work: An examination of the use of human bone in prehistoric Guam. In G. F. M. Rakita, J. E. Buikstra, L. A. Beck, & S. R. Williams (Eds.), *Interacting with the dead: Perspectives on mortuary archaeology for the new millennium* (pp. 305–315). Gainesville, FL: University Press of Florida.

Moore, D. R. (2012). What's new and what's cooking in the latte period pots. *Micronesica, 42*, 121–147.

Morrison, K. D. (1994). The intensification of production: Archaeological approaches. *Journal of Archaeological Method and Theory, 1*, 111–159.

Nunn, P. D. (2000). Environmental catastrophe in the Pacific Islands around A.D. 1300. *Geoarchaeology, 15*, 715–740.

Nunn, P. D. (2013). The end of the Pacific? Effects of sea level rise on Pacific Island livelihoods. *Singapore Journal of Tropical Geography, 34*, 143–171.

Nunn, P. D., Hunter-Anderson, R. L., Carson, M. T., Thomas, F., Ulm, S., & Rowland, M. J. (2007). Times of plenty, times of less: Last-millennium societal disruption in the Pacific Basin. *Human Ecology, 35*, 385–401.

Petersen, G. (2006). Micronesia's breadfruit revolution and the evolution of a culture area. *Archaeology in Oceania, 41*, 82–92.

Petersen, G. (2009) *Traditional Micronesian societies: Adaptation, integration, and political organization in the Central Pacific*. Honolulu: University of Hawaii Press.

Peterson, J. A. (2012). Latte villages in Guam and the Marianas: Monumentality or monumenterity? *Micronesica, 42*, 183–208.

Pregill, G. K., & Steadman, D. W. (2009). The prehistory and biogeography of terrestrial vertebrates in Guam, Mariana Islands. *Diversity and Distributions, 15*, 983–996.

Reed, E. (1952). *General report on archaeology and history of Guam* (Report prepared for Honorable Carlton Skinner, Governor of Guam). Santa Fe, NM: U.S. National Park Service.

Reinman, F. R. (1977). *An archaeological survey and preliminary test excavations on the Island of Guam, Mariana Islands, 1965–1966* (Miscellaneous Publication No. 1). Mangilao: Micronesian Area Research Center, University of Guam.

Russell, S. (1998). *Tiempon i Manmofo'na: Ancient Chamorro Culture and History of the Northern Mariana Islands* (Micronesian Archaeological Survey Rep. No. 32). Saipan: Commonwealth of the Northern Mariana Islands Division of Historic Preservation.

Spoehr, A. (1957). *Marianas prehistory: Archaeological survey and excavations on Saipan, Tinian and Rota. Fieldiana: Anthropology, Vol. 48*. Chicago: Chicago Natural History Museum.

Sudo, K.-I. (1996). Rank, hierarchy and routes of migration: Chieftainship in the Central Caroline Islands of Micronesia. In J. J. Fox & C. Sather (Eds.), *Origins, ancestry and alliance: Explorations in Austronesian ethnography* (pp. 57–72). Canberra: Department of Anthropology, Research School of Pacific Studies, Australian National University.

Thompson, L. (1940). The function of latte in the Marianas. *Journal of the Polynesian Society, 49*, 447–465.

Wickler, S. K. (2002). Terraces and villages: Transformation of the cultural landscape in Palau. In T. N. Ladefoged & M. W. Graves (Eds.), *Pacific landscapes: Archaeological approaches* (pp. 63–96). Los Osos, CA: Bearsville Press.

Wickler, S. K. (2004). Modelling colonisation and migration in Micronesia from a zooarchaeological perspective. In M. Mondini, S. Munoz, & S. K. Wickler (Eds.), *Colonisation, migration and marginal areas: A zooarchaeological approach* (pp. 28–40). Oxford: Oxbow Books.

Yawata, I. (1945). Peculiar forms of the stone-piles on the Mariana Islands: Mariana sekichuretsu iko no tokushu keishiki. *Zinruigaku Zasshi, 59*, 418–424 (in Japanese).

Zerega, N. J. C., Ragone, D., & Motely, T. J. (2004). Complex origins of breadfruit (*Artocarpus altilis*, Moraceae): Implications for human migration in Oceania. *American Journal of Botany, 91*, 760–766.

Zerega, N. J. C., Ragone, D., & Motely, T. J. (2006). Breadfruit origins, diversity, and human-facilitated distribution. In T. J. Motely & N. J. C. Zerega (Eds.), *Darwin's harvest: Origins, evolution, and conservation of crop plants* (pp. 213–238). New York: Columbia University Press.

Chapter 14
A.D. 1700–Present, Living with Colonialism and Globalisation

This book primarily is concerned with the archaeological record of landscape evolution in the Mariana Islands prior to A.D. 1700, but the present chapter highlights some of the profound transformations that have occurred within the historical and modern era. These last few hundred years have been chronicled extensively in written records, maps, illustrations, photographs, documentary archives, and various non-written traditions. These records add substantially to the knowledge of long-term evolutionary processes of human–environment interactions as outlined in the chapters of this book.

In addition to the material and written records of colonial impacts, foreign influences are reflected in the transformations of Chamorro language. The language now includes many Spanish loanwords, pervasively in multiple semantic fields including foods, colours, numbers, and kinship terms. Japanese and American English have contributed a few more loanwords and even have altered the patterns of speech and phrasing.

The use of writing has promoted some unfortunate errors in representing various aspects of Chamorro culture that were not originally intended to be communicated in text. Even the word *Chamorro* itself has unclear origins and may not have been used in reference to the people as a whole. The correct spelling should be *Tsamoru* in modern international conventions of phonetic orthography, but the established written word is unlikely to be altered at this point. Likewise, the island of Guam should be *Guahan*, and the island of Rota probably was once something similar to *Zarpan* as mentioned in the earlier Spanish notes and maps. If ever a single indigenous name applied to the islands as a whole group or to the people who lived there prior to the Spanish conquest, then this information has been lost. Among the efforts to develop a modern indigenous identity, *Taotato Tasi* refers to "people of the sea".

For each of the islands and many specific place names, the suffix "-an" is a widely recognised Malayo-Polynesian linguistic marker referring to a "place of something".

M.T. Carson, *Archaeological Landscape Evolution*,
DOI 10.1007/978-3-319-31400-6_14

In common usage, a place with the "-an" suffix applies not just to the land mass but rather to the surrounding ocean and the larger landscape context of real human experience. Very importantly, people do not live *on* an island such as Saipan, but rather they live *in* the realm of the natural and social context of this place.

Over the last few hundred years, islanders routinely have been told by outsiders that they live *on* a particular island. This fundamental shift in perception exemplifies many of the problems that people now face when attempting to understand island landscapes of Pacific Oceania. Living *on* an island implies that it must be a small and possibly insignificant dot lost amidst a vast ocean, contrary to the viewpoint of a "sea of islands" as a fully inhabited landscape and seascape (Hau'ofa 1994). Moreover, living *on* an island would imply a degree of detachment from the place, in contrast to the reality of living *in* a network of physical, biotic, cultural, and spiritual resources that all function together as a complex living system of a landscape.

Site Inventory and Dating

Within the last few centuries, population centres mainly have been concentrated in a few parts of the larger southern-arc islands. Current population records indicate 162,800 in Guam, 48,400 in Saipan, 3500 in Tinian, and 3200 in Rota. The northern-arc islands (collectively known as the Gani) are uninhabited at present due to volcanic activity and other safety issues, except for small numbers of adventurous visitors who sometimes stay for extended periods. The modern villages largely have followed the patterns of re-location of people due to the Spanish *reducción* program of the late 1600s, modified by gradually adding more settlements over time. Starting in the 1800s, given the low numbers of surviving Chamorro people, Carolinian Islanders from areas of Yap and Chuuk in Western Micronesia were encouraged to settle in the otherwise uninhabited parts of the Mariana Islands, leading to a long-term coexistence of Carolinian and Chamorro traditions still in effect today (Madrid 2006). The northern-arc islands supported surprisingly intensive commercial use of the land for sugarcane and other endeavours at different times, but currently they are abandoned except for very few people who live in Pagan and other islands on a periodic temporary basis.

Other than the historical and modern village areas that are intimately known today, the locations of battlefields, bridges, government facilities, and other historically significant places are remembered. These sites are commemorated in educational plaques and posters, rehabilitated structures, abandoned ruins, or sometimes nothing at all. These historical landscapes have influenced modern land developments and perceptions of Marianas history.

Historians justifiably may question this chapter's summary portrait of a single time period of A.D. 1700–Present, but this segment actually is the briefest of the full sequence as presented in this book. The proposed time period here, just like all others in the preceding chapters, should be understood to incorporate several internally variable contexts. Arguably, the society and landscape changed at a faster pace and at a larger scale during the centuries of foreign imperial engagements in the Marianas. On the other hand, the natural-cultural landscape at any moment in time has existed in its own context, and therefore the continually changing conditions must be acknowledged within every defined century, decade, or even a single year.

During the years A.D. 1700–Present, contexts of foreign rule resulted in radical shifts in the basic structure and functioning of life and landscape. Marianas history typically is described as a series of foreign-rule periods up through the present U.S.-allied arrangements in the Territory of Guam and the Commonwealth of the Northern Mariana Islands (see Fig. 5.2). Each foreign system imposed its own rules of governance, management of land and resources, and ongoing modifications of the Marianas landscape.

The earlier periods of Spanish rule probably were the most drastic in imposing fundamentally different worldviews and orchestrating all of the elements of supporting this new order. With a profoundly reduced native Chamorro population, foreign missionaries and militia in concert permeated all aspects of technological, economic, social, political, and religious life. Spanish perception of the region was illustrated in a symbolic map from 1761 that depicted the world as elements of the queen's body (Fig. 14.1), with Spain in the heart, New Spain of the Americas in the stomach, the Philippines in the feet, and the Mariana Islands not labelled but simply drawn as specks just above the ankles.

Later periods of Spanish, German, Japanese, and U.S. governments involved new elements of geopolitical strategy and industrial-based economy. Maps from different years concisely convey a sense of how the landscape transformed over time. Focusing on just one example in Hagatna of Guam (Fig. 14.2), graphic snapshots from 1914, 1954, and 2005 show development of more numerous housing areas and roads, as well as artificial filling of the seashore and the interior wetland. Traditional place names can be discerned, in some cases re-named or shifted in position, along with changing arrangements of the other features of an evolving landscape.

Landforms

During this period, cultural imprints greatly modified the landforms of the Mariana Islands, in some cases massively obscuring or even obliterating large sections of the natural landforms. Machine-powered clearing and levelling of lands created large-scale patterns of flattened terrain in support of modern housing, roads, and utility services. Additionally, masses of landfill material now constitute entirely artificial surfaces filling shallow shoreline zones and wetlands such as around Naval Base Guam, Cabras Island, and the Hagatna Basin in Guam.

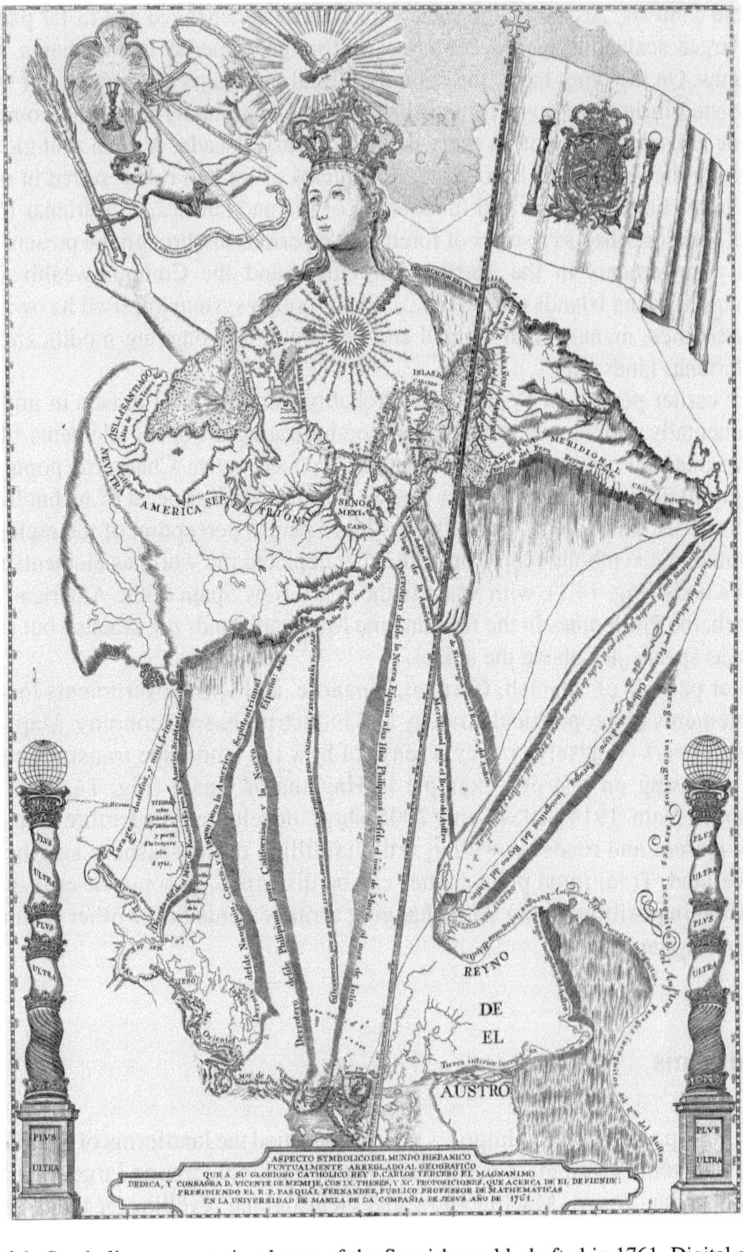

Fig. 14.1 Symbolic representational map of the Spanish world, drafted in 1761. Digital scan of original at the Micronesian Area Research Center, University of Guam

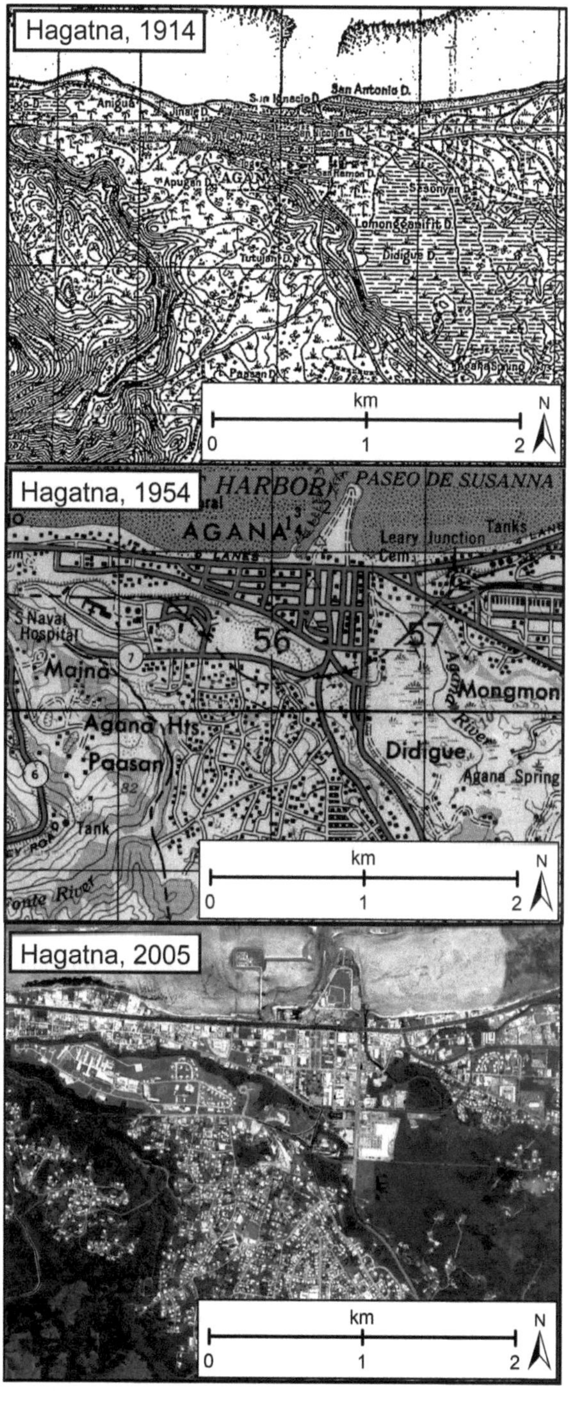

Fig. 14.2 Map series showing the same location of Hagatna in 1914, 1954, and 2005. Portions of maps were georeferenced from digital copies at the Micronesian Area Research Center, University of Guam. *Top* image: 1914 map by First Lieutenant C. L. Sturdevant, United States Army Corps of Engineers. *Middle* image: 1954 map by United States Army Engineer Base 64th Engineer Batallion. *Bottom* image: 2005 Quickbird satellite imagery

Resource Zones

Without any doubt, new technologies and lifestyles have altered the relations between people and resource zones. People have become habituated to obtain their daily necessary resources through rather impersonal and indirect means of supermarkets and government services. They meanwhile have learned how to live in residential housing areas, typically separate from their places of daily work and any other activities. Villages and natural landmarks have acquired new characteristics and traditions of Spanish and later contexts, sometimes even re-named accordingly.

Following A.D. 1700, human interactions with the Marianas landscape underwent severe transformations. The *latte* villages had been abandoned as residences, but they became respected as reminders of the past and as places of ancestors and spirits. Systems of managing lands and resource zones necessarily changed, no longer supporting the patterns of *latte* villages but instead supporting the new configuration of controlled population centres. People overall now experienced lessening opportunities to interact with the prior system of an inhabited landscape, to learn about its resources, and to situate place names and traditions within a living context of daily practice. Many people nonetheless maintained connections with the land, forest, and sea in different ways such as for hunting feral animals, tending to ranch lands, and gathering medicinal plants.

Foreign contacts brought an impressive roster of plants and animals to the Mariana Islands, thus creating new types of individual resources and functioning resource zones. Lands were used in entirely new ways to accommodate these foreign imports, in essence implementing a form of "ecological imperialism" in the Marianas. New crops included a mix of items from both Asia and the Americas, such as citrus fruits, chili peppers, maize, and other ingredients that quickly were adopted for local recipes. Pre-existing crops such as rice and sugarcane later were grown at industrial scales as part of politically controlled economic transformations of the region. New imported animals included pigs, dogs, chickens, horses, cattle, and water buffalo that massively transformed the local vegetation communities, soil properties, and overall character of the subsistence economy and patterns of land use.

In addition to the impacts of invasive species of plants and animals, the ways of managing these and other resources now became quite different from the patterns prior to A.D. 1700. With an emerging social and political order of foreign imperialism in the Marianas, lands were owned by the new rulers, by their agents, or by a select few privileged families. An overt imbalance of power forced people to enter a system of patronage with the recognised land owners or political controllers, in essence still functioning today. Most people needed to provide services and labour in exchange for the rights to use lands and resources, and some needed to pledge allegiance. Further developments followed the imposition of taxation and a money-based market economy, wherein people needed to obtain recognised monetary currency in order to survive.

Imports of plants and animals, industrial economies, and other factors of a modern world have re-structured the habitat zones of the Marianas. The landscape very strongly has been fashioned according to paved roads and personal vehicles that now are essential for connecting people with their daily resource needs. Most people over the last several decades have interacted with resource zones that are entirely artificial creations of office buildings, government institutions, and other trappings of a capitalist economy. Although they are clearly in a minority overall today, several people maintain small farms and gardens for providing at least some of their regular subsistence, and many capture seafood on a daily basis.

Material Culture

Within the last few hundred years, new technologies have changed the character of material culture. Pottery, stone, shell, and bone artefacts may be regarded as quaint reminders of a distant past. Many items of long-lasting durability have created increasingly larger volumes of a material record in concrete, metal, glass, and plastic. Additionally, a material-based capitalist economy has encouraged the accumulation of tangible and visible wealth of objects.

A full inventory of historical and modern material culture is outside the scope of this book. The key point here is to emphasise that material culture has been greatly transformed during this period, unlike anything that ever existed in the Mariana Islands previously. These transformations reflect the profound shifts in life and landscape at a fundamental systemic level.

Regional Context

The Mariana Islands historically have been a location of importance for military strategy, changing over time depending on international politics and also on war-related technologies. During the Spanish colonial period, the islands marginally supported external economic interests in the Americas and Island Southeast Asia. After maritime travel advanced with motorised vessels and allowed more sea-crossing traffic, the Mariana Islands were more reasonably within accessible range of growing numbers of international powers from different directions. Eventually with air travel and airborne missiles, the position of the Marianas has become critical in the middle of international relations from both east and west across the Pacific, of varying urgency according to how these polities relate with each other.

The current U.S. Government affiliations bring numerous benefits and well as frustrations in both Guam and the Commonwealth of the Northern Mariana Islands. Much the same can be said of any of the prior contexts of foreign rule. Discussions of political and economic independence are outside the scope of this book, but an awareness of the long-term natural and cultural landscape evolution very well may contribute to the development of truly sustainable solutions.

In a long-term view, the Marianas natural-cultural landscape has evolved through balancing traditions of looking inward and outward. Inward-looking self-reliance is essential for the meeting basic material needs of the population, while outward-looking external relations are necessary for securing a healthy role in international economics, politics, and society. The preceding chapters show that the Marianas landscape system has been strongest when it was inward-looking and weakest when it was forced to be outward-looking. Most certainly, a better balance will be needed for future sustainability of human–environment relations. These challenges inherently are exaggerated in an island setting such as the Marianas, but the lessons are instructive for all modern nations that have in a sense become artificial islands with persistently reified boundaries.

References

Hau'ofa, E. (1994). Our sea of islands. *Contemporary Pacific, 6*, 147–161.

Madrid, C. (2006). *Beyond distances: Governance, politics and deportation in the Mariana Islands from 1870 to 1877*. Saipan: Northern Mariana Islands Council for the Humanities.

Part III
Pursuing Research Questions

Chapter 15
First Inhabiting of a Landscape

Human settlement in the Mariana Islands by 1500 B.C. created a material record of a rarely captured event in world archaeology, when people initially inhabited a landscape for the first time. In other cases, archaeologists debate whether or not the oldest sites have been found or possibly have been missed in large continental settings such as reviewed in Chap. 2 concerning coastal China and California. Even in some neatly constrained island settings such as in the Hawaiian Archipelago, also mentioned in Chap. 2, the earliest periods of human presence still are ambiguous needing further investigation. Moreover in some places, archaeologists need to contend with a continuum of ancient hominid ancestors transitioning into modern humans, such as with *Homo erectus* and other species or subspecies in Africa, Asia, and Europe resulting in serious doubts about when the first 'human' experience with the landscape may have started. Given these uncertainties, archaeologists have been cautious about discussing what happens when people first encounter a genuinely untouched environment and begin the process of evolving or co-evolving with the landscape.

At some point in the murky distant past, probably at least 100,000 years ago, anatomically modern humans or their immediate ancestors ventured beyond their ancient biological home range in Africa and began interacting with previously uninhabited landscapes (Klein 2009). Today, all people inhabit landscapes that have been populated for innumerable generations. No part of the earth can be described genuinely as free of human influence, and no group of people can be described as unaffected by a preexisting landscape system. Although landscapes have evolved with people throughout the human experience worldwide, very little material evidence has been found about how this process started in any particular region.

The Mariana Islands attract special attention for their preserved archaeological records of when people first lived in this region of the world. In contrast, other areas typically offer windows into human encounters with landscapes where at least some people already had been living for an unknown length of time. In the case of the

M.T. Carson, *Archaeological Landscape Evolution*,
DOI 10.1007/978-3-319-31400-6_15

Remote Oceanic world, each of the many dozens of islands across the Pacific contains its own chronicle of first human presence, but the uniquely original instance in the region as a whole occurred when people made red-slipped pottery and lived in stilt-raised houses on the shores of the Mariana Islands. Some centuries following the initial settlement of the Marianas and continuing for several centuries, people incrementally explored and colonised the farther reaches of Remote Oceania and eventually inhabited a 'sea of islands', but these later events necessarily involved a much different natural–cultural landscape context than had existed during the first Marianas settlement period.

The oldest sites of the Mariana Islands hold invaluable material records for learning about how a landscape first became inhabited and began evolving as a complex natural–cultural system. The people who lived in this ancient landscape had no conceivable example to emulate or to avoid, so their experience was remarkably unique. Moreover, successful settlement in the Marianas required the longest ocean-crossing migration of its time, followed by the first ever human survival in the biologically foreign and socially unknown region of Remote Oceania.

The early-period Marianas sites potentially can change modern scientific understanding of how people and their landscapes first began to evolve together. In order to reach this goal, two major themes need to be discussed. First, how was the unprecedented sea-crossing voyage achieved in terms of technical practicalities, possible homeland sources, and motivating factors? Second, what processes were involved for installing and sustaining a human population in this new setting for the first time? Based on this information, the foundations of an inhabited landscape in the Marianas can be explored as a model for comprehending the essential core origins of landscapes as evolving natural–cultural systems.

Human Migration into a New Landscape

Marianas settlement at 1500 B.C. constituted the first human habitation of the Remote Oceanic world, necessarily involving a long-distance ocean-crossing migration that never before had been accomplished. These circumstances resulted in a unique case of a remotely isolated landscape system that has evolved with human presence throughout an extended sequence of changing environmental and social settings, beginning some centuries prior to any other example in the islands of Remote Oceania. The origins of natural–cultural landscape evolution thus can be examined here in ways that are not possible in any other case.

Before discussing the process of inhabiting the remote Marianas landscape, some practical issues must be addressed about how people managed their seaborne translocation across more than 2000 km in the first place. Without any doubt, the archaeological sites in themselves prove that people must have made the journey and survived with enough numbers to sustain a viable population. Nevertheless, no sailing crafts have been recovered from any pre-Spanish sites in the Marianas, so

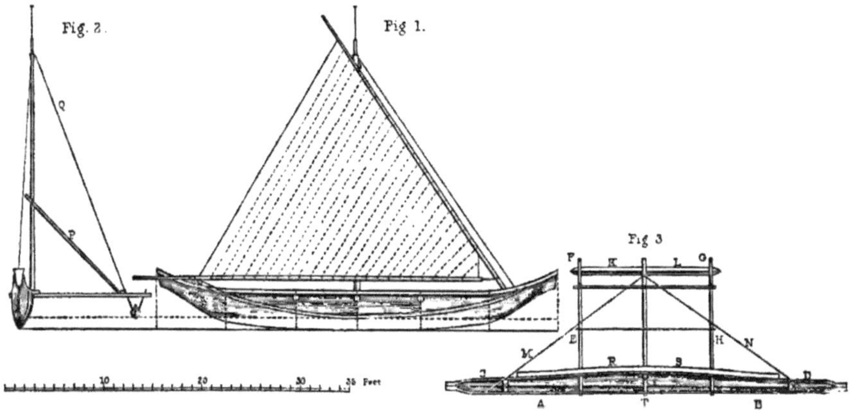

Fig. 15.1 Drawing of an outrigger sailboat in the Mariana Islands, drawn by Peircy Brett in 1742. Copy of image at the Micronesian Area Research Center, University of Guam

basic questions have arisen about the sailing technology, navigational knowledge, possible homeland origins, and conceivable motivations of the people who made this remote-distance migration.

Historical observations provide important information about the sailing crafts and skills of sailors during the 1500s through 1700s, serving as clues about more ancient periods. A single-outrigger sailboat apparently was manoeuvred with great skill in the Marianas and other islands of Micronesia over long distances and several continuous days at sea (Haddon and Hornell 1936–1938; Horridge 1995). No double-hulled canoes were reported in this region as had been described historically in Polynesia, so the older sailing traditions in the Marianas presumably did not include these features.

The first detailed sketch of a Marianas outrigger sailboat was drafted in 1742, by Peircy Brett during Lord Anson's visit in the islands (Fig. 15.1). At that time, Lord Anson marvelled at the ability of the small craft to sail very close into wind and with a speed greater than any other known vessel (in Barratt 1988). These characteristics very likely aided in shortening the number of days at sea for a long-distance voyage, critical when encountering circumstances of needing to travel against the prevailing winds and currents. A back-and-forth tacking or shunting strategy would allow progress against the wind, but the journey would become 2–3 times longer than a direct-line trip. When considering that a voyage from Island Southeast Asia to the Mariana Islands conceivably could require some weeks at sea, the abbreviation of the adventure by a few days would make a significant difference for the survival of the crew.

Prior to the 1742 documentation, reports of Marianas sailing canoes were sparse yet valuable. The oldest written description refers to Magellan's famous first encounter in 1521. According to Francisco Alvo's log book of the event on 6 March 1521 (in Stanley 1874:223): 'We saw a quantity of small sails coming to us, and they ran so, that they seemed to fly, and they had mat sails of a triangular shape, and they went both ways, for they made of the poop the prow, and of the prow the poop, as they wished'. This description likely was the source of today's popular reference to the 'flying proa' of the Marianas, highlighting the seemingly magical flying qualities of a canoe that could move quickly over the water and change direction at the whim of the sailors. The sailing direction could be changed by moving the sail, in an operation technically considered 'shunting' as distinctive from the operation of 'tacking'.

The word *proa* possibly relates to a generic word for a canoe in various parts of the Philippines and Indonesia, but the strongest linguistically attested word for a sailing canoe is *sakman* in Chamorro with direct cognates in the Malayo-Polynesian languages spoken throughout most of Island Southeast Asia (Pawley 2007). Additionally, the Chamorro language retains vocabulary for sailing terms and navigational knowledge, shared with other Malayo-Polynesian languages at a level indicative of an older shared ancestry prior to the independent developments of these languages in their separate areas (Pawley and Pawley 1994). A different vocabulary developed later among the Oceanic-speaking groups in Melanesia and Polynesia, referring to inventions of double-hulled designs, a fixed mast, and other technologies that evidently were not part of the *sakman* in the Mariana Islands or elsewhere in the farther western Asia-Pacific region (Pawley 2007).

According to the available historical and linguistic clues, something like a *sakman* probably carried the first migrants to the Mariana Islands. This vessel most likely was a single-outrigger sailing canoe, with a triangular sail that could be repositioned for redirecting the boat rapidly and efficiently. Moreover, the small and light vessels could be sailed at an impressive speed and at a close angle against the wind. The ancestral versions of *sakman* may not have been capable of storing much cargo other than a few people, but they could manage a long-distance voyage if guided by skilful and knowledgeable crew members. A successful voyage with a founding population may have included several of these vessels.

Sailing skills and knowledge very likely were known in Island Southeast Asia for several centuries prior to the eventual sea-crossing all the way to the Mariana Islands. In a setting where so many islands were inter-visible and accessible mostly in crossings of 200 km or less, Island Southeast Asia may have been a 'voyaging nursery' for people using simple watercraft and especially for people using single-outrigger sailing canoe prototypes of the Malayo-Polynesian *sakman* (Irwin 1992, 1998). In theory, this setting could nurture seafaring skills and knowledge among several widespread communities of the region, and sea voyages of short distances likely were experienced by many people with some frequency.

Historically, the seas of Island Southeast Asia were the home ranges of 'sea nomads' or 'sea gypsies', such as the Bajau and Orang Laut (literally 'sea people') who potentially contributed to long-distance seaborne migrations (Chen 2002; Sather 1995). These sea-oriented people periodically traded their sea catches and overseas goods with land-based communities for necessary access to plant foods, timbers for making their boats, and other resources. The mutually beneficial symbiotic relationships presumably involved more than exchange of a few basic materials, and conceivably these contexts facilitated long-distance transmittal of news and information, as well as general knowledge of remote lands and seas.

Sea nomads in the strictest definition would not have been solely responsible for establishing residential sites on the shores of the Mariana Islands or anywhere else, but they may have cooperated with others. Sea nomads very likely knew about overseas islands and brought this information back to others, but they would not be equipped by themselves to install permanent sedentary villages with a full repertoire of stilt-raised houses, growth of tree crops and root-tuber crops, and decorative pottery-making traditions. People who live most of their lives in boats at sea are the least likely groups to produce the distinctive forms of red-slipped pottery with the very rare dentate-stamped zone-filling designs found in the earliest Marianas sites as well as in potential homeland sites of Island Southeast Asia. These kinds of outcomes would have required at least some people who were familiar with land-based sedentary lifestyles and cultural traditions of Island Southeast Asia prior to 1500 B.C. to settle in the Marianas and re-enact the same traditions there.

A homeland source for the first Marianas settlers must have involved Island Southeast Asia, specifically in a place where people made the diagnostic early-period pottery with red-slipped and finely decorated surfaces. The additional trait of using white lime-infill in the pottery further indicates a homeland where people were familiar with chewing betelnut quids and preparing slaked lime as a necessary ingredient for the narcotic effect of the quids. As discussed in Chap. 8, this unique combination of traits occurred for the first time in the northern and central Philippines, as early as 2000–1800 B.C.

An origin of the oldest Marianas pottery-making traditions so far can be traced to the northern through central Philippines, but other places in Island Southeast Asia cannot be completely dismissed as potential contributing sources. The distinguishing use of white lime-infill to highlight dentate-stamped zone-filling over red-slipped vessels did not occur by accident, and so far it has been verified in only a few sites of the Philippines prior to 1500 B.C. that could be candidates of a Marianas homeland. However, a generic red-slipped pottery has been found in a much wider distribution throughout the Philippines and many parts of Indonesia by 1500 B.C., in a few rare cases with a coarser or bolder type of dentate-stamping occasionally with lime-infill of uncertain dating. Given the amount of archaeological research now accomplished in Island Southeast Asia by several local and international teams, the likelihood is rather small of discovering a previously unknown source of early-dated decorated pottery (Bellwood 1997; Hung 2008; Simanjuntak 2008; Tanudirjo 2001). The available evidence indicates an overall north-to-south spread of pottery-making traditions, beginning in the far northern Philippines as early as 2200 B.C. (Bellwood and Dizon 2013) and then proceeding southwards through other portions of the Philippines and eventually into Indonesia by 1500 B.C. in some cases but continuing through 1000 B.C. in others (Bulbeck and Nasruddin 2002; Spriggs 1999, 2007).

The homeland region as indicated by pottery traditions accords well with independent lines of evidence from genetics and linguistics, discussed in Chap. 5 but summarised briefly here. Modern DNA lineages indicate that the Chamorro people today descended from ancestors different from those responsible for other Remote Oceanic island settlements, instead matching more closely with populations now living in scattered areas of the Philippines and Indonesia (Vilar et al. 2013). Similarly, the Chamorro language very clearly had originated from a source in Island Southeast Asia, probably in the Philippines but close to the time when Proto Malayo-Polynesian languages were diversifying in the Philippines and perhaps in Indonesia as well, definitely prior to the developments of the Oceanic branch of Malayo-Polynesian languages spoken everywhere else in Remote Oceania (Blust 2000; Reid 2002; Zobel 2002). Additionally, Robert Blust (2009) observed that the Chamorro language retains features of people who always lived within a true typhoon zone such as in the Marianas and the Philippines.

Conceivably, the founding Marianas population included people from multiple villages, and possibly sea nomads were among the group. A homeland most probably was in the northern to central Philippines in terms of the contributions of pottery-making traditions and language, while the genetics lineages may have come from the same region or more broadly in Island Southeast Asia. In any case, Marianas settlement occurred during a context of rapid and widespread dispersals of pottery-making traditions, residential settlement patterns, Malayo-Polynesian languages, and presumably a number of genetics lineages into different areas of Island Southeast Asia. The patterns seen today in language and genetics reflect this broad geographic distribution of populations all related with the first Marianas settlers, their ancestors, and their descendants. In this situation, the most reasonable way of refining the search for a Marianas homeland is through archaeological dating of specific pottery assemblages and other materials.

Although occurring around the same time as other population movements in Island Southeast Asia, Marianas settlement was uniquely different because it involved a remote-distance ocean-crossing and installation of a population in an otherwise uninhabited place. In all other cases, the spread of red-slipped pottery-making traditions consistently stayed within the pre-inhabited region of Island Southeast Asia and within this manageable 'voyaging nursery'. The novel event in the Marianas was strikingly different from the established patterns of its time, and it invites much modern scholarly speculation about its technical practicalities and theoretical motivations.

Settlement of the Mariana Islands from a source in Island Southeast Asia generally would require sailing from west to east, against the predominant winds and currents that flowed from east to west, but this voyage evidently must have been possible. Sailors could take advantage of seasonal shifts in wind patterns, or they could navigate on long manoeuvers at angles against the wind. Geoffrey Irwin (1992, 1998) has proposed that ancient exploratory voyaging of the Pacific Ocean very likely followed these strategies, and moreover most of the major colonising episodes in the Pacific indeed were accomplished against the dominant wind directions. In theory, exploration in long back-and-forth tacking or shunting against the wind could provide maximum scouting coverage of a region, slowly moving from west to east, while simultaneously providing assurance of a quick return voyage running with the wind back to a known point of origin in the west whenever desired.

For sailors following this kind of an upwind-moving strategy, the Mariana Islands would present a wide 'target' stretched over more than 750 km north to south, more easily discovered than a narrow target of a few islands clustered closer together. Furthermore, advance scouting information would be invaluable for planning a successful colonising migration voyage with a skilled crew, enough provisions, and motivated people for establishing a new colony in a remote location.

Modern computer simulations of major wind and wave patterns so far cannot confirm a voyaging route from the Philippines to the Mariana Islands during any time of the year's seasonal wind shifts (Fitzpatrick and Callaghan 2013), but such voyages did occur in recorded history. As was mentioned in the end of Chap. 13, at least one historically known Chinese merchant (named Choco in Guam) was lost at sea when traveling from Manila (in the Philippines) to Ternate (in Indonesia), yet he somehow brought his small and presumably damaged craft to Guam in the 1640s (Garcia 2004:190). Also as noted in Chap. 13, the Spanish galleons from 1565 through 1815 followed a route from Manila northwards around the north end of the Mariana Islands on the way to the Americas. These ships potentially could stop in the northern islands of the Marianas or even follow the long north–south island arcs. Many ships were guided successfully along these routes, and a few were wrecked in the Marianas. Gonzalo de Vigo, originally from Magellan's expedition, deserted the *Trinidad* in 1522 and stayed at Maug in the far northern end of the Marianas, later collected during Salazar de Urdaneta's visit in 1526 (in Barratt 2003:41). Notably, the earlier European reporters admired the swiftness and direction-changing capabilities of the sailing canoes in the Marianas, likely more suitable for quick sea-crossing voyages than the large galleons that nonetheless managed.

Other possible voyaging routes to the Marianas may have passed through 'stepping stones' or 'staging areas' of other islands, such as in northeastern Indonesia or possibly in Palau and Yap. So far, neither Palau nor Yap has produced any archaeological traces as old as the first Marianas sites, so their role in supporting early Marianas settlement is questionable. The oldest known sites in Palau are dated about 1100–1000 B.C. (Fitzpatrick 2003; Fitzpatrick and Nelson 2011), and the oldest sites in Yap are unclear but probably no earlier than the findings in Palau (Intoh 1997). Islands of northeastern Indonesia contain a number of sites with generic red-slipped pottery, without the distinctive decorations of the Philippines and Marianas collections, dating as early as 1500 B.C. in Halmahera and the Talaud Islands (Bellwood 1997; Spriggs 2007; Tanudirjo 2001). Potentially these islands were known by expert seafarers who voyaged to the Marianas, and a voyaging route through these areas may have afforded greater chance of success in reaching the Mariana Islands.

According to the current archaeological evidence, nobody from any place in Island Southeast Asia made a successful sea-crossing migration into the realm of Remote Oceania prior to the first Marianas settlement about 1500 B.C. People potentially explored the ocean and knew about distant islands at an earlier date, but they did not make formal colonising migrations with material archaeological signatures until later. In any case, the successful sea-crossing to the Marianas must be understood as a novel development, for the first time accommodating a movement of people to live in a remote and isolated setting of small islands in the western Pacific.

The motivations presently are unknown for a remote-distance colonisation of the Mariana Islands, but several hypothetical scenarios can be proposed. The individual migrants did not need to agree on a single motivating factor, but rather they needed only to participate in the same voyage. Possible social contexts involved escalating population pressure in the coastal niches of Island Southeast Asia, negative impacts on the preferred natural resource supplies in the homeland region, or incentives for junior-ranked siblings to seek distant lands where they could establish their own lineages (Bellwood 2013:191–197).

The motivations for Marianas settlement may have entailed more extreme forms of the same reasons for any kind of migration, justifying the exceptional effort of the long journey, unprecedented isolation, and absence of any other people in the Marianas. These considerations could suggest that people followed drastic measures to escape from threats of warfare, religious or other persecution, famine, or natural disaster. Additional possibilities could relate to the need for a land-based outpost and supply station supporting sea nomads, forward planning of hopeful trading merchants, or others.

Initial Inhabiting of a Landscape

Whenever people migrate from one region to another, they bring with them a set of cultural practices, beliefs, language, and ways of interacting with the world. Immigrant populations create new aspects of an inhabited landscape wherever they settle. Remote islands are especially important for accessing archaeological and palaeo-environmental information about how these processes first began in places that previously lacked human habitation.

Unlike in the Marianas and other remote Pacific Islands, the archaeological records of human migrations in most other parts of the world refer to people moving into territories that already have been inhabited, perhaps for several millennia (Anthony 2007; Bellwood 2013). In these situations, new landscape systems developed through blending of the traditions of immigrants and indigenous groups, with variable outcomes of intrusion, innovation, and integration of the different traditions (Green 2000). These compound processes occurred when immigrant groups established sedentary residential settlements in Island Southeast Asia and presumably interacted with the preexisting mobile hunter-gatherer communities around 2000 B.C. (Bellwood 1997). Their later descendants, however, engaged in different contexts of inhabiting the landscapes of the remote Mariana Islands.

The processes of inhabiting the Mariana Islands, much like in any other remote island setting, involved three logically interlinked but qualitatively different steps of island discovery, sea-crossing migration, and finally the actual settlement (Graves and Addison 1994). In most cases, the archaeological record has preserved only the last step in this process, when people lived in a landscape and generated substantial archaeological materials attesting to their presence. The oldest Marianas sites indicate that settlement began by 1500 B.C., thus hinting at an unknown earlier timing of island discovery and the initial sea-crossing migratory voyage of the first successful settlers.

Marianas settlement was successful not only because of the discovery of the islands and making the unprecedented sea-crossing, but rather it was successful ultimately because people were able to establish a viable community in this place for the first time. In this case, a viable community of course involved having sufficient numbers of males and females for biological reproduction, but further it involved having people with the knowledge and skills to enact a full cultural system. The archaeological record reveals that people lived in stilt-raised villages, engaged in shoreline-oriented economies, shared red-slipped pottery and other traditions, and accordingly must have sustained these activities through social, political, ideological, and other factors of their cultural system.

The inhabited landscape system in the Marianas resulted from a set of influences from traditions in Island Southeast Asia, as well as new modifications or inventions locally specific to the Mariana Islands. In many ways, people acted on their prior notions of how to inhabit certain ecological niches and how to interact with the environment, based on the preexisting traditions in Island Southeast Asia prior to 1500 B.C. In other ways, people needed to adjust their modes of life or invent new strategies for coping with the specific conditions of the small and isolated Mariana Islands.

The first Marianas settlers installed their residential sites in certain shoreline niches that offered the most productive resources in familiar settings, mirroring the 'floating villages' on seashores, lakeshores, and riverbanks of Island Southeast Asia. These settings in the Marianas all were situated on narrow beaches with access to coral reefs, mangroves or similar shallow-water marshy zones, and forested island interiors. Houses were raised on wooden posts or cut tree trunks very close to the ancient shorelines, where shellfish comprised the vast bulk of protein in the local diets.

Necessary nutrition from plant foods may have been problematic in the Marianas, much like anywhere else in the islands of Remote Oceania that lacked the natural occurrence of plants and animals known in Southeast Asia. Of particular importance were starchy plant foods, such as the preferred varieties of taro and yam that needed to be imported. Bananas must have been imported as well, because they could not propagate without human mediation. Assorted sago-type palms, cycads, and an endemic seeded breadfruit (*Artocarpus mariannensis*) may have been major attractions for people to live in these islands, rather than any others of Remote Oceania where no such supplies of starchy plant foods existed prior to human occupation (Zerega et al. 2004, 2006). In particular, the non-seeded breadfruit (*A. altilis*) needed to be imported from overseas, so the existence of its seeded form naturally in the Marianas provided a preexisting source of dietary sustenance that otherwise was not available in other islands of Remote Oceania.

First human settlement in the Marianas coincided with a sudden change in the local vegetation, but the evidence has been limited in terms of the roster of taxonomically identifiable plants that must have been imported by people. The native forests certainly began to suffer the impacts of burning and clearing, coincident with notable increase of coconuts and the first appearance of both betelnut palms and ironwood trees in these islands (Athens and Ward 2004, 2006). Coconuts had occurred naturally in the Marianas ecosystem prior to human arrival, but people began to plant more of these useful trees immediately with the first settlement sites. The betelnut palm was valued for providing a nut as one of the three key ingredients of a chewable narcotic stimulant, combined with slaked lime and the leaf of *Piper betle* shrub (Zumbroich 2008). Ironwood trees (*Casuarina equisetifolia*) are known for their hard and dense wood, in fact so dense that a log will sink in water. Other likely imported items were bananas, varieties of taro and yam, and rice as attested at the time of Spanish contact but not yet confirmed in the earliest Marianas sites. Preserved starches are present on the most ancient stone tools and pottery, as well as in the sedimentary matrices, but they have been difficult to match with known reference collections.

One of the key challenges of early Marianas settlement likely involved the limited cargo space in the sailing canoes similar to the descriptions of the *sakman*. Each vessel probably could hold only a few passengers with very little capacity to carry other supplies. Some centuries later, the double-hulled canoes of Polynesia accommodated larger crews and heavy loads of supplies. More recently, the Spanish galleons may have been slow and awkward, but their key advantage was in the cargo capacity and durability at sea for several weeks. By comparison to these later examples, the first canoes reaching the shores of the Mariana Islands almost certainly carried far fewer people and overseas supplies, but island settlement nonetheless must have been successful in order to account for the continued human presence.

The small cargo capacity and long distance of sea-crossing to the Marianas necessarily restricted the inventory of supplies that could be transported from an overseas homeland. These practical factors very likely account for the absence of translocated animals in the early Marianas sites, in contrast to the usual import of pigs, dogs, chickens, and rats with the first peopling of other Remote Oceanic islands (Wickler 2004). Perhaps only a few plant foods were imported and subsequently grew successfully, but the natural occurrence of the seeded breadfruit very well may have allowed human populations to survive during the initial years of remote island settlement in the Marianas. Nonetheless, people definitely imported betelnut palms and ironwood trees, neither of which provided basic nutrition, so the first settlers most likely brought additional plants that were useful for subsistence and other purposes.

The most essential overseas imports were the people who inhabited the Marianas landscape for the first time. These people began installing residential housing in certain shoreline niches, directly altering these zones and further creating signatures in the vegetation communities, shellfish populations, and other aspects of the general environment. They presumably assigned names to specific places where they lived, as well as to the many other places of their surroundings.

First human settlement entailed the transformation of the Mariana Islands into an 'inhabited landscape' that could be incorporated into comprehensible terms of human experience. This concept conveys a sense of the natural landscape supporting or accommodating human life, while it equally reflects a sense of human social life imbued into the physical landscape. Similar viewpoints are expressed as 'dwelling' or 'being with the world' (Ingold 2000). These values may be expressed as living *in* an island rather than *on* an island, referring to the island as an inhabited space that encompasses landforms, coral reefs, ocean waters, plants and animals, soils, villages, and individual people in terms of how they relate to human experience.

The core concept of an 'inhabited landscape' is broadly shared across the Asia-Pacific region, and similar concepts may apply in other regions. Nonetheless, specific landscapes developed individually in diverse cases, as seen in the unique qualities of the Marianas landscape system when compared with other places of the region. The evident differences in human settlement dates, physical geography, and cultural settings all contributed to variable outcomes. These factors influenced the development of particular landscapes, and their combination may have produced synergistic effects.

While people definitely altered the natural setting of the Marianas and rendered the landscape into cultural terms, they of course were affected by the natural ecosystem in several important ways. People found only limited choices of where they could live and what kinds of resources they could use. For the early settlers seeking specific kinds of shoreline niches in support of a 'floating village' type of settlement, very few options were available in dispersed areas of the larger southern-arc islands. Additional limitations were imposed by the accessibility of sources of fresh water. Furthermore, the characteristics of settlement and land use largely were defined by the natural compositions of plant and animal communities, with few transported taxa at this early period.

In addition to the challenges of inherently limited natural resources in the Mariana Islands, the first settlers needed to contend with a degree of social isolation otherwise unknown in a homeland region of Island Southeast Asia. These conditions required more self-reliance among the first generations of island settlers than typically was the case in traditional Southeast Asian societies. These qualities potentially may have been preferred by people seeking new opportunities without competition in a distant land, but they likely altered the manners of social relations, kinship structure, and other aspects of society at least during the first generations of Marianas settlement.

Origins of Landscape Evolution

Landscapes evolved for millions of years without human beings, but human beings throughout their evolutionary history have lived with landscapes as a fundamental part of their experience. The earth's natural history has been deeply intertwined with human cultural history for several thousands of years, manifest in the landscapes that all people inhabit today. In this view, a landscape at any specific point in time embodies the results of the preceding long-term processes of natural and cultural histories. These processes continually evolve or co-evolve together, but rarely have any material records been preserved about how these processes first began to interact and develop into complex functioning systems.

In order to learn about the basic mechanisms of natural–cultural landscape evolution, this book has focused on how people inhabited the small and remote Mariana Islands. The islands of Remote Oceania were the last liveable places on earth to receive human populations, beginning about 1500 B.C. in the Marianas. A few centuries later, groups settled in dozens of islands across the Western Pacific, massively de-populating the native birds and other wildlife, replacing old forests with imported species, and transforming the land itself through a veritable siege of ecological imperialism. Before Remote Oceania as a whole became a socially active landscape of a 'sea of islands' however, the first human interaction with the Remote Oceanic landscape already had occurred in the Mariana Islands and had developed there for at least a few centuries.

Human settlement of Remote Oceania has produced an archaeological record of humanity's last moments of living in a pristine landscape where no other people previously had lived. Today, no place on earth genuinely can be described as free of human thought or action, and no people on earth can be described as free of the influence of their inhabited landscapes. In this sense, humankind's 'last look of Eden' has been captured in the first settlement sites of the Mariana Islands, referring to the first encounters between human beings and the Remote Oceanic environment.

The Marianas landscape already had existed as a product of natural history long prior to human arrival, but a fully functioning natural–cultural landscape system developed rather rapidly around 1500 B.C. The new developments in this case involved a joining of natural and cultural histories, mutually affecting each other but perhaps best understood as a unified complex system. People definitely needed to adapt to the facts of remote and small island contexts, as well as the naturally existing geological landforms, sea level, climate, raw materials of stones and clays, and native biological species in the Marianas. Some of the outcomes were preserved in material records, such as the installation of stilt-raised 'floating villages' in specific shoreline-oriented ecological niches, impacts on the shellfish and other nearshore resources, and manipulation of native forest composition. Other outcomes have been more difficult to document, such as place names and traditions about the landscape that involved cognitive processing of the natural and social environment.

In general terms, whenever people inhabit a landscape, they enter into a complex and dynamic relationship with this particular part of the world, and moreover they engage in a way of relating with the larger world as a whole. The experience of landscape in one place in theory can be transferred to other places, often requiring modification according to the circumstances of the new setting, such as when people from the long-populated and large land masses of Island Southeast Asia first settled in the remote and small islands of the Marianas where no other human beings previously had been living. The first Marianas settlers engaged with the landscape primarily in ways that they and their ancestors had been doing for some time previously in Island Southeast Asia, but the natural and social conditions in the Mariana

Islands in certain aspects differed from any conceivable prior experience. The plants, animals, weather patterns, and geological formations were only partially familiar, thus requiring exploration, experimentation, and new ways of integrating these elements into cultural perceptions of the landscape. Moreover, people encountered a landscape in the Marianas that did not yet support human communities, so individual places did not yet have names, stories, and associations with social relations, histories, or myths.

The initial inhabiting of the Marianas landscape may serve as an example of the general process of people engaging with new environmental and social settings. The first Marianas settlers evidently targeted ecological niches where they could enact or re-enact lifestyles most familiar to them, although they certainly needed to adapt to the unique local conditions. More drastic change would occur for people moving from coastal areas to land-locked inland zones, from constantly humid tropics to seasonally varied climate, or from riverbank to mountaintop settings. Further change would be necessary for groups moving into places where other communities already had been living, speaking a foreign language, and engaging with the landscape in a different way.

The first Marianas settlers found opportunities to inhabit their preferred niches in an otherwise uninhabited landscape, but most other immigrant communities in the world have not been so fortunate. In the Polynesian Outliers, for example, groups coming from West Polynesia settled in pre-inhabited areas of Micronesia and Melanesia after A.D. 1000, where they were forced into whatever niches happened to be unutilised or underutilised by the indigenous residents of the region (Carson 2012). These niches typically were in marginal ecological settings that were not at all reflections of the preferred habitats of a homeland region in West Polynesia, yet people lived there and applied place-names evocative of their homeland. In some cases, the Polynesian Outlier communities traded with their new neighbours, gaining reputations for providing craft-products such as woven mats and shell ornaments that were genuine cultural expressions yet not necessarily reproduced from unique ancestral homeland traditions.

The Polynesian Outlier communities, like many immigrant groups in the world, engaged with their new natural–cultural landscapes in ways that became emblematic of their identities as immigrants yet not necessarily reflecting their ancestral heritage. In modern cities worldwide, immigrants often become identified with jobs that otherwise are undesirable or unfilled. Accepting work as a gardener, cleaner, or taxi driver surely does not relate to ancestral homeland traditions of the immigrants. Likewise, finding affordable housing typically does not allow people to re-enact their idealised notions of an inhabited landscape. In some cases, immigrants provide their host communities with foods, medicinal treatment, or music and entertainment that can allow continued expression and strengthening of cultural heritage, effectively re-animating certain aspects of a homeland landscape in the new setting. Nonetheless, ancestral traditions inevitably need to be adapted in accordance with local raw materials, social standards, and other factors. Even in the clearest cases of naming a place after a homeland, such as in New England or New York, the 'new' instance is understood as having its own unique identity as a natural environment and as a place of cultural activities notably different from the homeland context.

The Marianas settlers did not need to engage with pre-established cultural groups, but rather their experience in creating a new landscape system involved probably one of the world's most extreme cases of re-creating a homeland setting in a new place. These people successfully inhabited ecological niches largely similar to the places where their predecessors and contemporaries had lived in their homeland region, and moreover they consciously reproduced very specific traditions of red-slipped pottery with rare decorative elements. The natural and cultural landscape in the Marianas thus reflected much of the setting of an overseas homeland in Island Southeast Asia, except that here people lived in a remote and small island setting for the first time. Additionally, certain parts of the landscape system needed to be adapted as suitable for the locally available stones, clays, plants, and animals with only limited overseas imports of a few plants. The overall result was a bottleneck subset of the homeland's more diverse traditions and ways of interacting with the landscape.

When people transformed the Marianas landscape into a culturally experienced place, at least three key steps already had occurred toward making this transformation possible. First was a change in perception of the homeland region, now accepted as a place that could be exited and possibly with no future return. Second was a change in conceptualising the ocean, formerly viewed as a natural resource zone and perhaps as a territorial boundary, now additionally viewed as a conduit for overseas migration. Third was a change in thinking of the Mariana Islands not as a vague overseas location but rather as a specific landscape that could be inhabited.

The three conceptual transformations of a homeland, gateway, and receiving zone are essential for any human migration (Carson and Hung 2014), and moreover they form the basis for comprehending how people create new landscape systems. In the Marianas case, the components of an Island Southeast Asian homeland, gateway conduit across the ocean, and receiving zone in the Mariana Islands all had existed previously, but they had not been activated culturally in ways that would allow a sea-crossing migration or the creation of a new landscape system until after a certain point in time. The cultural transformations of these three components may have originated asynchronously and for unrelated reasons, but eventually they were activated in unison. In other words, the reasons for exiting a homeland may have been brewing for some time before options of migration to the Mariana Islands became available. Moreover, the technologies and skills of remote-distance ocean voyaging may have developed independently of other concerns. Finally, the ability for people to sustain a landscape system in a remote and small island setting such as in the Marianas was in many ways different from the event of ocean-crossing migration.

By the time when people were living in the Mariana Islands, the landscape was perceived as a place that could support cultural life, as noted in the concept of an 'inhabited landscape'. Rather than a vague notion of a set of faraway islands, the islands now could be named individually and understood in relation to real people and experience. Specific landforms became landmarks of common reference, including areas of residential communities but also the many non-residential areas of concern in the ocean, mountains, and other elements of the landscape. People must have developed names for these places, leading to notions of both literal and figurative meanings of those place names.

The landscape system of the first Marianas settlement period did not last forever, but rather it underwent a series of substantial transformations. After a few centuries, people needed to adapt to lowering sea level and compounded impacts on the near-shore resources precisely in their targeted settlement habitats, and meanwhile a growing population required larger habitable space, more natural resources, and increasing interactions with the landscape as a whole. The initially inhabited shoreline niches later were stranded far inland and buried beneath thick layers of sediments of today's broad sandy beaches. The physical landscape was changing, and cultural interactions with the landscape needed to change accordingly.

Landscapes constantly are in a state of change, due to both natural forces and human agency, so the initial landscape system of the Marianas should not be expected to have survived completely intact. One of the oldest settlements of Unai Bapot in Saipan originally was a narrow and somewhat marshy fringe at the base of a limestone cliff at 1500 B.C., and only much later did this place obtain the characteristics of *unai* ('sandy beach') or an association with *bapot* ('steamboat'). Nevertheless, the contexts of the earliest settlement period created the essential foundations of the next several centuries of long-term natural–cultural landscape evolution that will be considered in the next chapter.

References

Anthony, D. W. (2007). *The horse, the wheel, and language: How bronze-age riders from the Eurasian steppes shaped the modern world*. Princeton, NJ: Princeton University Press.

Athens, J. S., & Ward, J. V. (2004). Holocene vegetation, savannah origins and human settlement of Guam. In V. Attenbrow & R. Fullagar (Eds.), *A Pacific odyssey: Archaeology and anthropology in the western Pacific—Papers in honour of Jim Specht* (pp. 15–30). Sydney: Australian Museum. Records of the Australian Museum, Supplement 29.

Athens, J. S., & Ward, J. V. (2006). *Holocene Paleoenvironment of Saipan: Analysis of a core from Lake Susupe*. Micronesian Archaeological Survey Report Number 35. Commonwealth of the Northern Mariana Islands Division of Historic Preservation, Saipan.

Barratt, G. (1988). *H.M.S. Centurion at Tinian, 1742: The ethnographic and historic records*. Micronesian Archaeological Survey Report Number 26. Commonwealth of the Northern Mariana Islands Division of Historic Preservation, Saipan.

Barratt, G. (2003). *The Chamorros of the Mariana Islands: Early European Records, 1521–1721*. Occasional Historical Papers Series, Number 10. Commonwealth of the Northern Mariana Islands Division of Historic Preservation, Saipan.

Bellwood, P. (1997). *Prehistory of the Indo-Malaysian Archipelago* (Rev. ed.). Honolulu: University of Hawai'i Press.

Bellwood, P. (2013). *First migrants: Ancient migration in global perspective*. Malden, MA: Wiley Blackwell.

Bellwood, P., & Dizon E. (Eds.). (2013). *4000 years of migration and cultural exchange: The archaeology of the Batanes Islands, Northern Philippines*. Terra Australis 40. Canberra: Australian National University E Press.

Blust, R. (2000). Chamorro historical phonology. *Oceanic Linguistics, 39*, 83–122.
Blust, R. (2009). The historical value of single words. In B. Evans (Ed.), *Discovering history through language: Papers in honour of Malcolm Ross* (Pacific Linguistics, Vol. 605, pp. 61–71). Canberra: Research School of Pacific and Asian Studies, Australian National University.
Bulbeck, F. D., & Nasruddin. (2002). Recent insights on the chronology and ceramics of the Kalumpang site complex, south Sulawesi, Indonesia. *Bulletin of the Indo-Pacific Prehistory Association, 22*, 83–99.
Carson, M. T. (2012). Recent developments in prehistory: Perspectives on settlement chronology, inter-community relations, and identity formation. In R. Feinberg & R. Scaglion (Eds.), *The Polynesian outliers: State of the art* (pp. 27–48). Pittsburgh: University of Pittsburgh. Ethnology Monographs Number 21.
Carson, M. T., & Hung, H.-c. (2014). Semiconductor theory in migration: Population receivers, homelands and gateways in Taiwan and Island Southeast Asia. *World Archaeology, 46*, 502–515.
Chen, C.-y. (2002). Sea nomads in prehistory on the southeast coast of China. *Bulletin of the Indo-Pacific Prehistory Association, 22*, 51–62.
Fitzpatrick, S. M. (2003). Early human burials in the western Pacific: Evidence for a c. 3000-year-old occupation on Palau. *Antiquity, 77*, 719–731.
Fitzpatrick, S. M., & Callaghan, R. T. (2013). Estimating trajectories of colonisation to the Mariana Islands, western Pacific. *Antiquity, 87*, 840–853.
Fitzpatrick, S. M., & Nelson, G. C. (2011). Purposeful commingling of adult and child cranial elements from the Chelechol ra Orrak Cemetery, Palau. *International Journal of Osteoarchaeology, 21*, 360–366.
Garcia, F. (2004). *The Life and Martyrdom of the Venerable Father Diego Luis de San Vitores of the Society of Jesus* (J. A. McDonough Ed. & M. M. Higgins, F. Plaza & J. M. H. Ledesma, Trans.). Translated and edited version of original published in 1684. Monograph Series Number 3. Micronesian Area Research Center, University of Guam, Mangilao.
Graves, M. W., & Addison, D. J. (1994). The Polynesian settlement of the Hawaiian Archipelago: Integrating models and methods in archaeological interpretation. *World Archaeology, 26*, 380–399.
Green, R. C. (2000). Lapita and the cultural model for intrusion, integration and innovation. In A. Anderson & T. Murray (Eds.), *Australian archaeologist: Collected papers in honour of Jim Allen* (pp. 372–392). Canberra: Coombs Academic Publishing, Australian National University.
Haddon, A. C., & Hornell, J. (1936–1938). *Canoes of Oceania*. Three volumes. Bishop Museum Special Publication Numbers 27, 28 and 29. Honolulu: Bishop Museum Press.
Horridge, A. (1995). The Austronesian conquest of the sea—Upwind. In P. Bellwood, J. J. Fox, & D. Tryon (Eds.), *The Austronesians: Historical and comparative approaches* (pp. 134–151). Canberra: Department of Anthropology, Research School of Pacific and Asian Studies, Australian National University.
Hung, H.-C. (2008). *Migration and cultural interaction in southern coastal China, Taiwan and the northern Philippines, 3000 BC to AD 100: The early history of Austronesian-speaking populations*. Ph.D. dissertation. Australian National University, Canberra.
Ingold, T. (2000). *The perception of the environment: Essays on livelihood, dwelling and skill*. London: Routledge.
Intoh, M. (1997). Human dispersals into Micronesia. *Anthropological Science, 105*, 15–28.
Irwin, G. J. (1992). *The prehistoric exploration and colonisation of the Pacific*. Cambridge: Cambridge University Press.
Irwin, G. J. (1998). The colonisation of the Pacific plate: Chronological, navigational and social issues. *Journal of the Polynesian Society, 107*, 111–143.
Klein, R. (2009). *The human career* (3rd ed.). Chicago: University of Chicago Press.
Pawley, A. (2007). The origins of early Lapita culture: The testimony of historical linguistics. In S. Bedford, C. Sand, and S. P. Connaughton (Eds.) *Oceanic explorations: Lapita and western Pacific settlement* (pp. 17–49). Terra Australis 26. Canberra: Australian National University E Press.

Pawley, A., & Pawley, M. (1994). Early Austronesian terms for canoe parts and seafaring. In A. K. Pawley, & M. D. Ross (Eds.) *Austronesian terminologies: Continuity and change* (pp. 329–361). Pacific Linguistics Series C-127. Department of Linguistics, Research School of Pacific and Asian Studies, Australian National University, Canberra.

Reid, L. A. (2002). Morphosyntactic evidence for the position of Chamorro in the Austronesian language family. In R. S. Bauer (Ed.), *Collected papers and Southeast Asian and Pacific languages* (Pacific Linguistics, Vol. 530, pp. 63–94). Canberra: Department of Linguistics, Research School of Pacific and Asian Studies, Australian National University.

Sather, C. (1995). Sea nomads and rainforest hunter-gatherers: Foraging adaptations in the Indo-Malaysian Archipelago. In P. Bellwood, J. J. Fox, & D. Tryon (Eds.), *The Austronesians: Historical and comparative approaches* (pp. 229–268). Canberra: Department of Anthropology, Research School of Pacific and Asian Studies, Australian National University.

Simanjuntak, T. (Ed.). (2008). *Austronesian in Sulawesi*. Yogyakarta, Indonesia: Center for Prehistoric and Austronesian Studies.

Spriggs, M. (1999). Archaeological dates and linguistic sub-groups in the settlement of the Island Southeast Asian-Pacific region. *Bulletin of the Indo-Pacific Prehistory Association, 18*, 17–24.

Spriggs, M. (2007). The Neolithic and Austronesian expansion within Island Southeast Asia and into the Pacific. In S. Chiu & C. Sand (Eds.), *From Southeast Asia to the Pacific: Archaeological perspectives on the Austronesian expansion and the Lapita cultural complex* (pp. 104–140). Taipei: Center for Archaeological Studies, Research Center for Humanities and Social Sciences, Academia Sinica.

Stanley, H. E. J. (Ed.). (1874). *The first voyage round the world by Magellan, Translated from the Accounts of Pigafetta and other contemporary writers*. London: The Hakluyt Society.

Tanudirjo, D. (2001). *Islands in between: Prehistory of the northeastern Indonesian Archipelago*. Ph.D. dissertation. Australian National University, Canberra.

Vilar, M. G., Chan, C. W., Santos, D. R., Lynch, D., Spathis, R., Garruto, R. M., & Koji Lum, J. (2013). The origins and genetic distinctiveness of the Chamorros of the Mariana Islands: An mtDNA perspective. *American Journal of Human Biology, 25*, 116–122.

Wickler, S. K. (2004). Modelling colonisation and migration in Micronesia from a zooarchaeological perspective. In M. Mondini, S. Munoz, & S. K. Wickler (Eds.), *Colonisation, migration and marginal areas: A zooarchaeological approach* (pp. 28–40). Oxford: Oxbow Books.

Zerega, N. J. C., Ragone, D., & Motely, T. J. (2004). Complex origins of breadfruit (*Artocarpus altilis*, Moraceae): Implications for human migration in Oceania. *American Journal of Botany, 91*, 760–766.

Zerega, N. J. C., Ragone, D., & Motely, T. J. (2006). Breadfruit origins, diversity, and human-facilitated distribution. In T. J. Motely & N. J. C. Zerega (Eds.), *Darwin's harvest: Origins, evolution, and conservation of crop plants* (pp. 213–238). New York: Columbia University Press.

Zobel, E. (2002). The position of Chamorro and Palauan in the Austronesian family tree: Evidence from verb morphosyntax. In F. Wouk & M. D. Ross (Eds.), *History and typology of Western Austronesian voice systems* (Pacific Linguistics, pp. 405–434). Canberra: Department of Linguistics, Research School of Pacific and Asian Studies, Australian National University.

Zumbroich, T. J. (2008). The origin and diffusion of betel chewing: A synthesis of evidence from South Asia, Southeast Asia and beyond. *Electronic Journal of Indian Medicine, 1*, 63–116.

Chapter 16
Long-Term Human–Environment Relations

A case study in the Mariana Islands describes how a landscape has varied across geographic space and through time, but further questions yet can be pursued about how and why these patterns and trends came to exist. Two ways of thinking refer to: (a) proximate accounts of how landscape systems have operated and have involved variable factors leading to change; and (b) ultimate explanations of why these systems should have existed and evolved. In most scientific philosophies, understanding *how* a system functions in a proximate descriptive sense is essential toward explaining *why* it should exist in an ultimate explanatory sense, but the two modes of thought ideally can guide and strengthen each other.

This book does not advocate any particular interpretive or explanatory theory, but rather it aims to study archaeological records of landscapes in the broadest sense, as a means to learn about long-term human–environment relations. Toward this goal, the preceding chapters provided a substantive description of how landscape systems functioned and how they varied in the case of a 3500-year record in the Mariana Islands, and a few examples in Chap. 2 illustrated the general applicability of landscape studies in other regions. In any case, multiple interpretations and hypotheses are possible in regards to the proximate or ultimate causes of changing forms of subsistence economy, management of forests, distributions of habitations and resource-use patterns, and other chronologically traceable variables. These changing interrelated components can be interpreted through any number of theories about ecologies, such as landscape ecology, evolutionary ecology, historical ecology, or behavioural ecology just as well as through other theories such as human or non-human agencies, post-modern self-reflexivity, experiential phenomenology, cognitive systems theory, notions of materiality, and other perspectives (Hodder 2012).

The present study refers to landscape evolution, but opinions differ about what may or may not be entailed in an evolutionary process. A liberal view accepts evolution as any change through time, thus accommodating many different ways of studying chronological variation as in the case of multiple components of a natural–

M.T. Carson, *Archaeological Landscape Evolution*,
DOI 10.1007/978-3-319-31400-6_16

cultural landscape system. A stricter definition in biology refers to random variation in the genetic coding of each reproductive generation that may or may not later contribute to further reproductive success of the members of a species who happen to retain these genetic traits. The biological definition works perfectly for its intended purpose of explaining biological generational descent with potential modification, but a different explanation is needed for cultural systems that can change significantly within the lifetimes of individual people and sometimes through deliberate intent. Similarly, chronologically changing characteristics in geological structure, climate, or sea level certainly do not follow the principles of biological evolution, although the evident chronologies could be described as evolutionary.

As complex systems, landscapes incorporate multiple components that interact and change continually, sometimes through different mechanisms and often at different rates or scales of time. Concurrent rhythms of change in multi-component landscapes may be understood as similar to Braudel's (1949) depiction of human history as simultaneously involving both slow and rapid paces of change in different aspects of a society. Some elements of a landscape can change only very slowly or appear to be mostly stable over very long periods of time, while others appear to undergo rapid change. Geological landforms are shaped through processes over tens of thousands of years or longer, while the forests growing in these landforms can change more quickly and potentially within a century or less through generational reproduction and human mediation. Conditions of global and regional climate evidently have fluctuated in multiple cycles of years, decades, centuries, millennia, and longer with occasional co-occurring points in some of these cycles.

Natural–cultural landscape evolution is most productively examined through long-term chronologies and multiple lines of evidence. In practical terms, longer chronologies allow more opportunities for observing changing conditions. Within whatever time scale is available, more lines of evidence enable better comprehension of complex landscape systems. This approach provides a reasonable means to examine the multiple concurrent components and rhythms of landscape evolution.

The Marianas example in this book extended over 3500 years, during which time changing conditions were evident in geology and landforms, climate, sea level and coastal ecology, water sources, plant and animal populations, patterns of residence and resource use, and material culture. In some cases, the chronological trends in these different factors coincided and created more profound system-wide effects than any of them could have produced alone. The preceding chapters already outlined the chronological sequence in a holistic illustration of each time interval. Here the major components of the landscape system each are reviewed in terms of their primary conditions, with attention to both continuity and change through time.

Geology and Landforms

The basic geological structure and landforms of the Mariana Islands may be viewed as the necessary foundations of terrain where people interacted with the landscape. Most geological processes have occurred at time scales far beyond the span of cultural presence in the Marianas, but a few notable transformations were evident in the configurations of landforms. Formations of limestone plateau terrain and volcanic hills with ridges and valleys did not change significantly in their basic geological structure or overall appearance, but they exhibited degrees of erosion and re-deposition of material in layers near the surface. The most profoundly changing conditions occurred in coastal landforms, as well as in other low-lying terrain with vulnerability to the fluctuations in sea level and the effects of surrounding slope erosion-deposition patterns.

Overall long-term stability has been witnessed in basic formations of limestone plateau terrain and volcanic hills. In principle, human perceptions and interactions with these landforms could have been stable over time, but the archaeological evidence shows very little cultural usage of these particular landforms until later in the chronological sequence. Prior to A.D. 1000, few sites have been preserved in these kinds of inland or upland terrain settings, possibly due to the limited scope of human activities outside coastal zones until later, as well as possibly due to the very thin sedimentary layers with limited possibilities of site preservation. Nonetheless, a person viewing a mountaintop or a limestone cliff today would see much the same basic landform structure that other people would have seen hundreds or even thousands of years ago. The major difference through time has been not in the shapes of these landforms but rather in the kinds of evidence of human activities that have been preserved there.

Slope erosion and re-deposition have altered the surfaces of nearly all terrain in the Mariana Islands, most often just minimally but in some cases more noticeably. Most of the terrain surfaces in higher elevations have lost small amounts of topsoil and occasionally portions of subsoil and rocky substrate. These materials were re-deposited in places where they settled in beds farther downslope, typically at bases of slopes, along valley floors, or near the low-elevation coastlines.

Slope erosion-deposition patterns were more or less constant through time, periodically enhanced by certain conditions such as loss of forest cover, increased storminess, or fluctuating sea level that sometimes coincided for greater total effects. When people manipulated the native forests through burning and clearing, soil surfaces were more vulnerable to erosion, first traceable in very limited fashion starting about 1500 B.C. and thereafter continuing variably. Additional soil vulnerability could occur during times of more punctuated storminess and sudden water flow with runoff, especially effective after periods of drought, so far not showing any major chronological patterns in the Marianas other than an increased frequency of storminess during the Little Ice Age about A.D. 1300 through 1850. Compounding with these factors, overall lowering sea level slowly allowed more of the slope-eroded material to accumulate over the increasingly stable lowland zones that became exposed higher above sea level in valley floors, wetlands, and coastal areas, magnified continually over time but with discernible incremental steps after 1100 B.C., A.D. 100, and A.D. 1000.

Coastal zones in the Marianas have been most vulnerable to changing conditions, and moreover they were the scenes of some of the most important cultural activities that needed to adjust with the dynamic natures of coastlines. The oldest Marianas settlement sites around 1500 B.C. were situated in unstable shoreline settings that soon were transformed by lowering sea level and new conditions of coastal landforms and ecologies, progressively altered after 1100 B.C. Coastal landforms gained greater stability after A.D. 100, but the broad coastal plains as seen today were stabilised closer to A.D. 1000. While coastal life consistently was a key defining factor in the Mariana Islands, the precise modes of living in coastal habitats needed to adapt in accordance with the ever-changing coastal morphologies and attendant ecologies.

Climate

Humid tropical climate, monsoonal weather pattern, and frequent typhoons have characterised the Marianas region for several thousands of years. These natural characteristics greatly have shaped the native vegetation, enabled luxurious plant growth, and guided cultural perceptions of annual and seasonal cycles. These aspects of the local environment would have been familiar to people coming from Island Southeast Asia, especially in the northern to central Philippines.

Steady hot temperatures and large amounts of annual rainfall have sustained minor fluctuations without any deep effects on the overall qualities of a humid tropical climate. Nonetheless climatic fluctuations certainly have occurred, perhaps with most significant outcomes relating to the predictability of rainfall patterns. Prolonged periods either without rain or else with unusually large amounts of rain could cause problems in planning for crop growth and general management of natural resources. The mostly stable and predictable conditions of the Little Climatic Optimum likely supported more reliable as well as abundant crop growth in support of larger populations during the centuries A.D. 1000 through 1300, followed by unpredictable unsteadiness of the Little Ice Age during the years A.D. 1300 through 1850, deserving of more research in relation to *latte* period societies and landscapes of the Marianas.

Sea Level and Coastal Ecology

Changing sea level directly affected coastal landforms and ecologies, of special significance for people living closely with seas and seashores of the Mariana Islands. The overall trend was a drawdown of sea level, starting about 1100 B.C. and continuing until reaching near modern conditions about A.D. 1000. Within the prolonged period of drawdown, at least one temporarily stable period occurred about A.D. 100–200, and others may yet be discerned. More recently, sea level has been rising, of great concern for many people living in small islands of Pacific Oceania.

Drawdown of sea level contributed to enlarged coastal plain landforms, infilling of low-lying wetland zones, and lowering of the absolute elevation of a freshwater aquifer lens floating over the sea level inside island land masses. Additional effects included the loss of habitat for supporting mangroves and certain kinds of shellfish such as *Anadara* sp. that otherwise would have provided important food supplies. Marshy shoreline zones transitioned into saltwater and brackish swamps, later to freshwater wetlands, and eventually to infilled dry landforms. Meanwhile, new habitats were created, better suited for different kinds of shellfish than in older periods, and eventually new coral reefs and nearshore ecosystems developed with healthy productivity.

The first Marianas settlers about 1500 B.C. evidently targeted a narrow range of shoreline niches, and these niches remained stable for at least a few centuries before sea level drawdown after 1100 B.C. In this context, the targeted shoreline niches provided stable and reliable supplies of shellfish and other resources. The only major transformative effects in these habitats prior to 1100 B.C. were caused by people who harvested certain resources. This kind of harvesting likely caused localised depression of favoured shellfish around the major habitation sites, later compounded with the effects of lowering sea level for potentially disastrous results.

By 700 B.C., people no longer could sustain the older patterns of shoreline-oriented settlements in the Marianas, and instead they needed to shift their lifestyles to include stronger focus on land-based resource zones. Coastal life by no means became lost, but rather the prior narrow specialisation in certain types of shorelines was no longer sustainable. People needed to broaden their use of resource zones and engage quite differently with the landscape than had been the case during earlier centuries.

Increasingly stable coastlines and thus more reliably stable coastal ecologies existed by A.D. 100 and especially after A.D. 1000, quite similar to modern conditions as observable today in the Mariana Islands. Coastal stability could support long-term and large-scale use of these resource zones, as reflected in the archaeological records of these later periods. By this time, however, people had developed much stronger usage of inland and upland terrain and habitats of the island interiors. A diversified broad-spectrum landscape system had been developing at least since 700 B.C. and probably earlier, and it became most extensive and formalised after A.D. 1000.

Water Sources

Water sources have changed over time in the Marianas in terms of their accessibility to human populations, primarily due to lowering sea level and related alterations of low-lying landforms where water could be accessed. As sea level lowered, a freshwater lens accordingly lowered in its absolute elevation floating over the saltwater base, where it could be accessed inside deep caves, in lowland basins, and in seeps just above sea level. Other sources of fresh water have been more or less steady without significant chronological change in rivers and streams, although they were restricted in number, primarily in southern Guam but also including a few in Saipan. Additionally, rainfall provided opportunity for water catchment with naturally periodic fluctuations in reliability.

Curiously, the oldest known settlement sites in the Marianas have not been found along riverbanks, but rather they have been found on ancient seashores where freshwater sources must have been obtained through rainfall catchment, inside deep cave pools, dripping from cave ceilings, or in coastal seeps. Prior to 1100 B.C., the higher sea level would have made pools of fresh water accessible inside a greater number of caves than is the case today with a lower level of a freshwater lens in the islands. While the freshwater sources inside caves steadily diminished over time, others developed in basins such as Susupe in Saipan and Hagatna in Guam, where lowering sea level eventually created saltwater marshes and later freshwater wetlands. Freshwater wetland habitat developed by A.D. 500 at Susupe and around A.D. 1000 at Hagatna.

Catchment of rainwater very likely was essential for human populations in the Mariana Islands, and accordingly at least some of the pottery vessels appear to have been used for water storage. The large-sized vessels after A.D. 1000 were especially suitable for containing large quantities of water, coincident with the greatest population size and settlement of the northern-arc islands lacking accessible potable groundwater. Coincidentally, the overall warmer and wetter conditions of the Little Climatic Optimum would have made rainwater catchment more reliable and productive during the years A.D. 1000 through 1300, but more punctuated storminess and occasional rainless periods after A.D. 1300 may have created less stability in rainwater collection.

Plant and Animal Populations

Biotic communities (plant and animal populations) famously are vulnerable to human interventions, but they equally must be recognised as products of natural history. Throughout the world, people have depopulated certain ranges of plants and animals, increased the numbers of some, imported others, and generally interfered with the composition of species biomass and diversity. People cannot have been responsible, however, for the geographic isolation, geological formations, and climate of the Mariana Islands that all influenced the kinds of plants and animals living here. They similarly could not have been responsible for the natural existence of megafauna in the American Continents, although they certainly adapted with technologies specifically suited for hunting these animals. People certainly adjusted their behaviours in accordance with the given environment and available resources, but those facts of natural history already had existed prior to human presence.

Human influences on biotic communities apparently have been exaggerated in places like the Mariana Islands, where fragile ecosystems had evolved for several thousands of years prior to a sudden invasion by human beings. Many of the native species had evolved without natural defence mechanisms and behaviours against predators. They likewise had not adapted to compete with other species invading or disturbing their preferred habitats.

In the Mariana Islands, an initial anthropogenic impact horizon has been very clear in palaeo-botanical evidence and only just now beginning to be detected in the archaeo-faunal records. The palaeo-botanical archives show thousands of years of undisturbed native forest growth, followed by abrupt burning and clearing of the forests, decline of native taxa coincident with increased growth of grasses and disturbance-related taxa, more growth of economically useful plants like coconut palms, and the arrival of betelnut palms, ironwood trees, and presumably other taxa imported from overseas by 1500 B.C. or slightly earlier. The archaeo-faunal records show decline in certain nearshore shellfish taxa and perhaps in beach-nesting turtles by 1100 B.C., but so far no major impacts on native birds have been documented until much later and especially after A.D. 1000 with the arrival of rats.

Prior to the Spanish occupation in the Mariana Islands, domesticated animals and formal agricultural fields were not parts of the natural–cultural landscape system. People certainly influenced the compositions of plant and animal resources, but they did not enter into deep interdependencies with livestock or crops. Despite periodic contacts with external groups who relied on domesticated animals and intensive agricultural field systems elsewhere in the Asia-Pacific region, these customs did not emerge in the Mariana Islands.

Patterns of Residence and Resource Use

In a large-scale view over the last few thousands of years, people continually lived in larger communities, in greater numbers of sites, and in more diverse types of settings. Some of this change was due to simple population growth and increasing density within the limited space and resources of the Mariana Islands. Other factors involved changing configurations of the available landforms and resource zones, caused by people and by various forces of nature.

In the Mariana Islands, patterns of residence and resource use became increasingly generalised and diversified over time. The first settlers by 1500 B.C. lived in narrowly defined shoreline-oriented niche habitats, but this practice eventually proved unsustainable due to changing coastal ecologies after 1100 B.C. and crossed a threshold of tolerable limits by 700 B.C. By this time, people already had started to rely more on land-based resource zones outside the shoreline niches, and this trend continued along with population growth. After A.D. 1000, large numbers of people lived in diverse settings of all of the Mariana Islands, where they engaged in generalised broad-spectrum management of habitats and natural resources in island interiors, coastlines, and oceans.

Population size cannot be concluded precisely from the Marianas archaeological remains, but approximate parameters can be based on the number of house structures or the total spatial extent of habitation layers. Numbers of houses or households are most easily estimated in the surface-visible sites post-dating A.D. 1000, especially where *latte* stone ruins are discernible as the remains of individual housing structures, but even these surface-visible cases include cultural deposits extending beyond the footprints of definite houses. In the deeper subsurface findings, the overall spatial extent of a residential complex needs to be estimated from limited windows of test-sampled portions that presumably once were connected as a whole.

The overall size, number, and distribution of residential sites increased steadily throughout the Marianas chronological sequence, with one apparent disruption in the trend. By 700 B.C., the preceding pattern of shoreline-oriented housing was no longer sustainable in elongate communities parallel with seashores, and instead people shifted to live in a greater number of smaller-sized clusters in slightly landward settings removed from the active shorefronts. These habitations after 700 B.C. each appeared to be smaller than in prior centuries, but they were more numerous in total. By A.D. 1, these habitations were increasing in size, and several new places were being used for residential and other activities.

The disruption around 700 B.C. signalled a system-wide change in the natural–cultural landscape, already underway since 1100 B.C. but reaching a threshold value about 700 B.C. forcing a major change. This turning point involved falling sea level and changing coastal ecologies, compounded with human-caused effects on shellfish and other nearshore resources. The physical landform structure was changing, along with composition of shellfish and other nearshore resources. People adapted by shifting their residences to slightly landward settings, where they adjusted their subsistence economy to draw on a broader range of resources other than the disappearing shoreline niches. Pottery forms and other artefacts meanwhile expressed less investment in decorative pieces and more concern with simple production and basic utilitarian demands.

The next profound change in the landscape system occurred around A.D. 1000, when people began living in villages formalised in stone-pillar *latte* structures and soon established these formal village systems throughout the islands. By this time, coastal zones had stabilised with new conditions and productive coral reef ecosystems, and coastal living remained quite strongly part of the Marianas communities. Meanwhile, land-based resources had become increasingly important over time and by now supplied much of the nutritional food base for an evidently large population.

Material Culture

The archaeological assemblages of the Marianas exhibit several changing components over time, in some cases linked with shifts in residential patterns, use of resource zones, and other factors of the landscape system. Abundant earthenware pottery provides the most robust reference for tracing chronological changes, but additional evidence is found in shell ornaments and other artefacts. In most cases, the chronological transitions were gradual over the course of some centuries each, although a few points were more abrupt.

Pottery traditions and shell ornaments followed similar trends of diminishing artistic output, apparently coinciding with increasing demands of subsistence economy in support of a continually growing population. The most finely made pottery of the early settlement period exhibited a definite change toward thicker, coarser, and less decorated varieties by 1100 B.C. when coastlines were transforming, followed by more pronounced change after 700 B.C. when people significantly changed their patterns of residence and resource use. A parallel trend was witnessed in shell beads and other ornaments. By A.D. 1000, pottery was large-sized, thick-walled, coarsely made, and minimally decorated while people lived in large numbers of formalised village complexes throughout the islands.

After A.D. 1000, the material signatures in the Marianas reflect a formalisation of social practice, likely related to an ordering of social, economic, and political relations among an evidently large number of people who needed continual multi-generational access to specifically defined territories. Large-sized and long-lasting objects were produced in the forms of stone-pillar *latte* house supports, stone mortars or grinding basins, large and thick pottery, and other material output at an impressively large number of sites. Moreover at this time, people were buried at their individual houses, suggesting an intended permanence of links between lineages and their fixed territories. Diversified economies relied on access to variable resource zones, but increasing roles of trees and root-tuber crops necessarily made people land-dependant within the necessarily limited island masses.

Continuity and Change

Both continuity and change have characterised the Marianas landscape over more than three millennia. The earliest and latest periods were remarkably different from one another in almost every aspect of the natural cultural landscape system, yet they were joined by a continuous sequence of changing conditions. Continuity is clear in the unbroken occupation since the time of first human presence in the region, with only one major disjuncture around A.D. 1700 due to the Spanish conquest. Prior to this historical event, the multiple components of the landscape system changed at different concurrent rates over time, with synergistic coinciding factors on rare occasions.

Long-term system-wide continuity clearly did not persist in the Marianas, or else the oldest and youngest time periods would have yielded much the same evidence of landform structure, natural resources, forms and distributions of housing, land-use patterns, and types of artefacts. All of these factors changed through time, mostly through a series of gradual transitions but rarely through points of punctuated transformation. One significant point of systemic transformation involved a loss of the originally targeted shoreline niches after 1100 B.C., and another involved the emergence of formalised village complexes and widespread territorial land-use patterns after A.D. 1000.

People did not uniformly abandon all of their shoreline habitations in a single event precisely at 1100 B.C., but rather several factors changed over the course of some centuries. By 1100 B.C., initial effects were evident in lowering sea level, changing shorelines, progradation of coastal landforms, declining amounts of preferred shellfish foods, enlargement of habitation zones, and transitions to less refined pottery and shell ornaments. In the context of these changing conditions, people continued to live in the shoreline niches until this lifestyle was no longer sustainable. After 700 B.C., the original shoreline niche settlement pattern no longer existed, but instead people had shifted into a new form of mixed coastal and inland patterns of residence and resource use. A fuller broad-spectrum pattern of settlement and resource use, however, would develop over another several centuries.

The emergence of formalised and widespread villages and land use after A.D. 1000 must be understood as the complex result of several factors that happened to coincide after many centuries of ongoing change. A shift toward broad-spectrum patterns of settlement and land use had been underway since at least 700 B.C. and arguably since 1100 B.C. if not earlier. Meanwhile, population size presumably had been growing steadily, with limited relief of density due to expansion into some inland zones but eventually requiring a larger scale of expansion. The largest inland and coastal settlements appeared after A.D. 1000, coincident with the formal *latte* stone-pillar housing and expansion of habitations into the smaller northern-arc islands. These factors combined with a result after A.D. 1000 of using stone instead of wood for house posts, burial of individuals at these houses, the first appearance of stone grinding basins, and broadly shared patterns of village layouts all indicative of deliberate investment in long-lasting material culture fixed in the landscape.

The Marianas example has shown that the landscape system at any point in time can be attributed to different proximate and ultimate causes. In some cases, fluctuation in sea level was the primary cause of change in multiple aspects of natural and cultural history. In other cases, population growth exerted a strong role in the imprint of people on the landscape, as well as in shaping social and economic relationships. Additional influencing factors apparently brought less dramatic results or were extended over longer periods of time, for example as natural ecosystems and human behavioural patterns adjusted to periods of climatic stability versus instability, changing forest compositions, and coral reef growth.

Regardless of proximate and ultimate causation, elements of change can be identified at variable time scales that continually overlap in their different rhythms. These qualities recall Braudel's (1949) notion of multiple currents of change in human society, often portrayed as consisting of long-lasting durable structures of history (known as *longue durée*) in contrast to rapid but superficially changing conditions (known as *histoire événementielle*) (following Simiand 1903). Multiple concurrent paces of change have been attested in the Marianas case, over time resulting in a completely different natural–cultural landscape system, and the complex coevolving dynamics deserve close examination.

Relativity of time scales must be remembered when working with concepts such as *longue durée* and *histoire événementielle*. Very little if any components of the ancient cultural landscape in the Mariana Islands were sustained intact throughout the entire 3500-year sequence. On the other hand, many aspects endured for considerable periods of some centuries or longer. In this case, the elements of deep structure in human culture can be appreciated as having changed over time, so that the profoundly defining elements during one time interval may have changed later. For instance, the founding populations in the Mariana Islands maintained ancestral practices of living in shoreline-oriented 'floating villages' and producing very finely decorated pottery traditions, clearly important enough to be replicated more than 2000 km from an overseas homeland, but these apparently deep-rooted cultural qualities were abandoned some centuries later. A second example in the Marianas refers to the later emergence of a structured system of villages and land use, coincident with formalisation of house architecture and other aspects of society, effectively overwriting the preceding traditions.

A change in a long-held cultural practice may signal a crisis event that demanded a radical shift in human behaviour, perhaps even more radical if it involved simultaneous change in several related components. As noted, at least two such systemic transformations occurred in the Marianas sequence, but both of these examples unfolded over the course of some centuries. Partly, the extended periods of change can be attributed to the innate complexities of social–ecological systems, wherein sudden shifts are improbable if not impossible. Nonetheless, archaeologically detected 'sudden' change may be described as having occurred within an interval of a few centuries. Additionally, some of the world's societies have created quick and profound change through imperial regimes and government policies, as happened during the Spanish conquest in the Mariana Islands. These observations may serve well as the modern international world faces challenges of policy-making about human response to environmental change and planning for sustainable futures.

References

Braudel, F. (1949). *La Méditerranée et la Monde Méditerranéan à l'Époche de Philippe II*. Three volumes. Paris: Armand Collin.

Hodder, I. (Ed.). (2012). *Archaeological theory today* (2nd ed.). Cambridge: Polity Press.

Simiand, F. (1903). Méthode historique et science social. *Revue de Synthèse Historique, 16*, 113–137.

Chapter 17
Future Directions

This book aims to facilitate discussions about landscape evolution in any part of the world, using an illustrative example of Mariana Islands archaeology. Building on the body of knowledge as presented in prior chapters, a few key points may be considered toward a sustainable future of archaeology and the different ways of learning about landscapes. Although presented only briefly, these issues are extremely important for making the present study relevant to modern and future concerns.

This study has accommodated a broad view of landscapes, finding resonance with other approaches in studying landscapes as complex systems. Following the premise that landscapes, however defined, are fundamental to human experience, this book has avoided any single overarching explanatory theory for comprehending why complex natural–cultural landscapes should exist or why they should evolve. Rather, the approach here has examined multiple fields of primary datasets as a basis to illustrate how landscape systems have functioned and changed over time in variable ways, at different concurrent paces or rhythms, and according to a number of interrelated mechanisms of proximate and ultimate causation.

Complex systems such as landscapes potentially can be understood through illustrative case studies and ideal 'model systems' that elegantly capture the essence of the otherwise daunting complexities. This book has provided one such richly illustrative example in the Mariana Islands, and its utility as a model system deserves more discussion. In one point of view, no single case study can serve as an accurate model of something so complex as the world's natural–cultural landscape systems. Although this kind of extremist stance never can be satisfied, the Marianas example has provided as much relevance as possible for general-global studies through drawing on several lines of evidence that each transcended periods of chronological change such as in landform configurations, sea level and coastal ecology, plant and animal communities, cultural patterns of residence and use of natural resources, and various aspects of material culture.

M.T. Carson, *Archaeological Landscape Evolution*,
DOI 10.1007/978-3-319-31400-6_17

Human interactions with landscapes undeniably have changed throughout evolutionary history, but the pace and scale both have increased notably within the last several decades of the modern industrialised and globalised world. People continue to experience landscapes, yet the mode of experience has become overall homogenised and impersonalised, giving way to a new standard of detached objectification of landscapes. This trend follows a general loss of linguistic diversity and cultural traditions worldwide, wherein people steadily are losing their opportunities to engage with their landscapes in the long-held traditional manners of their ancestors. In the Mariana Islands, for instance, foreign imperialism has interrupted the indigenous Chamorro landscape systems for more than 300 years now, yet the Marianas landscapes continue to be respected as invaluable heritage resources. In the Marianas and elsewhere worldwide, people at a large scale are learning to live in the modern globalised world that emphasises private ownership, capitalistic gains, and other ideologies that often conflict with traditional views of landscapes.

Living with landscapes in the modern world may need to develop with a duality of experience or a form of bilingualism, so that people can relate with their landscapes in different ways depending on context. Standard conventions of national and international laws, economics, and resource management provide a commonly understood language for discourse about landscapes as definable entities or properties. These conventions are useful for necessary communications with government policy-makers, but they often overlook or underemphasise the nuances of traditional cultural knowledge about landscapes. By bringing native languages and traditions more clearly into the discourse, landscapes can be appreciated more fully as complex natural–cultural systems. These different worldviews have much to offer one another, yet cross-communication has been problematic in most cases.

When considering different worldviews, landscapes must be remembered as dynamic and not at all fixed through time. Too often, the past is imagined as a singular monolithic entity, but in fact the dynamic characteristics of landscapes have been defining parts of the human experience in a long-term perspective. Deep chronological records of natural–cultural landscape systems are crucial for learning about how human societies can adapt to the continually changing conditions of their landscapes.

This book provides one thorough example in the Mariana Islands, useful toward general-global understanding of how the various components of a landscape can change through time and create conditions of further systemic evolution. Several similar case studies already exist or potentially can be synthesised from existing records in different regions, but new efforts are encouraged for facilitating cross-regional comparisons of the relevant datasets. Only through these kinds of efforts can enough solid information be marshalled for understanding how people relate with their landscapes in a long-term perspective transcending multiple instances of evolving conditions at different paces and scales through time.

Studies of ancient landscapes are particularly urgent right now in the Marianas, and possibly the same can be said in any place in our world where urbanisation and globalisation are forcing major change in the ways people relate with landscapes. Knowledge of the past has become essential in pursuing cultural heritage identities, and most people can recognise the importance of detailed scientific knowledge toward these goals. In a related issue, struggles for independent recognition have intensified over the past decades for the Chamorro people as an indigenous group, further complicated by a political separation between Guam and the Commonwealth of the Northern Mariana Islands. Continuing urban developments, private economic investments, military build-up, and public infrastructure vastly have altered the islands while also altering the roles of people in managing their environments and engaging with their landscapes. Meanwhile, at least some of the reigning government decision-makers and policy-makers tend to be more concerned with short-term economic gains, foreign aid, and taxable incomes and less concerned with investing in internal programs of long-term sustainable landscape systems.

Marianas archaeology and landscape studies in principle are significant for various stakeholders of cultural practitioners, archaeologists, biologists, geologists, resource managers, government leaders, and countless others who may or may not be able to work together productively. The multiple perspectives may be appreciated for emphasising the central role of landscapes in human societies, and in this sense the different perspectives all should be embraced as strengthening each other synergistically. Unfortunately, people inevitably miscommunicate and become drawn into hostile debates about which perspective is more correct or authoritative, perhaps symptomatic of an innate problem of how human beings relate with each other.

Despite its ongoing challenges, Marianas archaeology now has made research contributions of global significance. Enough information has been gained about long-term chronological sequences to enable new questions about the past, especially about how people interacted with changing natural–cultural landscapes. Archaeologists in the Marianas are beginning to conduct larger-format excavations, fine detail of recovery, inter-disciplinary analysis, and cross-regional comparisons. These new trends already have secured a strong role for Marianas archaeology in a larger Asia-Pacific context that had been overlooked for many decades, and they bear truly exciting potential to address research themes of global significance, such as landscape evolution considered in this book.

Landscapes may be interpreted in multiple ways, and this book intentionally has followed a liberal approach that allows diverse appreciation of landscapes as heritage resources in their own right. The notion of long-term evolving natural–cultural landscapes as heritage resources can be useful in a practical sense for cross-communication of different viewpoints and especially for developing sustainable roles for landscapes in resource management and government policies. The deep chronologies of evolving landscapes can be incorporated into public education, interpretive programs at parks and other preserved ecosystems, and decision-making about how to manage the many elements of the world's continually changing landscapes in the broadest sense. A necessary further step, though, involves integrating these kinds of perspectives of landscapes into daily practice and policies for long-term sustainability, for example in parks, landmarks, and natural refuge complexes ideally at large scales of cohesive ecosystems where long-term records can be experienced and appreciated.

A study of long-term natural–cultural landscape evolution has substantiated the pages of this book, but more research needs to be fulfilled in at least two facets. First, ongoing research surely will refine the details of the Marianas case and discover more of this region's archaeological potential. Second, investigations in other regions may build comparable case studies for global application. A model system in the Marianas serves as one useful example for learning about the processes involved in landscape evolution, but further studies will broaden and strengthen this research. In these ways, the present book may mark a beginning of future developments.

Index

M.T. Carson, *Archaeological Landscape Evolution*,
DOI 10.1007/978-3-319-31400-6

MIX
Papier aus verantwortungsvollen Quellen
Paper from responsible sources
FSC® C105338

If you have any concerns about our products,
you can contact us on
ProductSafety@springernature.com

In case Publisher is established outside the EU,
the EU authorized representative is:
Springer Nature Customer Service Center GmbH
Europaplatz 3, 69115 Heidelberg, Germany

Printed by Libri Plureos GmbH
in Hamburg, Germany